Annals of Mathematics Studies

Number 106

# TOPICS IN TRANSCENDENTAL ALGEBRAIC GEOMETRY

EDITED BY

PHILLIP GRIFFITHS

PRINCETON UNIVERSITY PRESS

PRINCETON, NEW JERSEY

1984

ISBN 0-691-08335-5 (cloth)
ISBN 0-691-08339-8 (paper)



Printed in the United States of America
by Princeton University Press, 41 William Street
Princeton, New Jersey

☆



Clothbound editions of Princeton University Press books are printed on acid-free paper, and binding materials are chosen for strength and durability. Paperbacks, while satisfactory for personal collections, are not usually suitable for library rebinding.

Library of Congress Cataloging in Publication data will be found on the last printed page of this book.

Table of Contents

## INTRODUCTION

During 1981-1982 the Institute for Advanced Study held a special year on algebraic geometry. Naturally there were a number of seminars, and this volume is essentially the proceedings of one of these.

The motif of the seminar was to explore the ways in which the recent developments in formal Hodge theory might be applied to problems in algebraic geometry. Accordingly, there were first a number of background lectures (Chapters I, II, IV; cf. also Chapter VII) dealing with classical Hodge theory, variations of Hodge structure, and mixed Hodge structures. There was also one talk (Chapter V) discussing some important recent developments in formal Hodge theory. Next, there were several talks explaining one of the first applications of the general theory; namely, the remarkable properties of the monodromy or Picard-Lefschetz transformation and the topology of degenerations (Chapters VI and VII). In one talk (Chapter III) some of the recent theory of infinitesimal variation of Hodge structure, which seeks to provide a direct link between formal Hodge theory and geometry by introducing a surrogate for the classical theta divisor in the Jacobian variety of curves, was presented.

One of the main results in classical Hodge theory is the Torelli theorem for curves, and in recent years there has been considerable interest in extending this theorem to higher dimensions. Accordingly there were quite a few talks devoted to this question (Chapters VIII, X, XII). Moreover, during the course of the year a number of new Torelli-type theorems were proved and these are also reported on in this volume (Chapters IX, XI, and XIII). It seems reasonable to hope that there is currently some substantial movement in the general Torelli problem.

One of the most important aspects of Hodge theory is the problem of algebraic cycles. One may argue that most of the progress to date consists in demolishing the intuition that one had built up by analogy with the case of divisors, and the theory of normal functions has played an important role in this development. For this reason there were a number of talks on this topic. In Chapter XIV the classical theory of normal functions is explained, and in Chapter XVII a possible connection between normal functions and infinitesimal variation of Hodge structure is explored. In Chapter XVI (which was not a talk in the seminar) Herb Clemens has explained some uses of normal functions to study homological, algebraic, and rational equivalence of higher codimensional cycles leading up to his recent example of non-finite generation. Finally, in Chapter XV El Zein and Zucker give a new result concerning the extension of normal functions across a singularity.

Needless to say, many important topics related to Hodge theory and/or algebraic cycles are not discussed in this book. Moreover, the aim of the talks in the seminar was to give a clear and hopefully intuitive picture of what is known and of some of the available techniques, rather than to attempt a complete-with-proofs presentation of the theory. Hopefully, this spirit is also communicated in these proceedings, and is supplemented by sufficient references so that the reader is able to track down the missing details and proofs.

Finally, I should like to personally acknowledge the role of Dr. Loring Tu in preparing this volume. Whatever clarity of exposition is present in Chapters I-IV and XII is due in large measure to his efforts. It is a pleasure to thank him together with all the contributors to this volume and participants in the seminar for their interest and efforts.

PHILLIP GRIFFITHS

Topics in

Transcendental

Algebraic Geometry

Chapter I

# VARIATION OF HODGE STRUCTURE

Phillip Griffiths
Written by Loring Tu

## §1. *Hodge structures*

Let $X$ be a compact Kähler manifold (e.g., a smooth projective variety). A $C^\infty$ form on $X$ decomposes into $(p,q)$-components according to the number of $dz$'s and $d\bar{z}$'s. Denoting the $C^\infty$ n-forms and the $C^\infty$ $(p,q)$-forms on $X$ by $A^n(X)$ and $A^{p,q}(X)$ respectively, we have the decomposition

$$A^n(X) = \bigoplus_{p+q=n} A^{p,q}(X) .$$

The cohomology $H^{p,q}(X)$ is defined to be

$$\begin{aligned} H^{p,q}(X) &= \{\text{closed } (p,q)\text{-forms}\}/\{\text{exact } (p,q)\text{-forms}\} \\ &= \{\phi \in A^{p,q}(X) : d\phi = 0\}/dA^{n-1}(X) \cap A^{p,q}(X) . \end{aligned}$$

THEOREM 1 (Hodge Decomposition Theorem). *Let* $X$ *be a compact Kähler manifold. Then in each dimension* $n$ *the complex de Rham cohomology of* $X$ *can be written as a direct sum*

$$H^n_{DR}(X,C) = \bigoplus_{p+q=n} H^{p,q}(X) .$$

REMARK 2. One can define a decreasing filtration on $A^n(X)$ by

$$F^pA^n(X) = A^{n,0}(X) \oplus \cdots \oplus A^{p,n-p}(X)$$

and a decreasing filtration on $H^n_{DR}(X)$ by

$$F^pH^n_{DR}(X) = H^{n,0}(X) \oplus \cdots \oplus H^{p,n-p}(X) .$$

The group $F^pH^n_{DR}(X)$ may also be described as

$$F^pH^n_{DR}(X) = \{\phi \in F^pA^n(X) : d\phi = 0\}/dA^{n-1} \cap F^pA^n(X) .$$

It has been found useful to extract the contents of the Hodge decomposition theorem into a definition, for it is not only the complex cohomology of a compact Kähler manifold that possesses such a decomposition.

DEFINITION 3. A *Hodge structure of weight* $n$, denoted $\{H_{\mathbf{Z}}, H^{p,q}\}$, is given by a lattice $H_{\mathbf{Z}}$ of finite rank together with a decomposition on its complexification $H = H_{\mathbf{Z}} \otimes \mathbf{C}$ :

$$H = \bigoplus_{p+q=n} H^{p,q}$$

such that

$$H^{p,q} = \overline{H^{q,p}} .$$

Here by a lattice of finite rank we mean simply a finitely generated free Abelian group.

Alternatively a Hodge structure of weight $n$ can be given by a lattice $H_{\mathbf{Z}}$ of finite rank together with a decreasing filtration on its complexification $H = H_{\mathbf{Z}} \otimes \mathbf{C}$ :

$$H = F^0 \supset F^1 \supset \cdots \supset F^n$$

such that

$$H \simeq F^p \oplus \overline{F}^{n-p+1}$$

The two definitions are equivalent, for given a decomposition $H = \oplus H^{p,q}$, one defines the filtration by

$$F^p = H^{n,0} \oplus \cdots \oplus H^{p,n-p} ,$$

and given a filtration $\{F^p\}_{p=0,\cdots,n}$, one defines the decomposition by

$$H^{p,q} = F^p \cap \bar{F}^q .$$

It is not difficult to check that these constructions satisfy the requisite properties. We may therefore denote a Hodge structure of weight $n$ either by $\{H_{\mathbf{Z}}, H^{p,q}\}$ or by $\{H_{\mathbf{Z}}, F^p\}$. The $H^{p,q}$ are called the *Hodge components of* $H$ and the filtration $\{F^p\}$ the *Hodge filtration* of $H$.

REMARK 4. By thinking heuristically of $F^p$ as forms possessing at least $p$ dz's, the various superscripts become more intelligible. For example, $\bar{F}^q$ would be forms possessing at least $q$ $d\bar{z}$'s. Since the total weight is $p+q=n$, $F^p \cap \bar{F}^q$ consists of forms having *precisely* $p$ dz's and $q$ $d\bar{z}$'s. Similarly, $\bar{F}^{n-p+1}$ consists of forms having at least $n-(p-1)$ $d\bar{z}$'s, or equivalently at most $p-1$ dz's; consequently, $F^p \oplus \bar{F}^{n-p+1}$ encompasses all $n$-forms.

EXAMPLES OF HODGE STRUCTURES. (a) (Hodge [8]). Let $X$ be a compact Kähler manifold. For any integer $n$ take

$$H_{\mathbf{Z}} = H^n(X, \mathbf{Z})/\text{torsion} .$$

Then

$$H = H_{\mathbf{Z}} \otimes \mathbf{C} = H^n_{DR}(X, \mathbf{C}) = \bigoplus_{p+q=n} H^{p,q}(X) ,$$

and $\{H_{\mathbf{Z}}, H^{p,q}(X)\}$ is a Hodge structure of weight $n$.
(b) (Deligne [3]). As in (a) but with $X$ any smooth complete abstract algebraic variety over $\mathbf{C}$. (Such a variety need not be Kähler, since it may admit no embedding into a projective space.)

A *polarized algebraic variety* is a pair $(X, \omega)$ consisting of an algebraic variety $X$ together with the first Chern class $\omega$ of a positive line bundle on $X$. Let

$$L : H^n(X, \mathbf{C}) \to H^{n+2}(X, \mathbf{C})$$

be multiplication by $\omega$. We recall below two fundamental theorems of Lefschetz.

THEOREM 5 (Hard Lefschetz Theorem). *On a polarized algebraic variety* $(X, \omega)$ *of dimension* $d$,

$$L^k : H^{d-k}(X,C) \to H^{d+k}(X,C)$$

*is an isomorphism for every positive integer* $k \leq d$.

Thus

$$L^{d-n} : H^n(X,C) \to H^{2d-n}(X,C)$$

is an isomorphism. The *primitive cohomology* $P^n(X,C)$ is defined to be the kernel of $L^{d-n+1}$. (For the geometric interpretation of this definition, see [6, p. 122].)

THEOREM 6 (Lefschetz Decomposition Theorem). *On a polarized algebraic variety* $(X, \omega)$, *we have for any integer* $n$ *the following decomposition*:

$$H^n(X,C) \simeq \bigoplus_{k=0}^{\left[\frac{n}{2}\right]} L^k P^{n-2k}(X,C) .$$

It follows that the primitive cohomology groups determine completely the full complex cohomology.

Let $(X, \omega)$ be a polarized algebraic variety. Define

$$H_Z = P^n(X,C) \cap H^n(X, Z)$$

and

$$H^{p,q} = P^n(X,C) \cap H^{p,q}(X) .$$

Then $\{H_Z, H^{p,q}\}$ is a Hodge structure of weight $n$. On this Hodge structure there is a bilinear form

$$Q : H_Z \times H_Z \to Z$$

given by

$$Q(\phi, \psi) = (-1)^{n(n-1)/2} \int_X \phi \wedge \psi \wedge \omega^{d-n} .$$

This bilinear form makes $\{H_Z, H^{p,q}\}$ into a *polarized Hodge structure* in the following sense.

DEFINITION 7. A *polarized Hodge structure of weight* $n$, denoted $\{H_Z, H^{p,q}, Q\}$ or $\{H_Z, F^p, Q\}$, is given by a Hodge structure of weight $n$ together with a bilinear form

$$Q : H_Z \times H_Z \to Z ,$$

which is symmetric for $n$ even and skew-symmetric for $n$ odd, satisfying the two *Hodge-Riemann bilinear relations*:

(8) $\qquad Q(H^{p,q}, H^{p',q'}) = 0$ unless $p' = n-p$ and $q' = n-q$ ,

(9) $\qquad (\sqrt{-1})^{p-q} Q(\psi, \bar{\psi}) > 0$ for any nonzero element $\psi$ in $H^{p,q}$ .

We define the *Weil operator* $C : H \to H$ by

$$C|_{H^{p,q}} = (\sqrt{-1})^{p-q} .$$

For example, $C(dz) = idz$ and $c(d\bar{z}) = -id\bar{z}$. In terms of the Hodge filtration $\{F^p\}$ the bilinear relations are

(10) $\qquad Q(F^p, F^{n-p+1}) = 0$ ,

(11) $\qquad Q(C\psi, \bar{\psi}) > 0$ for any nonzero element $\psi$ in $H$.

This bilinear form $Q$ is called a *polarization* on the Hodge structure.

*Two basic constructions and their relation with cycles*

We can associate to a Hodge structure $\{H_Z, H^{p,q}\}$ of weight $n$ one of two objects depending on whethei $n$ is even or odd.

(i) If $n = 2m$, then the *Hodge group* is

$$H_{\mathbf{Z}}^{m,m} = H_{\mathbf{Z}} \cap H^{m,m} .$$

An element of the Hodge group is called a *Hodge class*. The rank of the Hodge group $H_{\mathbf{Z}}^{1,1}$ is called the *Picard number* of $X$.

To motivate what is to follow, we recall the construction of the Jacobian of a curve. Let $C$ be a curve and $\gamma$ a 1-cycle on $C$. Integration of holomorphic 1-forms over $\gamma$

$$\omega \mapsto \int_{\gamma} \omega , \quad \omega \in H^0(C, \Omega^1) ,$$

defines a linear functional on $H^0(C, \Omega^1)$. Thus there is a map $H_1(C, \mathbf{Z}) \to (H^0(C, \Omega^1))^*$. The Jacobian of $C$ is

$$J(C) = \frac{(H^0(C, \Omega^1))^*}{H_1(C, \mathbf{Z})} .$$

Making the identifications

$$(H^0(C, \Omega^1))^* \simeq (H^{1,0})^* \simeq H^{0,1}$$

and

$$H_{\mathbf{Z}} \simeq H_1(C, \mathbf{Z}) ,$$

we can write

$$J(C) = H_{\mathbf{Z}} \backslash H^{0,1} .$$

Now we come to the construction of the intermediate Jacobian.

(ii) If $n = 2m-1$, then

$$H = \underbrace{H^{2m-1,0} \oplus \cdots \oplus H^{m,m-1}}_{F^m} \oplus H^{m-1,m} \oplus \cdots \oplus H^{0,2m-1}$$

and the *intermediate Jacobian* of $H$ is

$$J = H_{\mathbf{Z}} \backslash (\text{second half of Hodge decomposition})$$

$$= H_{\mathbf{Z}} \backslash H/F^m .$$

These two constructions, the Hodge group and the intermediate Jacobian, are closely related to the study of algebraic cycles on a smooth variety. Let $X$ be a smooth algebraic variety. Two cycles $Z_1$ and $Z_2$ on $X$ are *algebraically equivalent* if, roughly speaking, one can be deformed into the other via an algebraic family of cycles on $X$. To be more precise, there is an algebraic variety $S$ and an algebraic cycle $T$ in $S \times X$ such that $Z_1$ and $Z_2$ are the restrictions of $T$ to two fibers of the projection $\pi: S \times X \to S$ (see Figure 1). It may not be possible to choose $T$ effective even when $Z_1$ and $Z_2$ are.

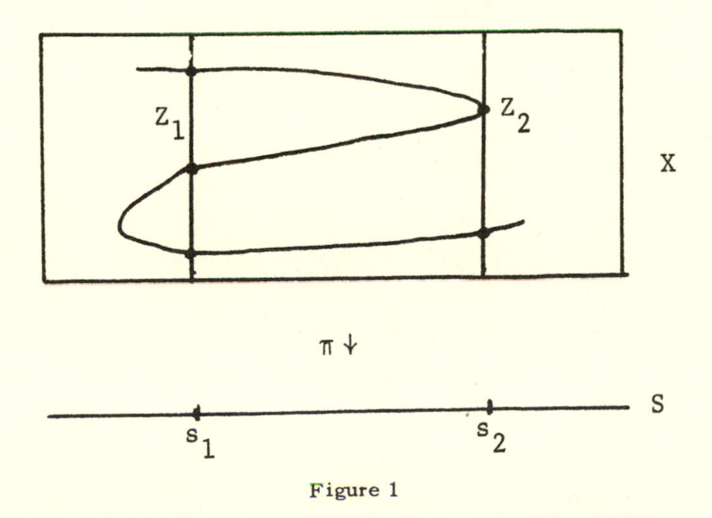


Figure 1

Two cycles are *rationally equivalent* if they are algebraically equivalent with a chain of $P^1$'s as the parameter space $S$ (Figure 2).

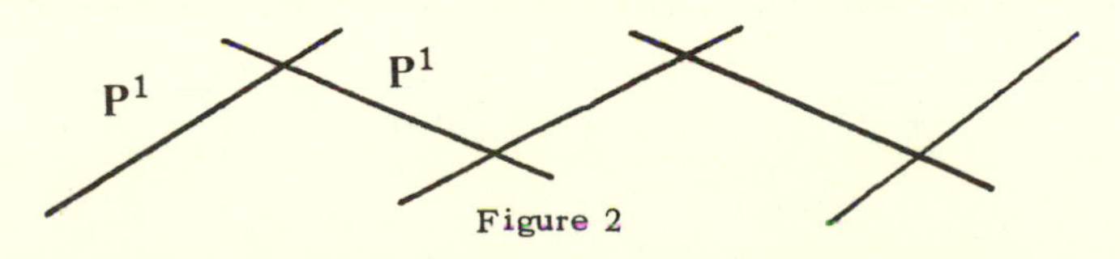


Figure 2

Let $\mathcal{Z}^m(X)$ be the Abelian group generated by the codimension m reduced and irreducible subvarieties of X. An element of $\mathcal{Z}^m(X)$ is a *codimension* m *algebraic cycle*. Denoting by $\mathcal{Z}_r^m(X)$, $\mathcal{Z}_a^m(X)$, and $\mathcal{Z}_h^m(X)$ the codimension m cycles rationally equivalent to zero, algebraically equivalent to zero, and homologically equivalent to zero respectively, there is a sequence of inclusions

$$\mathcal{Z}_r^m(X) \subset \mathcal{Z}_a^m(X) \subset \mathcal{Z}_h^m(X) \subset \mathcal{Z}^m(X) .$$

If Z is a cycle of codimension m on a variety X of dimension d, one can associate to the homology class of Z a fundamental class $f([Z]) \in H_{\mathbf{Z}}^{m,m}$ as follows. Let $Z_{reg}$ be the regular points of Z. Integration over $Z_{reg}$ defines a linear functional on $H^{2(d-m)}(X)$:

$$\psi \mapsto \int_{Z_{reg}} \psi , \qquad \psi \in A^{2(d-m)}(X) . \tag{12}$$

By Poincaré duality, this linear functional f(Z) may be identified with an element of $H^{2m}(X, \mathbf{Z})$. Because on a variety of dimension d a 2d-form must be of type (d,d), the form $\psi$ in (12) can be taken to be in $A^{d-m,d-m}(X)$. Hence

$$f(Z) \in (H^{d-m,d-m}(X))^* \simeq H^{m,m} .$$

If Z is homologous to zero, then by Stokes' theorem, the integral of a closed form over Z is zero. We have, therefore, a map

$$f : \mathcal{Z}^m(X)/\mathcal{Z}_h^m(X) \to H_{\mathbf{Z}}^{m,m} ,$$

called the *fundamental class map*.

Next we take up the relation between cycles and intermediate Jacobians. Given a smooth variety X of dimension d, the m^th^ intermediate Jacobian $J^m(X)$ is defined to be the intermediate Jacobian of the cohomology $H^{2m-1}(X)$:

$$J^m(X) = H^{m-1,m} \oplus \cdots \oplus H^{0,2m-1}(X)/H_{\mathbf{Z}} .$$

By Poincaré duality there is a canonical identification

$$J^m(X) = (F^{d-m+1} H^{2d-2m+1}(X))^*/\Lambda^* ,$$

where $\Lambda^*$ is the image of the map

$$\alpha : H_{2d-2m+1}(X, Z) \to (H^{2d-2m+1}(X))^*$$

given by integration. Our choice of the superscripts is dictated by the definition of the *Abel-Jacobi map* for codimension $m$ cycles:

$$u : \mathcal{Z}_h^m(X)/\mathcal{Z}_r^m(X) \to J^m(X) .$$

If $Z$ is a cycle homologous to zero in $X$, then $Z = \partial\Gamma$ for some chain $\Gamma$ of dimension $2(d-m)+1$. We define $u(Z)$ by

$$u(Z)(\psi) = \int_\Gamma \psi \quad \text{for all} \quad \psi \quad \text{in} \quad A^{2(d-m)+1}(X). \tag{13}$$

If $\Gamma$ is not a manifold, this integral is taken in the sense of currents. Liebermann ([9] and [10]) showed that all this makes sense.

An element of $H(X, Q) \cap H^{m,m}$ is called a *rational Hodge class*. By tensoring with $Q$, the fundamental class map can be defined over $Q$. The Hodge conjecture asks whether every rational Hodge class is the fundamental class of some algebraic cycle with rational coefficients. (There are counterexamples for torsion integral classes. See Atiyah and Hirzebruch, "Analytic cycles on complex manifolds," Topology 1 (1962), 25-45.)

The codimension one case of the Hodge conjecture is answered by the following theorem of Lefschetz.

THEOREM 1.14 (Lefschetz Theorem on (1,1)-Classes). *Let $X$ be a smooth projective variety. Then every integral (1,1)-class on $X$ is the fundamental class of a divisor on $X$.*

Before considering some low-dimensional examples of the Hodge group and the intermediate Jacobian, we want to identify two of the intermediate Jacobians. On a smooth variety $X$ of any dimension $d$ there is the exponential sequence

$$0 \longrightarrow Z \longrightarrow \mathcal{O} \xrightarrow{\exp} \mathcal{O}^* \longrightarrow 0 .$$

From the associated long exact sequence

$$\cdots \to H^1(X,Z) \to H^1(X,\mathcal{O}) \to H^1(X,\mathcal{O}^*) \to H^2(X,Z) \to \cdots ,$$

we see that $H^1(X,\mathcal{O})/H^1(X,Z)$ is the group of the isomorphism classes of the line bundles of first Chern class zero. This is by definition the *Picard variety* $\mathrm{Pic}^0(X)$ of $X$. Note that it is also the intermediate Jacobian $J^1(X)$:

$$J^1(X) \simeq \mathrm{Pic}^0(X) . \tag{15}$$

On the other hand, since

$$H^{2d-1}(X) = H^{d,d-1} \oplus H^{d-1,d} ,$$

the intermediate Jacobian

$$\begin{aligned} J^d(X) &= H^{d-1,d}/H_Z \\ &= H^0(X,\Omega^1)^*/H_1(X,Z) . \end{aligned}$$

This last group is by definition the *Albanese variety* $\mathrm{Alb}(X)$ of $X$. Hence

$$J^d(X) = \mathrm{Alb}(X) . \tag{16}$$

EXAMPLE 17. (a) For a smooth curve $C$ the intermediate Jacobian $J^1$ is the Jacobian of the curve, rational equivalence is linear equivalence, and the Abel-Jacobi map for $J^1$ is the usual Abel-Jacobi map

$$u : \mathrm{Div}^0(C) \to J(C) .$$

Note that in this case,

$$J^1 = J(C) = \mathrm{Pic}^0(C) = \mathrm{Alb}(C) .$$

(b) Let $F$ be a smooth surface.

For $H^1(F)$ the intermediate Jacobian is the Picard variety:

$$J^1(F) \simeq \mathrm{Pic}^0(F) .$$

For $H^2(F)$ the Lefschetz theorem on (1,1)-classes completely settles the nature of the Hodge classes.

For $H^3(F)$ the intermediate Jacobian is the Albanese variety:

$$J^2(F) \simeq \mathrm{Alb}(F) .$$

(c) Let $V$ be a threefold. For $H^1(V)$ and $H^5(V)$ the intermediate Jacobians $J^1(V)$ and $J^3(V)$ are again the Picard and the Albanese varieties respectively, of which we have some degree of understanding. The group $H^2(V)$ is taken care of by the Lefschetz theorem on (1,1)-classes. Since $H^4(V)$ is isomorphic to $H^2(V)$ by the Hard Lefschetz theorem, the Hodge conjecture holds for $H^{2,2}_{\mathbf{Z}}$. Indeed, if $\phi$ is an integral (2,2)-class, then $\phi = \omega \cdot \xi$ for some rational (1,1)-class $\xi$. By the Lefschetz theorem on (1,1)-classes, $\xi$ is a multiple of the fundamental class of a divisor $S$ in $V$. But then an integral multiple of $\phi$ is the fundamental class of a hyperplane section of $S$.

Thus the first mysterious group is $H^3(V)$ of a threefold $V$. If $h^{3,0}(V) = 0$, then

$$H^3(V) = H^{2,1} \oplus H^{1,2}$$

behaves very much like a Hodge structure of weight one. But if $h^{3,0}(V) \neq 0$, then the associated intermediate Jacobian $J^2(V)$ so far eludes understanding. We mention here one interesting special case.

SPECIAL CASE. Let $V \subset \mathbf{P}^4$ be a smooth quintic threefold. (Threefolds of lower degree all have $h^{3,0} = 0$.) Using Schubert calculus, for example, one can show that $V$ has 2875 lines, some counted with multiplicity. Because $H^2(V) = \mathbf{Z}$, the difference of any two lines is homologous to zero:

$$L_{ij} = L_i - L_j \in H^2(V) .$$

It is known that if $V$ is a general quintic threefold, then $u(L_{ij}) \neq 0$, where $u$ is the Abel-Jacobi map.

OPEN QUESTIONS. Are the only relations given by $u(L_{ij}) + u(L_{ji}) = 0$? Can the configuration $\{u(L_{ij})\} \subset J^2(V)$ be determined using infinitesimal variation of Hodge structure (cf. Chapters III, XII below)?

We remark that, by considering higher degree rational curves on $V$, Herb Clemens has drawn a very surprising conclusion (cf. Chapter XVI below).

Another question is to explicitly compute out the Abel-Jacobi mapping in one nontrivial example with $h^{3,0} \neq 0$; e.g., the threefold $\sum_{i=0}^{4} x_i^d = 0$ $(d \geqq 5)$ in $\mathbf{P}^4$.

(d) For a fourfold $X$, apart from $H^1$, $H^2$, $H^6$, and $H^7$, which we can take care of as before, virtually nothing is known about the rest.

PROBLEM. Try to understand the algebraic subvarieties of

$$X = F_1 \times F_2 \quad \text{(a product of two surfaces)}$$

and

$$X = \mathbf{C}^4/\text{lattice} \quad \text{(an Abelian variety)} .$$

§2. *Classifying spaces*

Let $H_{\mathbf{Z}}$ be a fixed lattice, $n$ an integer, $Q$ a bilinear form on $H_{\mathbf{Z}}$, which is symmetric if $n$ is even and skew-symmetric if $n$ is odd, and $\{h^{p,q}\}$ a collection of integers such that $p+q = n$ and $\Sigma h^{p,q} = \operatorname{rank} H_{\mathbf{Z}}$. As before, denote by $H$ the complexification $H_{\mathbf{Z}} \otimes \mathbf{C}$.

DEFINITION 18. With the notations above, the *classifying space* D for the polarized Hodge structures of type $\{H_{\mathbf{Z}}, Q, h^{p,q}\}$ is the set of all collections of subspaces $\{H^{p,q}\}$ of H such that

$$H = \bigoplus_{p+q=n} H^{p,q}, \quad \dim H^{p,q} = h^{p,q},$$

and on which Q satisfies the two bilinear relations (8) and (9).

Set $f^p = h^{n,o} + \cdots + h^{p,n-p}$. In terms of filtrations, D is the set of all filtrations

$$H = F^0 \supset F^1 \supset \cdots \supset F^n, \quad \dim F^p = f^p,$$

on which Q satisfies the bilinear relations (10) and (11).

A priori D is only a set, but in fact it can be given the structure of a complex manifold. The simplest way to do this is to represent D as an open subset of a homogeneous algebraic variety.

DEFINITION 19. The *compact dual* $\check{D}$ of D is the subspace of $\Pi_{p=0}^{n} G(f^p,H)$ consisting of the filtrations $\{F^p\}$ on H satisfying the first bilinear relation (10). Here $G(f^p,H)$ denotes the Grassmannian of $f^p$-dimensional subspaces of H.

Because the first bilinear relation (10) is a set of algebraic equations, the compact dual is clearly an algebraic variety. We will show below that it is in fact a complex manifold.

In connection with a polarized Hodge structure there are three basic Lie groups:

$$\begin{aligned} G_{\mathbf{Z}} &= \operatorname{Aut}(H_{\mathbf{Z}},Q) \\ &= \{g : H_{\mathbf{Z}} \to H_{\mathbf{Z}} \mid Q(g\phi,g\xi) = Q(\phi,\xi) \text{ for all } \phi, \xi \text{ in } H_{\mathbf{Z}}\}, \end{aligned}$$

$$G_{\mathbf{R}} = \operatorname{Aut}(H_{\mathbf{R}},Q),$$

$$G_{\mathbf{C}} = \operatorname{Aut}(H,Q).$$

EXERCISE 20. Show that $G_{\mathbf{C}}$ acts transitively on the compact dual $\check{D}$ and that $G_{\mathbf{R}}$ acts transitively on the classifying space $D$.

Because the group $G_{\mathbf{C}}$ acts transitively on $\check{D}$, the variety $\check{D}$ is smooth. Let $B$ be the stabilizer of a point in $\check{D}$. Then

$$\check{D} \simeq G_{\mathbf{C}}/B .$$

Since $D$ is an open subset of $\check{D}$, it is also a complex manifold. By Exercise 20,

$$D \simeq G_{\mathbf{R}}/V ,$$

where $V = G_{\mathbf{R}} \cap B$ is the stabilizer of a point in $D$.

We will find it useful to have the infinitesimal versions of these basic Lie groups. So let $\{H_0^{p,q}\}$ in $D$ be chosen as the base point, which we regard as the reference Hodge structure. The Lie algebra $\mathfrak{g}_{\mathbf{R}}$ of $G_{\mathbf{R}}$ has the following description:

$$\mathfrak{g}_{\mathbf{R}} = \{Y \in \operatorname{Hom}(H_{\mathbf{R}},H_{\mathbf{R}}) | Q(Y\psi,\eta) + Q(\psi,Y\eta) = 0 \text{ for all } \psi,\ \eta \text{ in } H_{\mathbf{R}}\} .$$

Analogous descriptions hold for the Lie algebras $\mathfrak{g}_{\mathbf{Z}}$ and $\mathfrak{g}_{\mathbf{C}}$ of $G_{\mathbf{Z}}$ and $G_{\mathbf{C}}$. Here $\mathfrak{g}_{\mathbf{Z}} = \mathfrak{g}_{\mathbf{C}} \cap \operatorname{Hom}(H_{\mathbf{Z}},H_{\mathbf{Z}})$; note that

$$\mathfrak{g}_{\mathbf{C}} = \mathfrak{g}_{\mathbf{Q}} \otimes \mathbf{C} .$$

We can give the space $\operatorname{Hom}(H,H)$ a Hodge structure of weight zero by setting

$$\operatorname{Hom}(H,H)^{r,-r} = \bigoplus_{p+q=n} \operatorname{Hom}(H^{p,q},H^{p+r,q-r})$$

and

$$\operatorname{Hom}(H,H)^{r,-r} = \overline{\operatorname{Hom}(H,H)^{-r,r}} .$$

Since $\mathfrak{g}_{\mathbf{C}}$ is a rationally defined subspace of $\operatorname{Hom}(H,H)$, it inherits a

Hodge structure of weight zero from Hom(H,H) :

$$\mathfrak{g}_{\mathbf{C}}^{r,-r} = \mathfrak{g}_{\mathbf{C}} \cap \mathrm{Hom}(H,H)^{r,-r} = \overline{\mathfrak{g}_{\mathbf{C}}^{-r,r}} .$$

If $\mathfrak{b}$ is the Lie algebra of the complex stability group $B$, then

$$\mathfrak{b} = \bigoplus_{r \geq 0} \mathfrak{g}_{\mathbf{C}}^{r,-r} ,$$

for these are precisely the infinitesimal automorphisms that leave the reference Hodge filtration fixed. If $\mathfrak{h}$ is the Lie algebra of the real stability group $H$, then

$$\mathfrak{h} = \mathfrak{b} \cap \mathfrak{g}_{\mathbf{R}} .$$

Over the compact dual $\check{D}$ we have the universal subbundles $\mathcal{F}^p \to \check{D}$. These are holomorphic vector bundles. Their quotient bundles

$$\mathcal{H}^{p,q} = \mathcal{F}^p/\mathcal{F}^{p+1}$$

are also holomorphic vector bundles. We call the restriction of these $\mathcal{H}^{p,q}$ to the classifying space $D$, the *Hodge bundles.* The Hodge bundle $\mathcal{H}^{p,q}$ is a vector bundle over $D$ whose fiber at the point $\{F^p\}$ is $F^p/F^{p+1} = H^{p,q}$. Note that $\mathcal{F}^p$ has the $C^\infty$ decomposition

$$\mathcal{F}^p = \mathcal{H}^{n,0} \oplus \cdots \oplus \mathcal{H}^{p,n-p} .$$

Because of the second bilinear relation, each Hodge bundle $\mathcal{H}^{p,q}$ has a $G_{\mathbf{R}}$-invariant metric, $(\sqrt{-1})^{p-q}Q(\ ,\ \bar{}\ )$, making it into a Hermitian vector bundle. As is well known, any Hermitian vector bundle has a canonical connection and therefore a curvature. We shall comment on the curvature of the Hodge bundles in the second lecture.

We now turn to the tangent bundle of $D$. First recall that over the Grassmannian $G(k,H)$ there is a canonical isomorphism

$$T(G(k,H)) \simeq \mathrm{Hom}(S,Q) ,$$

where $S$ and $Q$ are the universal subbundle and quotient bundle respectively. One way of giving this isomorphism is by the following recipe. Suppose $F(t)$ is an arc in $G(k,H)$ with initial point $F \in G(k,H)$ and initial vector $\xi \in T_F(G(k,H))$. For any $v$ in $F$, let $v(t)$ be a curve in $F(t)$ with $v(0) = v$. Then the homomorphism $\bar{\xi}: F \to H/F$ corresponding to the tangent vector $\xi$ is given by

$$\bar{\xi}(v) = \frac{d}{dt} v(t)\Big|_{t=0} \pmod{F}.$$

Denoting by $\mathcal{H}$ the trivial bundle with fiber $H$ over $\check{D}$, we therefore have

$$\begin{aligned} T(\check{D}) &\subset \bigoplus_{p=1}^{n} \operatorname{Hom}(\mathcal{F}^p, \mathcal{H}/\mathcal{F}^p) \\ &= \bigoplus_{p=1}^{n} \operatorname{Hom}(\mathcal{H}^{n,0} \oplus \dots \oplus \mathcal{H}^{p,n-p}, \mathcal{H}^{p-1,n-p+1} \oplus \dots \oplus \mathcal{H}^{0,n}) . \end{aligned}$$

Similarly, the tangent bundle $T(D)$ is also contained in this direct sum of homomorphisms of Hodge bundles. Because each Hodge bundle has a $G_{\mathbf{R}}$-invariant Hermitian metric, the classifying space $D$ also has a $G_{\mathbf{R}}$-invariant Hermitian metric, which we denote by $ds_D^2$.

REMARK 21. In fact, this $G_{\mathbf{R}}$-invariant metric on the classifying space $D$ is induced by the Killing form on the Lie algebra $\mathfrak{g}_{\mathbf{R}}$.

*Examples of classifying spaces*

EXAMPLE 22 (Weight One). For $n = 1$,

$$H = H^{1,0} \oplus H^{0,1}, \qquad H^{1,0} = \overline{H^{0,1}},$$

and $Q$ is a skew-symmetric bilinear form. Let $g = \dim H^{1,0}$. Then each filtration $H^{1,0} \subset H$ is an element of the Grassmannian $G(g,H)$. Relative to a suitable basis for $H$, the skew form $Q$ is represented by the matrix

$$\begin{bmatrix} 0 & I \\ -I & 0 \end{bmatrix} .$$

So D is the set of all g-dimensional isotropic subspaces of $\begin{bmatrix} 0 & I \\ -I & 0 \end{bmatrix}$ on $\mathbf{C}^{2g}$. (An *isotropic* subspace of a skew form J is a subspace V such that $J(V,V) \equiv 0$ .) These are precisely the *maximal* isotropic subspaces.

The classifying space D may be identified with the Siegel upper half space

$$\mathbf{H}_g = \{g \text{ by } g \text{ complex matrices } Z = X+iY \mid Z \text{ is symmetric and } Y \text{ is positive definite}\},$$

as follows. Each element of G(g,H) is represented by a 2g by g matrix $\Omega$, up to the equivalence relation

$$\Omega \sim B\Omega A, \quad B \in Sp(2g,\mathbf{Z}), \quad A \in GL(g,\mathbf{C}) .$$

The matrix $\Omega$ can be brought to the normal form

$$\Omega = \begin{bmatrix} I \\ Z \end{bmatrix} ,$$

where Z is a g by g complex matrix. By the first bilinear relation Z is symmetric. By the second bilinear relation Im Z is positive definite.

## §3. *Variation of Hodge structure*

Let $\pi : \mathfrak{X} \to S$ be a family of smooth polarized projective varieties. By this we mean that there is a commutative diagram

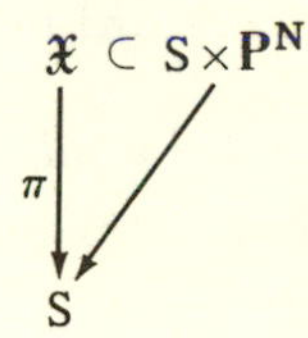


and that the fibers of $\pi : X \to S$ are smooth projective varieties. By associating to each fiber $X_s$ the Hodge decomposition of its primitive cohomology, we get locally maps from open subsets of $S$ into the classifying space $D$. Because one can identify the integer cohomology $H^n(X_s, Z)/\text{torsion}$ of a fiber with a fixed lattice $H_Z$ only up to the action of the monodromy group $\Gamma$, it is in general not possible to get a map from $S$ into $D$. Instead, what one has is a map into the quotient of $D$ by the action of the monodromy group

$$\phi : S \to \Gamma \backslash D ,$$

called the *period map*. It is shown in [4] that the period map is holomorphic.

A priori the differential of the period map

$$\phi_* : T(S) \to T(D) \subset \bigoplus_{p=1}^{n} \bigoplus_{r=1}^{p} (\mathrm{Hom}(\mathcal{H}^{p,q}, \mathcal{H}^{p-r,q+r}))$$

goes into the full tangent space of $D$ at each point. However, by [4] we do know that in fact $\phi_*$ shifts the Hodge filtration by only one; that is, $\phi_*$ maps into the subbundle $\bigoplus_{p=1}^{n} \mathrm{Hom}(\mathcal{H}^{p,q}, \mathcal{H}^{p-1,q+1})$.

To formalize the essential properties of the period map, we digress for a minute to discuss differential systems.

DEFINITION 23. A *differential system* on a complex manifold $X$ is given in one of two equivalent ways:

(i) either by a holomorphic subbundle $T_h(X)$ of the holomorphic tangent bundle $T(X)$, or

(ii) by an ideal $I$ in the complex $\Omega^*(X)$ of holomorphic forms on $X$ generated by a collection of 1-forms and their exterior derivatives.

An ideal such as in (ii) is called a *differential ideal*.

Given (i) we define $I$ to be the differential ideal generated by

$$\begin{aligned} \psi \in T_h(X)^{\perp} &= \text{annihilator of } T_h(X) \\ &= \{\psi \in \Omega^1(X) \mid \psi(v) = 0 \text{ for all } v \in T_h(X)\} \end{aligned}$$

and

$$d\psi \in \Omega^2(X) .$$

Conversely, given a differential ideal I we define the holomorphic subbundle $T_h(X)$ to be the kernel of all the 1-forms in I. (This requires a constant rank assumption.)

DEFINITION 24. An *integral manifold* of a differential ideal I on X is a holomorphic map of complex manifolds, $\phi : S \to X$, such that $\phi^* I = 0$.

This is equivalent to saying that

$$\phi_* T(S) \subset T_h(X) .$$

In checking integrality it suffices to check it on 1-forms, since if $\psi$ is a 1-form and $\phi^*\psi = 0$, then $\phi^* d\psi = d\phi^*\psi = 0$.

DEFINITION 25. The *horizontal differential system* on D is

$$T_h(D) = \{\xi \in T(D) \mid \xi(F^p) \subset F^{p-1}\} .$$

Because integrality is a local condition and because the period map $\phi$ is locally liftable to D, we may restate the horizontality of the period map in the following form: the period map of a family of polarized algebraic varieties is an integral manifold of the horizontal differential system on D.

DEFINITION 26. An *integral element* of a differential ideal I on a complex manifold X is a subspace $E \subset T_x(X)$ for some x in X such that

$$(*) \qquad \psi_\alpha|_E = 0$$

and

$$(**) \qquad d\psi_\alpha|_E = 0$$

for all generators $\psi_\alpha$ of degree 1 in I.

Note that (*) are linear equations and (**) are quadratic equations on a Grassmannian. In the theory of differential equations a system is said to be *involutive* if, roughly speaking, one cannot obtain new equations by differentiating the system. In other words, given (1) a system of differen-equations, one looks at (2) the set of all solutions, and then at (3) the system of all differential equations annihilating the solutions in (2). If (3) = (1), the system is said to be *involutive*.

OPEN QUESTION 27. Let $\phi: S \to \Gamma \backslash D$ be the period map of a family of polarized algebraic varieties and let I be the differential ideal determined by

$$\phi_*: T(S) \to T(\Gamma \backslash D) .$$

Are the differential equations of I an involutive system?

DEFINITION 28. A *variation of Hodge structure* is a map $\phi: S \to \Gamma \backslash D$, where S is a complex manifold and $\Gamma$ is a subgroup of $G_{\mathbf{Z}}$, such that $\phi$ is

(i) locally liftable,

(ii) holomorphic,

(iii) an integral manifold of the horizontal differential system $I_h$.

REMARK 29. Let $\tilde{S}$ be the universal covering of S. Then Condition (i) is equivalent to the existence of a map $\tilde{\phi}: \tilde{S} \to D$ which makes the following diagram commutative:

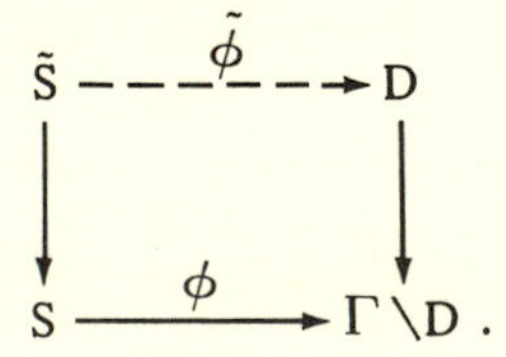

In terms of filtrations a variation of Hodge structure may be thought of as a family of variable Hodge filtrations, varying with s,

$$F_s^p = H_s^{n,0} \oplus \cdots \oplus H_s^{p,n-p}$$

on a fixed vector space H such that

(i) for each s the filtration is defined up to the action of an element of $\Gamma \subset G_{\mathbf{Z}}$,

(ii) $\partial F_s^p / \partial \bar{s} \subset F_s^p$,

(iii) $\partial F_s^p / \partial s \subset F_s^{p-1}$.

Let $\mathfrak{M}_g$ be the moduli space of smooth curves of genus g and $\Gamma_g = Sp(g, \mathbf{Z})$ the symplectic group. In Example 22, we noted that the classifying space for the polarized Hodge structures of weight 1 can be identified with the Siegel upper half space $\mathfrak{H}_g$. Hence, by associating to each curve its Hodge filtration $H^{1,0} \subset H^1$, we obtain a map

$$\phi : \mathfrak{M}_g \to \Gamma_g \backslash \mathfrak{H}_g .$$

However, because $\mathfrak{M}_g$ is not smooth and $\phi$ is not locally liftable around curves with automorphisms, it is not a variation of Hodge structure as we have defined the term. To take care of this situation, we introduce the notion of an *extended variation of Hodge structure*.

DEFINITION 30. Let S be any variety, possibly singular. A map $\phi : S \to \Gamma \backslash D$ is called an *extended variation of Hodge structure* if there is a smooth dense Zariski open set $S' \subset S$ such that $\phi|_{S'} : S' \to \Gamma \backslash D$ is a variation of Hodge structure.

By the local liftability property, given a variation of Hodge structure, we have the *monodromy representation*

$$\rho : \pi_1(S) \to \Gamma .$$

For an extended variation of Hodge structure, there is the monodromy representation

$$\rho : \pi_1(S') \to \Gamma ,$$

which, of course, depends on the open subset $S'$. But it turns out that the group $\rho(\pi_1(S')) \subset \Gamma$ is independent of $S' \subset S$. This is called the *monodromy group* of the extended variation of Hodge structure.

Speaking in general terms the global Torelli problem asks whether the Hodge decomposition of a variety determines the isomorphism class of the variety. The idea for proving such a statement would be to try to associate a geometric object to a Hodge structure, for example, the theta divisor of a Jacobian in the case of curves. In higher dimensions this cannot be done. What seems more amenable is the generic global Torelli theorem, which would say that the period map $\phi$ from some sort of moduli space $\mathfrak{M}$ onto its image $\phi(\mathfrak{M}) \subset \Gamma\backslash D$ has degree one; this means $\phi$ is generically one-to-one, but there could be points in $\Gamma\backslash D$ whose inverse images have more than one point. For an exposé of the recent progress in the pathology of the period map, see Chapter VIII of this volume. Generic global Torelli theorems are discussed in Chapters IX - XIII.

*The period map and Hodge bundles*

Let $\phi: S \to \Gamma\backslash D$ be a variation of Hodge structure with monodromy representation $\rho: \pi_1(S) \to \Gamma$. We view the universal covering $\tilde{S}$ of $S$ as a principal $\pi_1(S)$-bundle over $S$. Then the monodromy representation $\rho$ induces a lattice bundle $\mathcal{H}_{\mathbf{Z}}$ with group $\pi_1(S)$ over $S$, defined as the quotient of $\tilde{S} \times H_{\mathbf{Z}}$ by the equivalence relation

$$(\tilde{s}g, \psi) \sim (\tilde{s}, g\psi), \quad \text{where} \quad g \in \pi_1(S) .$$

The complexification of $\mathcal{H}_{\mathbf{Z}}$ is a complex vector bundle over $S$. Because $\mathcal{H}$ has the same transition functions as $\mathcal{H}_{\mathbf{Z}}$, it is locally constant and hence holomorphic. In the following, we will identify the holomorphic vector bundles over $S$ with the locally free sheaves of $\mathcal{O}_S$-modules. Associated to the locally constant sheaf $\mathcal{H}$ is a (1,0)-connection $\nabla$, relative to which the sections of $\mathcal{H}_{\mathbf{Z}}$ are locally constant.

At each point $s \in S$, the period map $\phi: S \to \Gamma\ D$ defines a filtration

$$\{0\} \subset \mathcal{F}^n_s \subset \cdots \subset \mathcal{F}^1_s \subset \mathcal{F}^0_s = \mathcal{H}_s ,$$

giving rise to a sequence of holomorphic subbundles $\mathcal{F}^p$ of $\mathcal{H}$. The

*Hodge bundle* $\mathcal{H}^{p,q}$ is defined to be the quotient bundle

$$\mathcal{H}^{p,q} = \mathcal{F}^p/\mathcal{F}^{p+1} .$$

There is a $C^\infty$ (*not* holomorphic) decomposition

$$\mathcal{H} = \bigoplus_{p+q=n} \mathcal{H}^{p,q} , \qquad \mathcal{H}^{p,q} = \overline{\mathcal{H}^{q,p}} .$$

In this context, the infinitesimal period relation becomes

$$\nabla \mathcal{F}^p \subset \mathcal{F}^{p-1} \otimes \Omega^1 .$$

In summary, a variation of Hodge structure $\phi : S \to \Gamma\backslash D$ gives the data $\{\mathcal{H}_Z, \mathcal{F}^p, \Delta, S\}$, where $\mathcal{H}_Z$ is a sheaf of lattices, $\mathcal{F}^p$ a filtration on $\mathcal{H} = \mathcal{H}_Z \otimes \mathcal{O}_S$, and $\nabla : \mathcal{H} \to \mathcal{H} \otimes \Omega^1$ the connection whose locally constant sections are $\mathcal{H}_Z \otimes C$ and which satisfies the infinitesimal period relation. Conversely, a set of data such as this determines a period map $\phi : S \to \Gamma\backslash D$. So a variation of Hodge structure can be given in either of these two equivalent ways.

EXAMPLE 31. Let $\pi : \mathfrak{X} \to S$ be a family of polarized algebraic varieties of dimension d all of whose fibers are smooth. Set

$$\mathcal{H}_Z = \mathcal{R}^n \pi_* Z ,$$

$$\mathcal{F}^p_s = F^p H^n_{DR}(X_s, C) .$$

Define $\nabla$ to be differentiation under the integral sign; this means,

$$\frac{d}{ds} \int_\gamma \omega(s) = \int_\gamma \nabla_{\frac{\partial}{\partial s}} \omega(s) ,$$

where $\gamma \in H_n(X_s, Z)$ is a family of cycles. Then $\{\mathcal{H}, \mathcal{F}^p, \nabla, S\}$ is a variation of Hodge structure.

*Normal functions*

Given a variation of Hodge structure $\{\mathcal{H}_{\mathbf{Z}}, \mathcal{H}^{p,q}, \nabla, S\}$, the connection induces by the infinitesimal period relations a map

$$(*) \qquad D : \mathcal{H}/\mathcal{F}^p \to (\mathcal{H}/\mathcal{F}^{p-1}) \otimes \Omega^1 .$$

A section $\nu$ of $\mathcal{H}/\mathcal{F}^p$ with $D\nu = 0$ is said to be *quasi-horizontal.* The quasi-horizontal sections of course include the horizontal ones (those for which $\nabla\nu = 0$ ).

In case the variation of Hodge structure has odd weight $n = 2m-1$, we can define a family of intermediate Jacobians by setting

$$\mathcal{J} = \mathcal{H}_{\mathbf{Z}} \backslash \mathcal{H}/\mathcal{F}^m \to S .$$

Because $\nabla\mathcal{H}_{\mathbf{Z}} = 0$, (*) induces a map on $\mathcal{J}$ :

$$D : \mathcal{J} \to (\mathcal{H}/\mathcal{F}^{m-1}) \otimes \Omega^1 .$$

The kernel of $D$,

$$\mathcal{J}_h = \{\nu \in \mathcal{J} | D\nu = 0\} ,$$

is called the *sheaf of normal functions.* The sections of $\mathcal{J}_h$ are the *normal functions.* These are discussed in Chapters XIV - XVII.

EXAMPLE 32. Let $\mathfrak{X} \to S$ be a family of algebraic varieties and $\mathfrak{Z} \subset \mathfrak{X}$ a codimension $m$ cycle such that each intersection

$$\mathfrak{Z} \cdot X_s = Z_s \in \mathcal{Z}_h^m(X_s)$$

is homologous to zero. Define

$$\nu(s) = u_{X_s}(Z_s) \in J(X_s)$$

where $u_{X_s} : \mathcal{Z}_h^m(X_s) \to J(X_s)$ is the Abel-Jacobi map on $X_s$. It is known that $\nu : S \to \mathcal{J}$ is holomorphic and assumes values in $\mathcal{J}_h$ (see [5]). It is called the *normal function associated to* $\mathcal{Z}$.

DEFINITION 33. A polarized Hodge structure $\{H_Z, H^{p,q}, Q\}$ is said to be *unimodular* if $\det Q = \pm 1$.

REMARK. The intermediate Jacobian of a unimodular polarized Hodge structure of odd weight is a principally polarized complex torus.

PROBLEM 34 (Beauville). Let $\phi$ be a unimodular polarized Hodge structure. Suppose $\phi$ can be written as a direct sum of unimodular polarized Hodge structures:

$$\phi = \oplus \phi_\nu .$$

Is this decomposition unique?

REMARK. The answer is no if the Hodge structures are not assumed unimodular.

## REFERENCES

[1] M. Cornalba and P. Griffiths, "Some transcendental aspects of algebraic geometry," in Proceedings of Symposia in Pure Mathematics, Vol. 29(1975), A.M.S., pp. 3-110.

[2] J. Carlson, M. Green, P. Griffiths, and J. Harris, "Infinitesimal variation of Hodge structures," to appear in Comp. Math.

[3] P. Deligne, "Théorie de Hodge III," Publ. Math. I.H.E.S. 44(1974), pp. 5-78.

[4] P. Griffiths, "Periods of integrals on algebraic manifolds I, II," Amer. J. Math. 90(1968), pp. 568-626, pp. 805-865.

[5] ________, "Periods of integrals on algebraic manifolds III," Publ. Math. I.H.E.S. 38(1970), pp. 125-180.

[6] P. Griffiths and J. Harris, "*Principles of Algebraic Geometry,*" John Wiley and Sons, New York, 1978.

[7] P. Griffiths and W. Schmid, "Recent developments in Hodge theory: a discussion of techniques and results," in *Discrete Subgroups of Lie Groups and Applications to Moduli,* Oxford University Press, 1973.

[8] W. V. D. Hodge, "*The Theory and Applications of Harmonic Integrals,*" second ed., Cambridge University Press, New York, 1952.

[9] D. Lieberman, "Higher Picard varieties," Amer. J. Math. 90 (1968), pp. 1165-1199.

[10] ________, "On the module of intermediate Jacobians," Amer. J. Math. 91 (1969), pp. 671-682.

[11] W. Schmid, "Variation of Hodge structure: the singularities of the period mapping," Inv. math. 22 (1973), pp. 211-319.

PHILLIP GRIFFITHS
MATHEMATICS DEPARTMENT
HARVARD UNIVERSITY
CAMBRIDGE, MA 02138

LORING TU
MATHEMATICS DEPARTMENT
JOHNS HOPKINS UNIVERSITY
BALTIMORE, MD 21218

## Chapter II

# CURVATURE PROPERTIES OF THE HODGE BUNDLES

Phillip Griffiths
Written by Loring Tu

We consider a polarized variation of Hodge structure $\phi : S \to \Gamma \backslash D$, which we think of locally as a variable polarized Hodge decomposition on a fixed vector space:

$$\begin{cases} H = \bigoplus_{p+q=n} H_s^{p,q} \\ F_s^p = H_s^{n,0} \oplus \cdots \oplus H_s^{p,n-p}, \end{cases}$$

where $s$ varies over the variety $S$. (To be strictly correct, $s$ should be in the universal covering $\tilde{S}$, for otherwise it may not be possible to have the fixed vector space $H$. Locally the description just given is fine.) We have

$$\frac{\partial H_s^{p,q}}{\partial \bar{s}} \subseteq H_s^{p+1,q-1} \oplus H_s^{p,q}$$

and by conjugation,

$$\frac{\partial H_s^{p,q}}{\partial s} \subseteq H_s^{p,q} \oplus H_s^{p-1,q+1};$$

or in terms of filtrations,

$$\frac{\partial F_s^p}{\partial \bar{s}} \subseteq F_s^p$$

and

$$\frac{\partial F_s^p}{\partial s} \subseteq F_s^{p-1} .$$

The differential $\phi_*$ of the period map assumes values in the horizontal subspace

$$\phi_* : T_s(S) \to \bigoplus_p \mathrm{Hom}(H_s^{p,q}, H_s^{p-1,q+1}) .$$

If $v \in T_s(S)$ is a tangent vector, we write

$$\phi_*(v) = \bigoplus_p v_p ,$$

where

$$v_p \in \mathrm{Hom}(H_s^{p,q}, H_s^{p-1,q+1}) ,$$

and we write

$$v_p^* \in \mathrm{Hom}(H_s^{p-1,q+1}, H_s^{p,q})$$

for the adjoint of $v_p$.

Since the variation of Hodge structure is polarized by the second bilinear relation, on the Hodge bundle $\mathcal{H}^{p,q}$ there is a Hermitian metric given by

$$<\psi,\eta> = (\sqrt{-1})^{p-q} Q(\psi,\overline{\eta}) ,$$

making the Hodge bundle $\mathcal{H}^{p,q}$ into a Hermitian vector bundle.

§1. *Connections and curvature*

We recall here the curvature formulas for subbundles and quotient bundles of a Hermitian vector bundle. These are worked out in Griffiths and Harris [6, pp. 71-79]; however, because we use a different subscript convention, we will sketch the computations below.

If $E$ is a Hermitian vector bundle with connection $D : \mathcal{A}^0(E) \to \mathcal{A}^1(E)$ and $e_1, \cdots, e_n$ is a unitary frame for $E$, then the connection matrix $\theta$

relative to this frame is the matrix of 1-forms given by

$$De_j = \Sigma \theta_{ij} e_i .$$

The curvature matrix $\Theta$ is the matrix of (1,1)-forms given by

$$D^2 e_j = \Sigma \Theta_{ij} e_i .$$

Since

$$\begin{aligned} D^2 e_j &= \Sigma\, d\theta_{ij} e_i - \Sigma\, \theta_{ij} \wedge \theta_{\ell i} e_\ell \\ &= \Sigma\, d\theta_{ij} e_i + \Sigma\, \theta_{\ell i} \wedge \theta_{ij} e_\ell \\ &= \Sigma\, (d\theta_{ij} + \theta_{ik} \wedge \theta_{kj}) e_i \quad \text{[replacing } \ell \text{ by } i \\ &\qquad\qquad\qquad\qquad\qquad \text{and } i \text{ by } k] , \end{aligned}$$

we have

$$\Theta = d\theta + \theta \wedge \theta .$$

Given an exact sequence

$$0 \to S \to E \to Q \to 0 ,$$

the *second fundamental form*

$$\sigma : \mathfrak{A}^0(S) \to \mathfrak{A}^1(Q)$$

is the composition of $D|_S$ followed by the projection to the quotient bundle $Q$. Using the Hermitian metric we may identify the quotient bundle $Q$ with $(S)^\perp$ and write $E = S \oplus Q$. Let $e_1, \cdots, e_r$ be a unitary frame for $S$ and $e_{r+1}, \cdots, e_n$ a unitary frame for $Q$. Then the connection matrix for $E$ is

$$\theta = \begin{bmatrix} \theta_S & -{}^t\overline{\sigma} \\ \sigma & \theta_Q \end{bmatrix}$$

and the curvature matrix is

$$\Theta = d\theta + \theta \wedge \theta = \begin{bmatrix} d\theta_S + \theta_S \wedge \theta_S - {}^t\bar{\sigma} \wedge \sigma & * \\ * & d\theta_Q + \theta_Q \wedge \theta_Q - \sigma \wedge {}^t\bar{\sigma} \end{bmatrix}.$$

Therefore, if $\Theta_S$ and $\Theta_Q$ are the curvature matrices of S and Q respectively, then

$$\Theta_S = \Theta|_S + {}^t\bar{\sigma} \wedge \sigma \tag{1}$$

and

$$\Theta_Q = \Theta|_Q + \sigma \wedge {}^t\bar{\sigma}. \tag{2}$$

Denote the metric on E by $< , >$. A connection D on E is the *metric connection* if

(a) $d<v,w> = <Dv,w> + <v,Dw>$, and

(b) the (0,1)-part of D is $\bar{\partial}$.

If $D = D' + D''$ is the decomposition of D into its (1,0)-part and (0,1)-part, then the metric connections on the subbundle S and the quotient bundle Q are

$$D_S = (D'|_S - \sigma) + \bar{\partial}$$

and

$$D_Q = D'|_Q + (\bar{\partial} + {}^t\bar{\sigma}).$$

PROPOSITION 3. *Let* $D = D' + D''$ *be the metric connection on a Hermitian vector bundle* E. *If* e *is a holomorphic section of* E, *then*

$$\partial\bar{\partial}<e,e> = <D'e,D'e> - <\Theta e,e>.$$

*Proof.* By comparing the types in Condition (a) of a metric connection, we have

$$\partial<v,w> = <D'v,w> + <v,D''w>,$$

$$\bar{\partial}<v,w> = <D''v,w> + <v,D'w>.$$

Therefore,

$$\partial\bar{\partial}\langle e,e\rangle = \partial\langle e,D'e\rangle$$
$$= \langle D'e,D'e\rangle + \langle e,D''D'e\rangle$$
$$= \langle D'e,D'e\rangle + \langle e,\Theta e\rangle \qquad \text{because } \Theta = D''D' \text{ for holomorphic sections}$$
$$= \langle D'e,D'e\rangle - \langle \Theta e,e\rangle \qquad \text{because } {}^t\bar{\Theta} = -\Theta\,.$$

q.e.d.

§2. *The curvature of Hodge bundles*

There are two metrics on the cohomology bundle $\mathcal{H}$: the Hodge metric $\langle\ ,\ \rangle$ and the nondegenerate indefinite Hermitian form $(-\sqrt{-1})^n Q(\ ,\ ^-)$, which we will call the *indefinite metric* and denote by $(\ ,\ )_i$. On the Hodge bundle $\mathcal{H}^{p,q}$ these two metrics differ by a sign

$$\langle\ ,\ \rangle = (-1)^p(\ ,\ )_i\,.$$

So the curvature of $\mathcal{H}^{p,q}$ is the same relative to either metric.

Let $\psi_p : \mathcal{F}^p \to \mathcal{H}/\mathcal{F}^p$ be the second fundamental form of $\mathcal{F}^p$ in $\mathcal{H}$ relative to the indefinite metric. By the horizontality of the Gauss-Manin connection, $\psi_p$ induces a map: $\mathcal{F}^p/\mathcal{F}^{p+1} \to \mathcal{F}^{p-1}/\mathcal{F}^p$, which we also denote by $\psi_p$. Thus the second fundamental form may be viewed as a map of Hodge bundles

$$\psi_p : \mathcal{H}^{p,q} \to \mathcal{H}^{p-1,q+1}\,.$$

A section $e$ of the Hodge bundle $\mathcal{H}^{p,q}$ is said to be *quasi-horizontal* if $\psi_p e = 0$.

PROPOSITION 4. *Let* $\langle\ ,\ \rangle$ *be the Hodge metric and* $e$ *and* $e'$ *two* $C^\infty$ *sections of the Hodge bundle* $\mathcal{H}^{p,q}$. *The curvature form* $\Theta$ *of* $\mathcal{H}^{p,q}$ *is given by*

$$\langle\Theta e,e'\rangle = \langle\psi_p e,\psi_p e'\rangle + \langle{}^t\bar{\psi}_{p+1}e,{}^t\bar{\psi}_{p+1}e'\rangle\,.$$

*Proof.* Since $\psi_p$ is defined relative to the indefinite metric, we will compute $\Theta_{\mathcal{H}^{p,q}}$ relative to the indefinite metric. Consider the defining exact sequences for the Hodge bundles:

$$0 \to \mathcal{F}^p \to \mathcal{H} \to \mathcal{H}/\mathcal{F}^p \to 0$$

$$0 \to \mathcal{F}^{p+1} \to \mathcal{F}^p \to \mathcal{H}^{p,q} \to 0\,.$$

By the curvature formulas for subbundles and quotient bundles ((1) and (2)),

$$\Theta_{\mathcal{F}^p} = {}^t\overline{\psi}_p \wedge \psi_p \quad \text{[because } \mathcal{H} \text{ is flat]}$$

$$\Theta_{\mathcal{H}^{p,q}} = {}^t\overline{\psi}_p \wedge \psi_p + \psi_{p+1} \wedge {}^t\overline{\psi}_{p+1}\,.$$

Therefore,

$$(\Theta_{\mathcal{H}^{p,q}}e,e)_i = -(\psi_p e, \psi_p e)_i - ({}^t\overline{\psi}_{p+1}e, {}^t\overline{\psi}_{p+1}e)_i\,.$$

In terms of the Hodge metric this formula is

$$(-1)^p\langle\Theta e,e\rangle = -(-1)^{p-1}\langle\psi_p e,\psi_p e\rangle - (-1)^{p+1}\langle{}^t\overline{\psi}_{p+1}e, {}^t\overline{\psi}_{p+1}e\rangle\,.$$

q.e.d.

A $(1,1)$-form

$$\omega = \frac{\sqrt{-1}}{2}\,\Sigma h_{ij}(z)\,dz_i \wedge d\overline{z}_j$$

is *positive* if $(h_{ij})$ is a positive definite Hermitian matrix. A matrix $A$ of $(1,1)$-forms is *positive* if $\langle Ae,e\rangle$ is a positive $(1,1)$-form for every vector $e$. The matrix $A$ is *negative* if $-A$ is positive. The notion of positive or negative semidefiniteness is defined analogously. A Hermitian vector bundle $E$ is positive or negative according as $i\Theta_E$ is positive or negative.

COROLLARY 5. *The Hodge bundles* $\mathcal{H}^{0,n}$ *are negative semidefinite.*

*Proof.* By the proposition

$$i\langle \Theta_{\mathcal{H}^{0,n}} e,e\rangle = i\langle {}^t\overline{\psi}_1 e, {}^t\overline{\psi}_1 e\rangle ,$$

which is a negative semidefinite (1,1)-form. q.e.d.

For the proof of the next proposition recall that a real-valued function $f$ is *plurisubharmonic* if its *Levi form* $i\partial\overline{\partial}f$ is positive semidefinite. By the maximum principle the only plurisubharmonic functions on a compact manifold are the constants.

PROPOSITION 6. *Over a compact base a holomorphic quasihorizontal section of* $\mathcal{H}^{p,q}$ *is holomorphic and flat as a section of* $\mathcal{H}$.

*Proof.* Combining the Levi form formula (Proposition 3) with the curvature formula (Proposition 4), we get

$$\partial\overline{\partial}\langle e,e\rangle = \langle D_p' e, D_p' e\rangle - \langle \psi_p e, \psi_p e\rangle - \langle {}^t\overline{\psi}_{p+1} e, {}^t\overline{\psi}_{p+1} e\rangle .$$

Because $e$ is quasihorizontal,

$$(*) \qquad i\partial\overline{\partial}\langle e,e\rangle = i\langle D_p' e, D_p' e\rangle - i\langle {}^t\overline{\psi}_{p+1} e, {}^t\overline{\psi}_{p+1} e\rangle .$$

This is positive semidefinite, since $D_p'$ has type (1,0) and ${}^t\overline{\psi}_{p+1}$ has type (0,1). So $\langle e,e\rangle$ is plurisubharmonic and hence constant. It then follows from (*) that $D_p' e = 0$ and ${}^t\overline{\psi}_{p+1} e = 0$. Since

$$\nabla e = D_p' e + \psi_p e = 0$$

and

$$\overline{\partial} e = D_p'' e \pm {}^t\overline{\psi}_{p+1} e = 0 ,$$

$e$ is flat and holomorphic as a section of $\mathcal{H}$. q.e.d.

APPLICATION 7 (Theorem of the Fixed Part). *Let* $\{\mathcal{H}_{\mathbf{Z}}, \mathcal{H}^{p,q}, \nabla, Q, S\}$ *be a polarized variation of Hodge structure over a compact base. If* $e$ *is a global holomorphic flat section of* $\mathcal{H}$, *then the* (p,q)*-components of* e *are also holomorphic and flat as sections of* $\mathcal{H}$.

*Proof.* Let $e = e_r + e_{r+1} + \cdots + e_n$ be the decomposition of e with $e_p \in \mathcal{H}^{p,q}$. Since e is flat,

$$0 = \nabla e = (D_r' + \psi_r)e_r + (D_{r+1}' + \psi_{r+1})e_{r+1} + \cdots .$$

By comparing types

$$\psi_r e_r = 0 .$$

Similarly, since e is holomorphic,

$$0 = D''e = (D_r'' \pm {}^t\overline{\psi}_{r+1})e_r + \cdots$$

so that

$$D_r'' e_r = 0 .$$

Therefore, $e_r$ is a holomorphic flat section of $\mathcal{H}^{r,n-r}$. By Proposition 6 it is holomorphic and flat as a section of $\mathcal{H}$. Next apply this argument to $e - e_r$, which we now know to be a global holomorphic flat section of $\mathcal{H}$. By induction all the (p,q)-components of e are holomorphic and flat as sections of $\mathcal{H}$. q.e.d.

For the next application, we note that if $\mathcal{V}$ and $\mathcal{V}'$ are two variations of Hodge structures over the same base S, then $\pi_1(S)$ acts on $\mathrm{Hom}(H_{\mathbf{C}}, H_{\mathbf{C}}')$ by

$$(g, \sigma) \mapsto g^{-1}\sigma g$$

for $g \in \pi_1(S)$ and $\sigma \in \mathrm{Hom}(H_{\mathbf{C}}, H_{\mathbf{C}}')$.

APPLICATION 8 (Rigidity Theorem). *Let* $\mathcal{V}$ *and* $\mathcal{V}'$ *be two variations of Hodge structures over a compact base space* S. *Suppose* $\mathcal{V}$ *and* $\mathcal{V}'$ *agree at one point* $s_0$ *via an isomorphism* $\mu_0 : H_{\mathbf{C}} \to H_{\mathbf{C}}'$ *and suppose the monodromy action is equivariant*:

$$\mu_0(ge) = g\mu_0(e) .$$

*Then* $\mathcal{V}$ *and* $\mathcal{V}'$ *are isomorphic everywhere.*

*Proof.* By parallel translation $\mu_0$ extends to a flat, possibly multivalued section $\mu$ of $\mathrm{Hom}(\mathcal{V}, \mathcal{V}')$, which is an isomorphism of vector spaces everywhere. The equivariance of the monodromy says precisely that this section $\mu$ is single-valued. By the theorem of the fixed part $\mu$ has type $(0,0)$ everywhere, so it preserves the Hodge filtrations everywhere. q.e.d.

APPLICATION 9 (Semi-simplicity). Deligne has proved that the monodromy representation of a variation of Hodge structure over a quasiprojective base is completely reducible over $\mathbb{Q}$ (Deligne [4, 4.2.6)]). To keep the presentation simple we will content ourselves with the following weaker statement.

THEOREM 9.1. *Let* $\mathcal{V} = \{\mathcal{H}_Z, \mathcal{H}^{p,q}, \nabla, Q, S\}$ *be a variation of Hodge structure over a compact base,* $\Gamma$ *the monodromy group, and* $H^\Gamma$ *the space of invariant cohomology classes. Then the restriction of the bilinear form* $Q$ *to* $H^\Gamma$ *is nondegenerate. It follows that if* $(H^\Gamma)^\perp$ *is the orthogonal complement of* $H^\Gamma$ *with respect to* $Q$ *, then*

$$H = H^\Gamma \oplus (H^\Gamma)^\perp$$

*as* $\Gamma$*-modules.*

*Proof.* Let $C: H \to H$ be the Weil operator. We claim that if $e = \Sigma\, e_p$ is an invariant cohomology class, then so are each Hodge component $e_p$ and $Ce$. Let $e(s)$ be the flat global section of the cohomology bundle $\mathcal{H}$ obtained by parallel translating $e$ and let $e(s) = \Sigma\, e_p(s)$ be its Hodge decomposition. By the theorem of the fixed part, each $e_p(s)$ is also flat. Therefore, $e_p(s)$ can be obtained from $e_p$ by parallel translation. Since $e_p(s)$ is a single-valued flat section, $e_p$ is an invariant cohomology class. If $\gamma \in \Gamma$, then

$$\begin{aligned}\gamma Ce &= \gamma\Sigma(\sqrt{-1})^{p-q}\, e_p \\ &= \Sigma(\sqrt{-1})^{p-q}\, e_p \\ &= Ce\ .\end{aligned}$$

This proves the claim. Since $Q(C\overline{e},e) > 0$ for $e \neq 0$, the bilinear form $Q$ is nondegenerate on $H^{\Gamma}$.

Because $\Gamma$ preserves $Q$, the orthogonal complement $(H^{\Gamma})^{\perp}$ is also invariant under $\Gamma$. Let $e \in H^{\Gamma} \cap (H^{\Gamma})^{\perp}$. Then $Q(C\overline{e},e) = 0$. So $e = 0$ and $H^{\Gamma} \cap (H^{\Gamma})^{\perp} = \{0\}$. Since $H^{\Gamma}$ and $(H^{\Gamma})^{\perp}$ have complementary dimensions in $H$, we get the direct sum decomposition

$$H = H^{\Gamma} \oplus (H^{\Gamma})^{\perp}\ . \qquad \text{q.e.d.}$$

§3. *Curvature of the Classifying Space*

Let $X$ be a Hermitian manifold. Denote by $T_x(X)$ the $(1,0)$-tangent space at $x$ in $X$. For a holomorphic tangent vector $\xi$ in $T_x(X)$ we define the *holomorphic sectional curvature* $K(\xi)$ as follows.

For the metric $ds^2 = h(z)dzd\overline{z}$ on the unit disk $\Delta$, the Gaussian curvature is defined to be

$$K = -\frac{1}{\pi}\frac{1}{h}\frac{\partial^2 \log h}{\partial z \partial \overline{z}}\ .$$

In general, if $X$ is a Hermitian manifold with metric $ds_X^2$ and $f : \Delta \to X$ is a holomorphic map such that

$$(*) \qquad f(0) = x \quad \text{and} \quad f_*(\partial/\partial z)_0 = \xi\ ,$$

then we set

$$K_f = \text{Gaussian curvature of } f^* ds_X^2 \text{ at the origin in } \Delta\ .$$

The *holomorphic sectional curvature* $K(\xi)$ is defined to be

$$K(\xi) = \sup_f K_f\ ,$$

where $f$ ranges over all holomorphic maps $f: \Delta \to X$ satisfying the initial conditions (*).

FACT 10 (Wu [12]). *If* $(R_{i\bar{j}k\bar{\ell}})$ *is the curvature tensor relative to an orthonormal frame near* $x$ *and* $\xi = (\xi^j)$ *is a unit vector relative to the same frame, then*

$$K(\xi) = \Sigma R_{i\bar{j}k\bar{\ell}} \xi^i \bar{\xi}^j \xi^k \bar{\xi}^\ell .$$

EXAMPLE 11. On the unit disk $\Delta$ the Poincaré metric

$$ds_\Delta^2 = \frac{2}{\pi} \frac{dz\, d\bar{z}}{(1 - |z|^2)^2}$$

has curvature $K \equiv -1$. On the upper half-plane $\mathfrak{h} = \{w = u + iv \mid v > 0\}$, which is conformally equivalent to $\Delta$, the Poincaré metric is given by

$$ds_{\mathfrak{h}}^2 = \frac{1}{2\pi} \frac{dw\, d\bar{w}}{v^2} .$$

THE AHLFORS LEMMA. *Let* $X$ *be a Hermitian manifold and* $T_h \subset T(X)$ *a holomorphic subbundle of the tangent bundle such that the holomorphic sectional curvature* $K(\xi) \leq -1$ *for all nonzero vectors* $\xi$ *in* $T_h$. *Then for any holomorphic map* $f: \Delta \to X$ *such that* $f_*(\partial/\partial z) \in (T_h)_{f(z)}$, *we have*

$$f^*(ds_X^2) \leq ds_\Delta^2 .$$

REMARK 12. The map $f$ in the Ahlfors Lemma is an integral manifold of $T_h$. The assertion is that such a map is *distance-decreasing*.

Recall that for the classifying space $D$, we have

$$\begin{array}{l} T(D) \subset \bigoplus_p \bigoplus_{r \geq 0} \mathrm{Hom}(\mathcal{H}^{p,q}, \mathcal{H}^{p-r,q+r}) \\ \cup \\ T_h(D) \subset \bigoplus_p \mathrm{Hom}(\mathcal{H}^{p,q}, \mathcal{H}^{p-1,q+1}) . \end{array}$$

Since each Hodge bundle $\mathcal{H}^{p,q}$ has a Hermitian metric, the tangent bundle T(D) inherits a Hermitian metric by functoriality.

BASIC COMPUTATION 13 ([3] and [7]). The holomorphic sectional curvature of the horizontal subbundle $T_h(D)$ is bounded above by a negative constant:

$$K(\xi) \leq -A < 0 \quad \text{for all} \quad \xi \in T_h(D)\,.$$

REMARK 14. This follows (nontrivially) from the curvature formula for the Hodge bundles (Proposition 4). It is convenient to normalize so that $A = 1$.

As a consequence of this computation and the Ahlfors Lemma, the classifying space D has all the function-theoretic properties of a bounded domain *relative to horizontal maps.*

A special case of the distance-decreasing property of the period map is the following.

PROPOSITION 15. *Let* $\phi : \Delta^* \to \Gamma\backslash D$ *be a period map and* $\phi_* : T\Delta^* \to TD = \oplus \mathrm{Hom}(\mathcal{F}^p, \mathcal{H}/\mathcal{F}^p)$ *its differential. Write* $\phi_* = \oplus(\phi_*)_p$. *If* e *is a section of* $\mathcal{F}^p$, *then*

$$\left\|(\phi_*)_p\left(\frac{\partial}{\partial r}\right)e\right\| \leq \frac{C\|e\|}{r \log \frac{1}{r}}$$

*for some constant* C.

*Proof.* Recall that the Poincaré metric on $\Delta^*$ is

$$\frac{2}{\pi}\,\frac{dr \otimes dr + r^2 d\theta \otimes d\theta}{\left(r \log \frac{1}{r}\right)^2}.$$

Hence,

$$\left\|\frac{\partial}{\partial r}\right\|_{\Delta^*} = \sqrt{\frac{2}{\pi}}\,\frac{1}{r \log \frac{1}{r}}\,.$$

By the distance-decreasing property of the period map,

$$\left\|\phi_* \frac{\partial}{\partial r}\right\| \leq \sqrt{\frac{2}{\pi}} \frac{1}{r \log(1/r)} .$$

Therefore,

$$\|(\phi_*)_p(\partial/\partial r) e\| \leq \|(\phi_*)_p(\partial/\partial r)\| \; \|e\| \leq \|\phi_* \partial/\partial r\| \; \|e\|$$

$$\leq \sqrt{\frac{2}{\pi}} \frac{\|e\|}{r \log(1/r)} . \qquad \text{q.e.d.}$$

APPLICATION 16 (Removable Singularity Theorem). *Let* $Z$ *be a subvariety of the algebraic variety* $S$ *and* $\phi : S - Z \to \Gamma \backslash D$ *a variation of Hodge structure such that the monodromy is locally finite around* $Z$. *Then* $\phi$ *extends to give an extended variation of Hodge structure* $\phi : S \to \Gamma \backslash D$.

For a proof see Griffiths [5, p. 156].

APPLICATION 17 (Monodromy Theorem). *Let* $\phi : \Delta^* \to \{T^n\} \backslash D$ *be a variation of Hodge structure over the punctured disk with*

$$\phi_*(\textit{generator of } \pi_1(\Delta^*)) = T .$$

*Then all the eigenvalues of* $T$ *are roots of unity.*

The proof that follows is due to Borel. We will need to quote the theorem of Kronecker that an algebraic integer all of whose conjugates have absolute value $1$ is a root of unity.

*Proof.* Since $T$ can be represented as an integer matrix, the eigenvalues of $T$ are algebraic integers. By Kronecker's theorem, it suffices to show that their conjugates are all of absolute value $1$.

Because the upper half-plane $\mathfrak{h}$ is simply connected, the map $\phi$ can be lifted as in the following commutative diagram

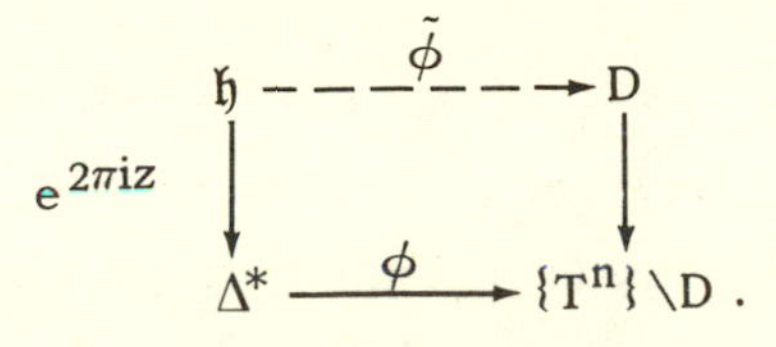

Note that $\tilde{\phi}(z+1) = T\tilde{\phi}(z)$. Let $\{z_n\}$ be a sequence of points in $\mathfrak{h}$ and $p_0$ the base point in $D$. Denote by $\rho_X(x,y)$ the distance between the points $x$ and $y$ on a Hermitian manifold $X$. Then

$$\rho_D(\tilde{\phi}(z_n+1), \tilde{\phi}(z_n))$$

$$= \rho_D(T\tilde{\phi}(z_n), \tilde{\phi}(z_n))$$

$$= \rho_D(Tg_n p_0, g_n p_0) \quad \text{for some } g_n \in G_R, \text{ because } G_R \text{ acts transitively on } D$$

$$= \rho_D(g_n^{-1}Tg_n p_0, p_0) \quad \text{by the } G_R\text{-invariance of the metric on } D.$$

On the other hand, by the distance-decreasing property of horizontal maps,

$$\rho_D(\tilde{\phi}(z_n+1), \tilde{\phi}(z_n)) \leq \rho_{\mathfrak{h}}(z_n, z_n+1) = \frac{1}{\operatorname{Im} z_n}.$$

So if $\{z_n\}$ is chosen so that $\operatorname{Im} z_n \to 0$, then

$$\rho_D((g_n^{-1} Tg_n)p_0, p_0) \to 0 \quad \text{as} \quad n \to \infty.$$

Because $D$ is a metric space,

$$\lim_{n\to\infty} (g_n^{-1} Tg_n)p_0 = p_0.$$

It follows that $g = \lim_{n\to\infty} g_n^{-1} Tg_n$ is in the stabilizer $H$ of $p_0$. Because $H$ is compact, the eigenvalues of $g$ and hence of $T$ all have absolute value $1$. q.e.d.

REMARK 18. A matrix $T$ satisfying

$$(T^N - I)^{k+1} = 0$$

for some integers $N$ and $k$ is said to be *quasi-unipotent.* The least such $k$ is called the *index* of quasi-unipotency. An equivalent formulation of the monodromy theorem is that $T$ is quasi-unipotent. Schmid has shown that for a variation of Hodge structure of weight $n$ over the punctured disk $\Delta^*$, the index of quasi-unipotency is at most $n$. See Schmid [11].

REMARK 19. In the geometric case, the monodromy theorem is due to Landman [10] and Katz [9].

## §4. *Algebraization and regular singular points*

Let $\mathcal{V} = \{\mathcal{H}_Z, \mathcal{H}^{p,q}, \nabla, Q, S\}$ be a variation of Hodge structure over an algebraic base $S$. A priori the Hodge bundles are only holomorphic bundles. Of course, if the base is a smooth compact algebraic variety, then by Serre's GAGA principle, the Hodge bundles are also algebraic bundles. It turns out that even when the base is noncompact, the Hodge bundles can be given an algebraic structure which is uniquely characterized by a growth condition on its sections. The proof of this algebraization theorem is based on a curvature computation which we will briefly explain.

First, some notations and terminologies. Let $S$ be an algebraic variety, not necessarily compact. A *smooth compactification* of $S$ is a smooth compact variety $\overline{S}$ such that $D = \overline{S} - S$ is a divisor with normal crossings. Thus the neighborhoods at infinity are punctured polycylinders $P^* = (\Delta^*)^k \times \Delta^{n-k}$. Such a smooth compactification exists by Hironaka. We let $\eta_{P^*}$ be the Poincaré metric on $P^*$. Recall that for a polarized Hodge structure $\{H_Z, H^{p,q}, Q\}$ a positive definite metric can be defined on $H^{p,q}$ by setting

$$<\psi,\eta>_p = (\sqrt{-1})^{p-q} Q(\psi,\overline{\eta}) .$$

The *Hodge length* of $v$ in $H$, $v = \sum_{p=1}^{n} v^{p,q}$, is then

$$\|v\| = \Big(\sum_{p=1}^{n} <v^{p,q},v^{p,q}>_p\Big)^{1/2} .$$

The algebraicity of the Hodge bundles is a consequence of the following general theorem.

THEOREM 20. *With the notations above, let* $E \to S$ *be a Hermitian vector bundle whose curvature satisfies*

$$-C\eta_{P^*} \leq \frac{\langle \theta e, e \rangle}{\langle e, e \rangle} \leq C\eta_{P^*}$$

*locally at infinity. Then there exists a unique algebraic structure on* $E$ *whose algebraic sections* $e(s)$ *are characterized by moderate growth, that is,*

$$\|e(s)\| = 0(|s|^{-a})$$

*for a local parameter* $s$ *on* $S$ *and for some positive integer* $a$.

The point is that by the Ahlfors Lemma the curvatures of the Hodge bundles satisfy the inequality of the theorem and so over an algebraic base, $\mathcal{H}^{p,q}$ has the prescribed algebraic structure. To get an algebraic structure with moderate growth on $\mathcal{F}^p$ one first shows that the extension class of

$$0 \to \mathcal{F}^{p+1} \to \mathcal{F}^p \to \mathcal{H}^{p,q} \to 0$$

is algebraic. By induction we may assume $\mathcal{F}^{p+1}$ algebraic. The algebraicity of $\mathcal{F}^p$ then follows from that of $\mathcal{H}^{p,q}$, $\mathcal{F}^{p+1}$, and the extension class.

REMARK 21. When the variation of Hodge structure comes from geometry, there is an a priori algebraic structure on the Hodge bundles $\mathcal{H}^{p,q}$. For if $\pi : \mathcal{X} \to S$ is a family of polarized algebraic varieties, then $\mathcal{H}^{p,q}$ can be identified with the direct image sheaf

$$\mathcal{H}^{p,q} \simeq \mathcal{R}^p \pi_* \Omega^p_{\mathcal{X}/S} ,$$

where $\Omega^p_{\mathcal{X}/S} = \Lambda^p \Omega^1_{\mathcal{X}/S}$ is the sheaf of algebraic p-forms along the fiber. However, it can be shown that this a priori algebraic structure agrees with the algebraic structure with moderate growth given by the theorem (see Cornalba and Griffiths [2]).

*Regular singular points*

To explain intuitively the theorem on regular singular points, we consider a family of smooth projective varieties $f: \mathfrak{X} \to \Delta^*$. Let $\omega(s) \epsilon H^k(X_s)$ be a smoothly varying collection of $C^\infty$ k-forms on the fibers of the family and $c(s_0)$ a k-cycle in $X_{s_0}$. Horizontally displace $c(s_0)$ to obtain k-cycles $c(s) \epsilon H_k(X_s, Z)$. Although $c(s)$ is in general a multivalued section of $\bigcup_s H_k(X_s, Z)$, over an angular sector $\Delta^*$ it has single-valued branches. The assertion of the theorem is that *if the Hodge length of* $\omega(s)$ *has moderate growth, i.e.,*

$$\|\omega(s)\| = 0(|s|^{-a}) \text{ for some positive integer } a,$$

*then over any angular sector the period of* $\omega(s)$ *relative to a single-valued branch of* $c(s)$ *also has moderate growth*:

$$\left|\int_{c(s)} \omega(s)\right| = 0(|s|^{-a}).$$

To formalize this, we first make a few definitions.

DEFINITION 22. Let $\pi: E \to \Delta^*$ be a Hermitian vector bundle. A holomorphic section $e$ of $E$ is said to be *meromorphic at the origin* if

$$\|e(s)\| = 0(|s|^{-a}) \text{ for some positive integer } a.$$

Such a section is also called a *meromorphic section* of $E$.

DEFINITION 23. Let $\pi: E \to \Delta^*$ be a holomorphic vector bundle with a connection $\nabla$, and let $\{v_1, \cdots, v_n\}$ be a (possibly multivalued) flat frame for $E$. The connection is said to have *regular singular point* at the origin (relative to the flat frame $\{v_1, \cdots, v_n\}$) if on any angular sector the coefficients of any meromorphic section $e$ relative to the flat frame $\{v_1,\cdots,v_n\}$ have at most poles at the origin; in other words, if $e(s) = \Sigma b_i(s)v_i$, then the $b_i(s)$ are multi-valued meromorphic functions at the origin.

THEOREM 24. *The Gauss-Manin connection of the cohomology bundle* $\pi : \mathcal{H} \to \Delta^*$ *of a polarized variation of Hodge structure* $\mathcal{V} = \{\mathcal{H}_Z, \mathcal{H}^{p,q}, \nabla, Q, \Delta^*\}$ *has a regular singular point at the origin.*

The proof of this theorem depends on an estimate for the length of a holomorphic flat section of $\pi : \mathcal{H} \to \Delta^*$. Here by a flat section we mean a flat section with respect to the Gauss-Manin connection $\nabla$. If $D$ is the metric connection on $\mathcal{H}$ relative to the indefinite metric $(-\sqrt{-1})^n Q(\ ,\ ^-)$. Then the Gauss-Manin connection is the $(1,0)$-part $D'$ of $D$. Thus a holomorphic flat section is also flat with respect to $D$ (but not with respect to the metric connection of the Hodge metric on $\mathcal{H}$).

Recall that the metric connection $D_p = D_p' + D_p''$ on the Hodge bundle $\mathcal{H}^{p,q}$ is related to the metric connection $D = D' + D''$ on $\mathcal{H}$ by

$$D_p' + \psi_p = D'|_{\mathcal{H}^{p,q}}$$

$$D_p'' \pm {}^t\psi_{p+1} = D''|_{\mathcal{H}^{p,q}} .$$

PROPOSITION 25. *On any angular sector of* $\Delta^*$ *the Hodge length of a holomorphic flat section* $e$ *of* $\mathcal{H} \to \Delta^*$ *satisfies*

$$C_1\left(\log \frac{1}{|s|}\right)^{-k} \le \|e(s)\|^2 \le C_2\left(\log \frac{1}{|s|}\right)^{k}$$

*where* $C_1$, $C_2$, $k$ *are positive constants.*

*Proof.* Let $e = \Sigma e_p$ be the Hodge decomposition of $e$, where $e_p$ is the $(p,n-p)$-component. Because $e$ is flat,

$$D_p' e_p = -\psi_{p+1} e_{p+1} .$$

Because e is holomorphic,

$$D''_p e_p = \pm {}^t\psi_{p-1} e_{p-1} .$$

Thus the radial derivative of the Hodge length is

$$\begin{aligned}\frac{\partial}{\partial r} <e,e> &= \Sigma \frac{\partial}{\partial r} <e_p, e_p> \\ &= 2 \text{ Re } \Sigma <D\left(\frac{\partial}{\partial r}\right) e_p, e_p> \\ &= -2 \text{ Re } \Sigma <\psi_{p+1}\left(\frac{\partial}{\partial r}\right) e_{p+1}, e_p> \pm <{}^t\psi_{p-1}\left(\frac{\partial}{\partial r}\right) e_{p-1}, e_p> .\end{aligned}$$

By the Schwarz inequality,

$$|<\psi_{p+1}\left(\frac{\partial}{\partial r}\right) e_{p+1}, e_p>| \leq \|\psi_{p+1}\left(\frac{\partial}{\partial r}\right) e_{p+1}\| \, \|e_p\| .$$

By the distance-decreasing property of the period map,

$$\left\|\psi_{p+1}\left(\frac{\partial}{\partial r}\right) e_{p+1}\right\| \leq \frac{C\|e_{p+1}\|}{r \log \frac{1}{r}} .$$

Therefore,

$$\left|<\psi_{p+1}\left(\frac{\partial}{\partial r}\right) e_{p+1}, e_p>\right| \leq \frac{C\|e_{p+1}\| \, \|e_p\|}{r \log \frac{1}{r}} \leq \frac{C\|e\|^2}{r \log \frac{1}{r}}$$

So the radial derivative satisfies

$$\frac{\left|\frac{\partial}{\partial r} <e,e>\right|}{<e,e>} \leq \frac{k}{r \log \frac{1}{r}}$$

for some constant $k$. Integrating this inequality with respect to $r$ yields the desired estimate. q.e.d.

## REFERENCES

[1] M. Cornalba and P. Griffiths, "Analytic cycles and vector bundles on noncompact algebraic varieties," Inv. math. 28 (1975), pp. 1-106.

[2] M. Cornalba and P. Griffiths, "Some transcendental aspects of algebraic geometry," in Proceedings of Symposia in Pure Mathematics, Vol. 29(1975), A.M.S., pp. 3-10.

[3] P. Deligne, "Le travaux de Griffiths," Séminaire Bourbaki Exp. 376(1970), Lecture Notes in Math. 180(1971), pp. 213-237.

[4] ———, "Théorie de Hodge II," Publ. Math. I.H.E.S. 40(1972), pp. 5-57.

[5] P. Griffiths, "Periods of integrals on algebraic manifolds III," Publ. Math. I.H.E.S. 38(1970), pp. 125-180.

[6] P. Griffiths and J. Harris, "*Principles of Algebraic Geometry,*" John Wiley and Sons, New York, 1978.

[7] P. Griffiths and W. Schmid, "Locally homogeneous complex manifolds," Acta Math. 123(1969), pp. 253-302.

[8] ———, "Recent developments in Hodge theory: a discussion of techniques and results," in Discrete Subgroups of Lie Groups and Applications to Moduli, Oxford University Press, 1973, pp. 31-127.

[9] N. Katz, "Nilpotent connections and the monodromy theorem: applications of a result of Turritin," Publ. Math. I.H.E.S. 39(1971), pp. 175-232.

[10] A. Landman, "On the Picard-Lefschetz transformation for algebraic manifolds acquiring general singularities," Transaction of the A.M.S. 181(1973), pp. 89-126.

[11] W. Schmid, "Variation of Hodge structure: the singularities of the period mapping," Inv. math. 22(1973), pp. 211-319.

[12] H. Wu, "A remark on holomorphic sectional curvature," Ind. Math. Jour., vol. 22(1973), pp. 1103-1108.

[13] A Remark on Holomorphic Sectional Curvature, Ind. Math. Jour., vol 22(1973), pp. 1103-1108.

REMARK ADDED IN PROOF: There is increasing evidence that the curvature properties of the Hodge bundles play an essential role in the classification theory of algebraic varieties. We give here a short bibliography to this interesting development:

[1] T. Fujita, On Kähler fibre spaces over curves, J. Math. Soc. Japan 30(1978), pp. 779-794.

[2] Y. Kawamata, Characterization of abelian varieties, Comp. Math. 43(1981), pp. 253-276.

[3] ———, Kodaira dimension of algebraic fibre spaces over curves, Inv. Math. (1982).

[4] E. Viehweg, Die additivität der Kodaira dimension für projektive Faserräume über Varietäten des allgemeinen Typs, Jour. reine & angew. Math. 330(1982), pp. 132-142.

[5] ————, Weak positivity and the additivity of the Kodaira dimension, II: the local Torelli map, manuscript.

PHILLIP GRIFFITHS
MATHEMATICS DEPARTMENT
HARVARD UNIVERSITY
CAMBRIDGE, MA 02138

LORING TU
MATHEMATICS DEPARTMENT
JOHNS HOPKINS UNIVERSITY
BALTIMORE, MD 21218

## Chapter III
# INFINITESIMAL VARIATION OF HODGE STRUCTURE

Phillip Griffiths
Written by Loring Tu

Giving a polarized Hodge structure $\{H_{\mathbf{Z}}, H^{p,q}, Q\}$ of weight one is equivalent to giving a polarized Abelian variety $(A,\omega)$, as follows. We set the complex torus $A$ to be $H^{0,1}/H_{\mathbf{Z}}$. Via the identification

$$\begin{aligned}\mathrm{Hom}(\Lambda^2 H_{\mathbf{Z}}, \mathbf{Z}) &\simeq \mathrm{Hom}(\Lambda^2 H_1(A,\mathbf{Z}),\mathbf{Z}) \\ &\simeq \mathrm{Hom}(H_2(A,\mathbf{Z}),\mathbf{Z}) \\ &\simeq H^2(A,\mathbf{Z}) \,,\end{aligned}$$

the polarization $Q$ corresponds to a class $[\omega] \in H^2(A,\mathbf{Z})$. It can be checked that as a consequence of the two bilinear relations, $[\omega]$ is a positive (1,1)-class. In case the polarization is unimodular, the class $[\omega]$ defines a divisor $\Theta$, unique up to translation, called the *theta divisor* of $A$. It is the geometric object $\Theta$ that plays the major role in classical Hodge theory.

Much of the formal aspects of classical Hodge theory has been extended to higher weights; for example, the asymptotic behavior of a family of Hodge structures. However, one may argue that the applications to geometry have fallen short of expectations, as evidenced by the lack of progress on higher codimensional cycles. We suggest that one reason for this is the following.

OBSERVATION 1. A general Hodge structure of weight $n \geq 2$ (where $h^{2,0} \geq 2$ if $n = 2$) does not come from geometry, so that there is no "natural" way of assigning a geometric object such as $\Theta$ to it.

EXPLANATION. Let $D \subset \check{D}$ be the classifying space for the polarized Hodge structures $\{H_{\mathbf{Z}}, H^{p,q}, Q\}$ and $T_h(\check{D})$ the horizontal subbundle, given by

$$\{\xi \in T(\check{D}) \mid \xi F^p \subset F^{p-1}\}.$$

Set

$$I^{(1)} = \mathcal{O}(T_h(\check{D})^{\perp}) \subset \Omega^1_{\check{D}}.$$

Define $I \subset \Omega^{\bullet}_{\check{D}}$ to be the sheaf of ideals generated by the 1-forms $\theta \in I^{(1)}$ and their exterior derivatives $d\theta$. Then $I$ is a $G_{\mathbf{C}}$-invariant differential system on $D$ and hence induces a differential system on $\Gamma\backslash D$, which we also denote by $I$. In this context the horizontality condition in the definition of a variation of Hodge structure $\phi: S \to \Gamma\backslash D$ amounts to requiring that $\phi$ be an integral manifold of the differential system $I$; that is,

$$\phi^*(\theta) = 0 \quad \text{for all} \quad \theta \quad \text{in} \quad I.$$

The point is that if the weight $n \geq 2$ (assuming $h^{2,0} \geq 2$ if $n = 2$), then $I \neq (0)$. Thus $\phi(S)$ can never contain an open subset of $\Gamma\backslash D$. q.e.d.

Since the differential system $I$ appears to be causing the trouble, we will try and use it to extract some geometry. This leads to the topic of today's talk, *the infinitesimal variation of Hodge structure*, a work which is still in the experimental stage.

DEFINITION 2. An *infinitesimal variation of Hodge structure* $V = \{H_{\mathbf{Z}}, H^{p,q}, Q, T, \delta\}$ of weight $n$ is given by a polarized Hodge structure $\{H_{\mathbf{Z}}, H^{p,q}, Q\}$ of weight $n$ together with a vector space $T$ and a linear map (here $q = n-p$)

$$\delta: T \to \bigoplus_{1 \leq p \leq n} \mathrm{Hom}(H^{p,q}, H^{p-1,q+1})$$

satisfying

(1) $\delta_{p-1}(\xi_1)\delta_p(\xi_2) = \delta_{p-1}(\xi_2)\delta_p(\xi_1)$ for $\xi_1, \xi_2 \in T$,

(2) $Q(\delta(\xi)\psi, \eta) + Q(\psi, \delta(\xi)\eta) = 0$ for $\psi \in H^{p,q}, \eta \in H^{q+1,p-1}$.

In particular, an infinitesimal variation of Hodge structure is an *integral element* of the horizontal differential system I on D.

EXAMPLE 3. If $\mathfrak{X} \to S$ is a family of polarized algebraic varieties and $\phi : S \to \Gamma \backslash D$ is its period map, then the differential of the period map

$$\phi_* : T_{s_0}(S) \to \bigoplus_{1 \le p \le n} \mathrm{Hom}(H^{p,q}, H^{p-1,q+1})$$

gives rise to an infinitesimal variation of Hodge structure in which $T = T_{s_0}(S)$ and $\delta = \phi_*$.

Whereas a polarized variation of Hodge structure has no algebraic invariants (because $G_{\mathbb{R}}$ acts transitively on the classifying space D), an infinitesimal variation of Hodge structure has too many. Of these, five have thus proved useful in geometric problems.

CONSTRUCTION #1. *The kernel of the* n-*th coiterate of the differential.*

For $\xi_1, \cdots, \xi_n$ in T consider the map

$$\delta(\xi_1) \cdots \delta(\xi_n) : H^{n,0} \to H^{0,n}$$

It follows from (1), (2), and the symmetry of Q that

$$\begin{aligned} Q(\delta(\xi_1) \cdots \delta(\xi_n)\psi, \eta) &= (-1)^n Q(\psi, \delta(\xi_n) \cdots \delta(\xi_1)\eta) \\ &= Q(\delta(\xi_1) \cdots \delta(\xi_n)\eta, \psi) . \end{aligned}$$

So in fact $\delta(\xi_1) \cdots \delta(\xi_n)$ is in $\mathrm{Hom}^{(s)}(H^{n,0}, H^{0,n})$, the symmetric maps from $H^{n,0}$ to $(H^{n,0})^*$. Define

$$\delta^n(\xi_1, \cdots, \xi_n) = \delta(\xi_1) \cdots \delta(\xi_n) .$$

By (1), $\delta^n$ is symmetric in its arguments and so induces a map

$$\delta^n : \mathrm{Sym}^n T \to \mathrm{Hom}^{(s)}(H^{n,0}, H^{0,n}) ,$$

called the n-th *iterate of the differential.* Note that

$$\mathrm{Hom}^{(s)}(H^{n,0},H^{0,n}) = \mathrm{Sym}^2(H^{n,0})^* .$$

The dual of $\delta^n$ is then

$$(\delta^n)^* : \mathrm{Sym}^2 H^{n,0} \to \mathrm{Sym}^n T^* .$$

Our first invariant is

$$\mathcal{Q}(V) = \ker(\delta^n)^* ;$$

it may be viewed as a linear system of quadrics on $PH^{0,n}$.

EXAMPLE 4. Let $C$ be a curve of genus $g$ and $T = H^1(C,\Theta)$. It is well known that this $T$ effectively functions as the tangent space to the moduli space $\mathfrak{M}_g$ at $C$. Let

$$\delta : T \to \mathrm{Hom}^{(s)}(H^{1,0},H^{0,1})$$

be the differential of the period map at $C$. This gives the *universal infinitesimal variation of Hodge structure* of $C$. Then

$$\delta^* : \mathrm{Sym}^2 H^{1,0} \to T^* .$$

Note that $H^{1,0} = H^0(C,K)$ and $T^* = H^0(C,K^2)$. Therefore,

$$\delta^* : \mathrm{Sym}^2 H^0(C,K) \to H^0(C,K^2)$$

is the obvious map and

$$\mathcal{Q}(V) = \text{quadrics through the canonical curve } \phi_K(C) .$$

COROLLARY 5. *A general curve of genus* $g \geq 5$ *can be reconstructed from its universal infinitesimal variation of Hodge structure.*

*Proof.* A general curve of genus $g \geq 5$ is not hyperelliptic, trigonal, or a plane quintic. By a theorem of Babbage, Enriques, and Petri the canonical image of such a curve is the intersection of all the quadrics through it. Hence, the base locus of $\mathcal{Q}(V)$ is the canonical curve of $C$. q.e.d.

COROLLARY 6 (Generic Global Torelli). *For* $g \geq 5$ *the extended period map*

$$\phi : \mathfrak{M}_g \to \phi(\mathfrak{M}_g) \subset \Gamma \backslash \mathfrak{H}_g$$

*has degree one.*

*Proof.* Suppose $\phi$ has degree $d \geq 1$. Let $Z \in \phi(\mathfrak{M}_g)$ be a regular value. Then $\phi$ is a local isomorphism around $\phi^{-1}(Z) = \{C_1, \cdots, C_d\}$. It follows that $C_1, \cdots, C_d$ all have the same universal infinitesimal variation of Hodge structure. By Corollary 5 all the $C_i$'s are equal. q.e.d.

Of course, this theorem is a consequence of the well-known global Torelli theorem for curves, but the virtue of this proof is that it does not use the theta divisor $\Theta$ and so has a chance to generalize. In general, speaking philosophically, if a variety can be reconstructed from its infinitesimal variation of Hodge structure, then the generic global Torelli theorem holds (cf. Chapters XII and XIII).

CONSTRUCTION #2. *Degeneracy loci of iterates of the differential.*

In this construction we consider the (n–2p)-th iterate $\delta^{n-2p}$ of the differential. As in the previous construction

$$\delta^{n-2p} : \mathrm{Sym}^{n-2p} T \to \mathrm{Hom}^{(s)}(H^{n-p,p}, H^{p,n-p}) .$$

This induces a map

$$f : T \to \mathrm{Hom}^{(s)}(H^{n-p,p}, H^{p,n-p})$$

$$f(\xi) = \delta^{n-2p}(\xi, \cdots, \xi) = \delta(\xi) \cdots \delta(\xi) .$$

Define

$$\Sigma_{p,k} = \{\xi \in \mathbf{P}T \mid \mathrm{rank}\, f(\xi) \leq k\} .$$

NOTATION.

$\Sigma_k = \Sigma_{0,k}$ = locus where the n-th iterate has rank $\leq k$.

$\Sigma = \Sigma_{h^{n,0}-1}$ = locus where the n-th iterate drops rank.

EXAMPLE 7. We keep the notations of Example 4. Fix $n = 1$ and $p = 0$. Then

$$f(\xi) = \delta(\xi) : H^0(C,K) \to H^1(C,\Theta) ,$$

and

$$\Sigma_{0,1} = \{\xi \in PH^1(C,\Theta) \mid \operatorname{rank} \delta(\xi) \leq 1\} .$$

PROPOSITION 8. *For a general curve* C *of genus* $g \geq 5$ *the rank one degeneracy locus is the bicanonical image*:

$$\Sigma_{0,1} = \phi_{2K}(C) .$$

EXPLANATION. For x in C we have $\xi_x = \phi_{2K}(x)$ in $PH^1(C,\Theta)$. It is an easy lemma that

$$\ker \delta(\xi_*) \supseteq H^0(C,K(-x)) .$$

Since $H^0(C,K(-x))$ is a hyperplane in $H^0(C,K)$,

$$\operatorname{rank} \delta(\xi_*) \leq 1 .$$

This gives a map

$$\phi_{2K}(C) \to \Sigma_{0,1} .$$

The remainder of the proof of the proposition may be found in [4, §Vc)]. From this result we get another proof of the generic global Torelli theorem for curves.

CONSTRUCTION #3. *Annihilator of a Hodge class.*

Given an infinitesimal variation of Hodge structure $V = \{H_{\mathbf{Z}}, H^{p,q}, Q, T, \delta\}$ of even weight $n = 2m$ and a Hodge class $\gamma \in H^{m,m}_{\mathbf{Z}}$, we set $H^{m+k,m-k}(-\gamma) = \{\psi \in H^{m+k,m-k} \mid Q(\delta^k(\xi)\psi,\gamma) = 0 \text{ for all } \xi \text{ in } T\}$.

EXAMPLE 9. Let S be a smooth surface of degree $d \geq 5$ in $\mathbf{P}^3$ and $\omega \in H^2(S,\mathbf{Z})$ its hyperplane class. A *Hodge line* is by definition a cohomology class $\lambda \in H^{1,1}(S) \cap H^2(S,\mathbf{Z})$ satisfying

$$\begin{cases} \lambda^2 = 2-d \\ \lambda \cdot \omega = 1 \, . \end{cases}$$

We will view $H^{2,0}(-\lambda)$ as a subspace of $H^0(S,\Omega^2)$, the holomorphic 2-forms on $S$. It is a theorem (cf. [4]) that the base locus of the forms in $H^{2,0}(-\lambda)$ is a line $\Lambda$ with fundamental class $\lambda$:

$$\bigcap_{\psi \in H^{2,0}(-\lambda)} (\psi) = \Lambda \, . \tag{9.1}$$

REMARK 10. This result generalizes to the following (*loc. cit.*). Let $C$ be a smooth curve of degree $d$ and genus $g$ in $\mathbf{P}^3$ such that $h^1(C,N_{C/\mathbf{P}^3}) = 0$ and let $S$ be a surface of sufficiently large degree $\geq m(d,g)$ containing $C$. If $V$ is the universal infinitesimal variation of Hodge structure of $S$ and $\gamma$ is the class of $C$, then

$$C = \bigcap_{\psi \in H^{2,0}(-\lambda)} (\psi) \, ;$$

that is, the curve $C$ may be reconstructed from $V$ and $\gamma$. A similar statement holds for a curve in $\mathbf{P}^r$.

As an application of (9.1) we will sketch in Chapter V a proof of the following theorem.

THEOREM 11. *Any smooth surface in* $\mathbf{P}^3$ *with the same universal infinitesimal variation of Hodge structure as the Fermat surface* $F_d = \{x_0^d + x_1^d + x_2^d + x_3^d = 0\}$ *of degree* $d \geq 5$ *is projectively equivalent to* $F_d$. *Furthermore,*

$$\mathrm{Aut}(V(F_d)) = \mathrm{Aut}(F_d) \, .$$

We will motivate and then give a result concerning Construction #1. Given a family of polarized algebraic varieties, which we think of as $\{X_s\}_{s \in S}$, there is the Kodaira-Spencer map at $s = s_0$

$$\rho : T_{s_0}(S) \to H^1(X,\Theta) \, , \qquad X = X_{s_0} \, .$$

There is also the cup product map

$$\kappa : H^1(X,\Theta) \to \bigoplus_{1 \leq p \leq n} \mathrm{Hom}(H^{p,q}(X), H^{p-1,q+1}(X)) .$$

It can be shown ([2]) that the differential of the period map is

$$\delta = \kappa \circ \rho : T_{s_0}(S) \to \bigoplus_{1 \leq p \leq n} \mathrm{Hom}(H^{p,q}(X), H^{p-1,q+1}(X)) .$$

Because $\mathfrak{X} \to S$ is a projective family, the hyperplane class $\omega$ is constant. Therefore,

$$0 = \delta(\xi)(\omega) = \rho(\xi) \wedge \omega \quad \text{for any} \quad \xi \in T_{s_0}(S) .$$

Recalling the definition of the primitive cohomology, it follows that $\delta(\xi)$ carries a primitive class to a primitive class. Setting $T = T_{s_0}(S)$, and $H^{p,q} = H^{p,q}_{\mathrm{prim}}(X)$, we then have

$$\delta = \kappa \circ \rho : T \to \bigoplus_{1 \leq p \leq n} \mathrm{Hom}(H^{p,q}, H^{p-1,q+1}) .$$

This infinitesimal variation of Hodge structure $V = \{H_{\mathbf{Z}}, H^{p,q}, Q, T, \delta\}$ is said to *arise from geometry*.

The motivation for Construction #1 is as follows. Except when $X$ is a curve, $H^1(X,\Theta)$ is in general not particularly geometric. However, there is a map

$$\rho^n : \mathrm{Sym}^n H^1(X,\Theta) \to H^n(X, \Lambda^n\Theta) .$$

The map $\rho^n$ is the composition

$$\otimes^n H^1(X,\Theta) \to H^n(X, \otimes^n\Theta) \to H^n(X, \Lambda^n\Theta)$$

of two alternating maps and so is symmetric. It is straightforward to verify that the following diagram is commutative:

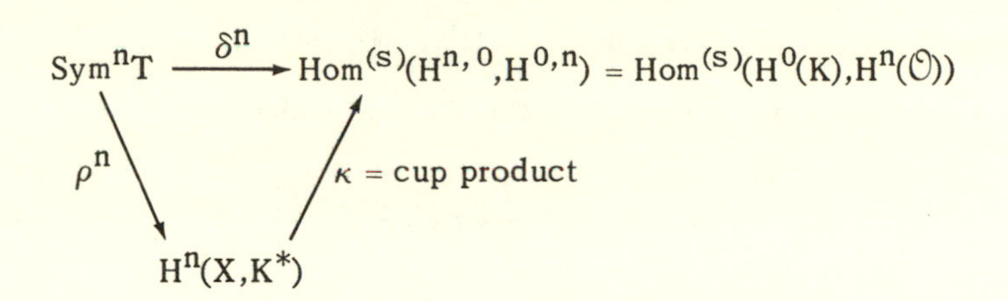


The dual of this is our *basic diagram*:

(12)

$$\begin{array}{ccc} \mathrm{Sym}^2 H^0(X,K) & \xrightarrow{(\delta^n)^*} & \mathrm{Sym}^n T^* \\ & {}_{\mu}\searrow \quad \nearrow_{\lambda} & \\ & H^0(X,K^2) \;, & \end{array}$$

where $\mu$ is the dual of the cup product and is given by multiplication. To get some feeling for what is going on, we assume that $\mu$ is onto. Then quite formally we get the exact sequence

(*) $$0 \to \ker \mu \to \ker \delta^{n*} \to \ker \lambda \to 0 \, .$$

Note that

$$\ker \mu = I_{\phi_K(X)}(2) = \{\text{quadrics through the canonical image } \phi_K(X)\} \, .$$

So (*) may be rewritten as

$$0 \to I_{\phi_K(X)}(2) \to \mathcal{Q}(V) \to \ker \lambda \to 0 \, .$$

When $n = 1$ and $\mathfrak{X} \to S$ is the Kuranishi family of curves, $\ker \lambda = (0)$ and we have

$$\mathcal{Q}(V) \simeq I_{\phi_K(X)}(2)$$

as in Example 4. In the general situation, to be able to interpret geometrically the *infinitesimal Schottky relations* $\mathcal{Q}(V)$ one must first understand $\ker \lambda$. We give a partial result in this direction.

For $X = X_{s_0} \subset P^r$, we take $T = H^0(X, N_{X/P^r})$. Consider the composition of the Gauss map followed by the Plücker embedding:

$$X \xrightarrow{\gamma} G(n+1, r+1) \xrightarrow{P} P(\Lambda^{n+1} C^{r+1}) = P^N .$$

If $X$ is locally given by $x(u_1, \cdots, u_n) \in C^{r+1} - \{0\}$, then

$$(P \circ \gamma)(u) = x(u) \wedge \frac{\partial x}{\partial u_1}(u) \wedge \cdots \wedge \frac{\partial x}{\partial u_n}(u) .$$

Since $x$ is a section of $L = \mathcal{O}_X(1)$,

$$(P \circ \gamma)^* \mathcal{O}_{P^N}(1) = KL^{n+1} .$$

We define $\Gamma$ to be the pullback of the hyperplane sections

$$\Gamma = (P \circ \gamma)^* H^0(P^N, \mathcal{O}(1)) \subset H^0(X, KL^{n+1})$$

and the *Gauss linear system* to be

$$\Gamma_{2K} = \text{image of } \Gamma \otimes H^0(X, KL^{-(n+1)}) \to H^0(X, K^2) .$$

THEOREM 13 (see [1]). $\Gamma_{2K} \subseteq \ker \lambda$.

It follows that $\mu^{-1}(\Gamma_{2K}) \subseteq \ker (\delta^n)^* = \mathcal{Q}(V)$; that is to say, *the Gauss linear system always gives infinitesimal Schottky relations*. This is consistent with the experimentally observed phenomenon that *a geometric understanding of Hodge theory frequently involves dual varieties* (e.g., Andreotti's proof of Torelli for curves, the intermediate Jacobian of the cubic and other Fano threefolds, the generic global Torelli theorem for hypersurfaces (cf. Chapters XII and XIII below)).

Since this talk was given (in November 1981) there has been progress in IVHS. Some of this occurred during the year and is reported on in Chapter XIII. Additional applications of the methods of IVHS by F. Catanese (On the period map of surfaces with $K^2 = \chi = 2$, preprint available from Dipartimento di Matematica, Universitá Degli Studi di Pisa), Donagi and Tu (work in progress), and M. Green (also work in progress) are encouraging signs for a stronger interplay between formal Hodge theory and geometry.

Finally, an excellent exposition [5] of IVHS in the setting of general moduli problems will soon appear.

## REFERENCES

[1] J. Carlson, M. Green, P. Griffiths, and J. Harris, Infinitesimal variation of Hodge structures, to appear in Comp. Math.

[2] P. Griffiths, Periods of integrals on algebraic manifolds, I, II, Amer. J. Math. 90(1968), 568-626, 805-865.

[3] ________, Infinitesimal variation of Hodge structures (III): determinantal varieties and the infinitesimal invariant of normal functions, to appear in Comp. Math.

[4] P. Griffiths and J. Harris, Infinitesimal variation of Hodge structures (II): an infinitesimal invariant of Hodge classes, to appear in Comp. Math.

[5] C. Peters and J. Steenbrink, Exposition of some aspects moduli theory and infinitesimal variations of Hodge structure, Proc. Katata Conf. (1982), to appear.

PHILLIP GRIFFITHS
MATHEMATICS DEPARTMENT
HARVARD UNIVERSITY
CAMBRIDGE, MA 02138

LORING TU
MATHEMATICS DEPARTMENT
JOHNS HOPKINS UNIVERSITY
BALTIMORE, MD 21218

Chapter IV

# ASYMPTOTIC BEHAVIOR OF A VARIATION OF HODGE STRUCTURE

Phillip Griffiths
Written by Loring Tu

## §1. *Nilpotent Orbit Theorem*

Consider a variation of Hodge structure over a punctured disk,

$$\phi : \Delta^* \to \{T^k\}\backslash D ,$$

where $T$ is the image under the monodromy representation of the generator of $\pi_1(\Delta^*)$. By the monodromy theorem $T$ is quasi-unipotent:

$$(T^N - I)^{m+1} = 0 \quad \text{for some} \quad N, \quad m \in \mathbf{Z} .$$

Let $s$ be the coordinate on $\Delta^*$. Replacing $s$ by $s^N$, we may assume that $T$ is unipotent. If the variation of Hodge structure arises from a degenerating family, this amounts to pulling the family back to an N-fold cover of $\Delta^*$. Because the upper half plane $\mathfrak{h}$ is simply connected, the local liftability of the period map $\phi$ implies its global liftability to $\mathfrak{h}$:

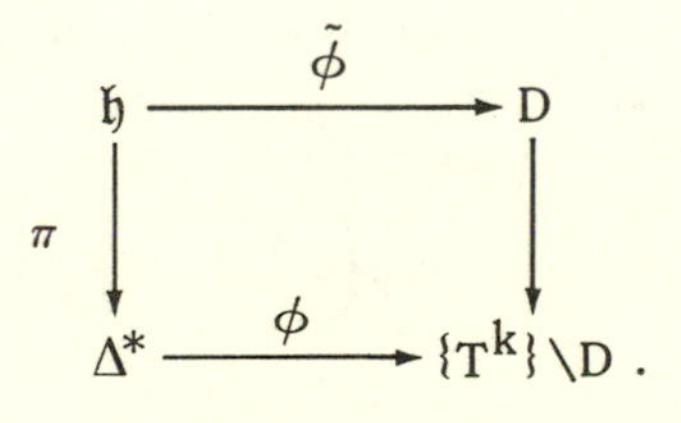


Here

$$\mathfrak{h} = \{\omega = u + iv \mid v > 0\}$$

and

$$s = \pi(\omega) = \exp(2\pi i\omega)\ .$$

Note that

$$\tilde{\phi}(\omega+1) = T\tilde{\phi}(\omega)\ .$$

In terms of the data $\mathcal{O} = \{\mathcal{H}_{\mathbf{Z}}, \mathcal{F}^p, \nabla, S = \Delta^*\}$ we think of $T$ as an action on the lattice $H_{\mathbf{Z}} = (\mathcal{H}_{\mathbf{Z}})_{s_0}$, which of course induces an action on $H_{\mathbf{C}} = (\mathcal{H}_{\mathbf{C}})_{s_0}$. Analytic continuation around $s = 0$ gives

$$H_{e^{2\pi i}s} = TH_s\ .$$

Because $T$ is unipotent, its logarithm can be defined as a finite sum:

$$N = \log T = (T-I) - \frac{(T-I)^2}{2} + \cdots + (-1)^{m+1}\frac{(T-I)^m}{m}\ .$$

It is an elementary fact that every holomorphic vector bundle over the punctured disk is trivial. Hence the cohomology bundle $\mathcal{H} \to \Delta^*$ is trivial. Each trivialization gives rise to an extension of a bundle over the disk $\Delta$. Of the many trivializations possible, we single out one, called the *privileged extension*, defined as follows.

Using the lattice bundle $\mathcal{H}_{\mathbf{Z}}$ inside $\mathcal{H}$, it is possible to speak of the *horizontal displacement* of an element $e$ in the fiber $\mathcal{H}_{s_0}$. By horizontally displacing $e$, we get a multi-valued flat global section $e(s)$ of $\mathcal{H}$ over $\Delta^*$; $e(s)$ is multi-valued because $e((\exp 2\pi i)s) = Te(s)$. Define

$$\sigma_e(s) = \exp\left(-\frac{\log s}{2\pi i}N\right)e(s)\ .$$

Because

$$\sigma_e((\exp 2\pi i)s) = \exp\left(-\frac{\log s}{2\pi i}N\right)T^{-1}Te(s) = \sigma_e(s)\ ,$$

$\sigma_e(s)$ is a *single-valued* holomorphic section of $\mathcal{H}$ over $\Delta^*$.

DEFINITION 1. The *privileged extension* of $\mathcal{H} \to \Delta^*$ to $\overline{\mathcal{H}} \to \Delta$ is given by taking $\{\sigma_e\}$ to be a holomorphic frame, as e ranges over a basis of the fiber $\mathcal{H}_{s_0}$.

Return now to the period map $\phi : \Delta^* \to D\backslash\Gamma$ of a degenerating family. As before, we have the diagram

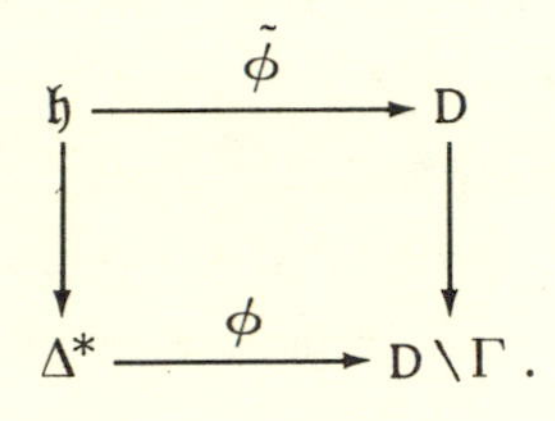

Set

$$\tilde{\psi}(w) = \exp(-wN)\tilde{\phi}(w) \in \check{D} .$$

Then

$$\tilde{\psi}(w+1) = \tilde{\psi}(w)$$

so that $\tilde{\psi}$ descends to a *single-valued* map $\psi : \Delta^* \to \check{D}$ given by

$$\begin{aligned} \psi(s) &= \tilde{\psi}(w) \\ &= \tilde{\psi}\left(\frac{\log s}{2\pi i}\right) \\ &= \exp\left(-\frac{\log s}{2\pi i} N\right)\phi(s) . \end{aligned}$$

THEOREM 2. *The map* $\psi : \Delta^* \to \check{D}$ extends across the origin to a map $\psi : \Delta \to \check{D}$.

For a proof see Cornalba and Griffiths [1, p. 89] or Griffiths and Schmid [2, p. 104]. The idea is as follows. We view $\psi$ as a map into a product of Grassmannians, represented by a matrix whose entries are the periods. Composing $\psi$ with the Plücker embedding P gives a map $P \circ \psi$ into a projective space. By the theorem on regular singular points, the periods have at most poles at the origin. Since $P \circ \psi$ is given by the minors of $\psi$, it is meromorphic at the origin. By factoring out the common factors of $P \circ \psi$, it follows that $P \circ \psi$ and hence $\psi$ can be extended across the origin.

DEFINITION 3. The filtration $\psi(0) \in \check{D}$ will be called the *limiting filtration* and will be denoted by $\{F^p_\infty\}$.

We remark that $\{F^p_\infty\}$ may well lie outside $\bar{D}$. It arises in two contexts; one is the nilpotent orbit theorem (to be discussed momentarily), and the other is as the Hodge filtration in the limiting mixed Hodge structure (to be discussed below).

DEFINITION 4. The *nilpotent orbit* of a degenerating family over $\Delta^*$ is the map $\mathcal{O} : h \to \check{D}$ given by

$$\mathcal{O}(w) = \exp(wN)\psi(0)$$

The nilpotent orbit satisfies

$$\mathcal{O}(w+1) = T\mathcal{O}(w) .$$

Schmid's nilpotent orbit theorem says that this nilpotent orbit is a very good approximation of the original period map.

THEOREM 5 (Nilpotent orbit theorem). a) *The nilpotent orbit is horizontal.* b) *For* $\operatorname{Im} w >> 0$, *the nilpotent orbit assumes values in* $D$. c) *The nilpotent orbit osculates to the period map to very high order; more precisely, there are constants* $A$ *and* $B$ *such that for* $\operatorname{Im} w \geq A > 0$,

$$\rho_D(\mathcal{O}(w), \tilde{\phi}(w)) \leq (\operatorname{Im} w)^B e^{-2\pi \operatorname{Im} w} .$$

For a proof see Cornalba and Griffiths [1, p. 90]. The original proof is in [3].

§2. *The* $SL_2$*-orbit Theorem*

The nilpotent orbit theorem is just the first step. For example, using it we still do not see that $N^{n+1} = 0$, where $n$ is the weight of the Hodge structure in question – this is the strong form of the monodromy theorem, giving $n+1$ as a bound on the Jordan blocks of the monodromy matrix. Moreover, we do not see from it the answer to the main question:

QUESTION. Is the Hodge length $\|e\|$ of an invariant cohomology class $e \in H$ bounded on $\Delta^*$ ?

The affirmative answer to this is what is needed to extend the theorem on the fixed part and its consequences to a variation of Hodge structure with an arbitrary algebraic base space. The point is that a *bounded* plurisubharmonic function on an algebraic variety (possibly noncomplete) is constant.

These results are consequences of the $SL_2$-orbit theorem of W. Schmid [3]. Roughly speaking, given a nilpotent orbit $\mathcal{O}: \mathfrak{h} \to D$, the $SL_2$-orbit theorem enables us to construct a variation of Hodge structure

$$\phi : \mathfrak{h} \to D$$

which lifts to a homomorphism of Lie groups

$$\psi : SL_2(\mathbf{R}) \to G_{\mathbf{R}}$$

such that

$$\rho_D(\mathcal{O}(w), \phi(w)) \leq C(\mathrm{Im}\ w)^{-1}$$

for some constant $C$. Thus the nilpotent and $SL_2$-orbits are asymptotic to each other. A stronger but more technical form of this asymptotic behavior is possible (see Schmid [3]). Rather than stating it, we will now discuss the consequences.

First, given a variation of Hodge structure of weight $n$ over $\Delta^*$, we get

$$N^{n+1} = 0 .$$

Furthermore, there exists a unique ascending filtration $\{W_\ell\}$ of $H_Q$, called the *monodromy filtration,*

$$0 \subset W_0 \subset \cdots \subset W_{2n} = H_Q ,$$

satisfying

$$N : W_i \to W_{i-2}$$

$$N^k : W_{n+k}/W_{n+k-1} \xrightarrow{\sim} W_{n-k}/W_{n-k-1} .$$

One should think of the monodromy operator $N$ as the analogue of the operator ''cup product with the Kähler class'' in the Hard Lefschetz theorem.

The monodromy filtration is uniquely characterized by these properties. For by taking $k = n$, we get

$$N^n : W_{2n}/W_{2n-1} \xrightarrow{\sim} W_0 ,$$

from which it follows that

$$W_{2n-1} = \ker N^n$$

and

$$W_0 = \operatorname{im} N^n .$$

The other terms of the monodromy filtration can now be defined by induction, as follows. Set

$$H'_Q = W_{2n-1}/W_0 .$$

Then $N$ induces an operator $N'$ on $H'_Q$ satisfying $(N')^n = 0$. The filtration on $H'_Q$ is

$$\begin{array}{ccccc} 0 \subset W'_0 & \subset \cdots \subset & W'_{2n-2} & = & H'_Q \\ \| & & \| & & \\ W_1/W_0 & & W_{2n-1}/W_0 & . & \end{array}$$

So $W_{2n-2}$ and $W_1$ are the inverse images of $W'_{2n-3}$ and $W'_0$ respectively under the projection $W_{2n-1} \to H'_{\mathbf{Q}}$. This process continues and uniquely constructs the monodromy filtration on $H_{\mathbf{Q}}$.

REMARK. Ron Donagi points out the following picture of the monodromy weight filtration: Set

$$N^{p,q} = \operatorname{im} N^p \cap \ker N^{n-q} .$$

These are the obvious spaces that can be constructed from the pair $(H,N)$. Note that $N^{p,q} \supset N^{p+1,q}$ and $N^{p,q} \supset N^{p,q+1}$. Then

$$W_q = \operatorname{span}\Big( \sum_{r+s \leqq n-q} N^{r,s} \Big) .$$

DEFINITION 6. A *mixed Hodge structure* $\{H_{\mathbf{Q}}, F^p, W_\ell\}$ of weight $n$ consists of an ascending weight filtration, defined over $\mathbf{Q}$,

$$0 \subset W_0 \subset \cdots \subset W_{2n} = H_{\mathbf{C}} ,$$

such that the Hodge filtration induces a pure Hodge structure of weight $m$ on the graded piece $Gr_m = W_m/W_{m-1}$ of the weight filtration for each $m = 0, \cdots, 2n$. The induced Hodge filtration on $Gr_m$ is

$$F^p(Gr_m) = (F^p \cap W_m)/(F^p \cap W_{m-1}) .$$

For $r \in \mathbf{Z}$, a *morphism of type* $(r,r)$ of mixed Hodge structures is a rationally defined map $f: H_{\mathbf{Q}} \to H'_{\mathbf{Q}}$ such that

$$f(W_\ell) \subset W'_{\ell+2r}$$

and

$$f(F^p) \subset (F')^{p+r} .$$

The main consequence of the proof of the $SL_2$-orbit theorem is that given a variation of Hodge structure over $\Delta^*$, the limiting filtration $\{F^p_\infty\}$

together with the monodromy weight filtration $\{W_\ell\}$ gives a mixed Hodge structure on the vector space $H_{\mathbf{Q}}$. This is called the *limiting mixed Hodge structure*. Relative to the limiting mixed Hodge structure $N$ is a morphism of type $(-1,-1)$.

We also get a characterization of the monodromy filtration in terms of the growth of the Hodge length, namely,

$$W_\ell = \left\{e \in H : \|e\| = 0\left(\left(\log \frac{1}{|s|}\right)^{\frac{\ell-n}{2}}\right)\right\}.$$

COROLLARY 7. *Every local invariant cohomology class has bounded Hodge length.*

*Proof.* Since

$$N = (T-I) - \frac{(T-I)^2}{2} + \cdots + (-1)^{n+1} \frac{(T-I)^n}{n}$$

and

$$T-I = N + \frac{N^2}{2!} + \cdots + \frac{N^n}{n!},$$

$T-I$ and $N$ have the same kernel. So the invariant cohomology classes are precisely $\ker N$. Since $\ker N \subset W_n$, the characterization of $W_n$ above proves the corollary. q.e.d.

REMARK. There is an interesting, and also *confusing*, point concerning the limiting mixed Hodge structure. At the risk of making matters worse, we shall attempt to clarify it.

Given a variation of Hodge structure over the punctured disk, we lift to the upper-half-plane and consider the VHS as a holomorphically varying filtration

$$\begin{cases} F_w^p \subset H & p = 0, \cdots, n \quad \text{and} \quad \operatorname{Im} w > 0 \\ F_{w+1}^p = TF_w^p & \end{cases}$$

on the fixed vector space $H$. Set $h_p = \dim F_w^p$ and $h = h^0 = \dim H$.

LEMMA 8. *There exists a holomorphic basis* $f_i(w) \in F_w^p$ *satisfying*

$$\begin{cases} \text{(i)} \quad f_i(w+1) = Tf_i(w) \\ \text{(ii)} \quad f_i(w) = \sum_{\alpha=0}^{k} f_{i\alpha}(w)w^\alpha, \text{ where} \\ \text{(iii)} \quad f_{i\alpha}(w+1) = f_{i\alpha}(w) \text{ and } f_{i\alpha}(w) = O(1) . \end{cases}$$

*Here,* $f_i(w)$ *and* $f_{i\alpha}(w)$ *are vectors in the fixed vector space* $H$.

*Proof.* Set $s = e^{2\pi i w}$ and

$$\tilde{F}_s^p = \exp(-wN) F_w^p .$$

Then $\{\tilde{F}_s^p\}_{s \in \Delta^*}$ gives a holomorphically varying and single-valued filtration on the vector space. By our discussion above, the holomorphic mapping $\Delta^* \xrightarrow{\psi} G(h_p, H)$ extends across $s = 0$; we let $\tilde{f}_i(s) \in \tilde{F}_s^p$ be a holomorphically varying basis. Then

$$f_i(w) = \exp(wN)\tilde{f}_i(e^{2\pi i w}) \in F_w^p \subset H$$

satisfies the requirements of the lemma. q.e.d.

We now consider the non-zero vector

$$\Lambda_p(w) = f_1(w) \wedge \cdots \wedge f_{h_p}(w) \in \Lambda^{h_p} H$$

Clearly we have

$$\begin{cases} \Lambda_p(w+1) = (\Lambda^p T) \cdot \Lambda_p(w) \\ \Lambda_p(w) = w^{h_p} \tilde{\Lambda}_p(w) + O(w^{h_p - 1}) \end{cases}$$

where $\tilde{\Lambda}_p(w+1) = \tilde{\Lambda}_p(w)$, and where $\tilde{\Lambda}_p\left(\frac{\log s}{2\pi i}\right)$ is a holomorphic and non-vanishing function for $|s| < \varepsilon$. Then

$$\begin{aligned}\hat{\Lambda}_p(w) &= w^{-h_p}\Lambda_p(w)\\ &= \tilde{\Lambda}_p(w) + O(w^{-1})\end{aligned}$$

gives the Plücker coordinate of $F_w^p$. In particular, $\hat{\Lambda}_p(i\infty) = \lim_{w\to i\infty} \hat{\Lambda}_p(w)$ exists as a point in $G(h_p, H) \subset P\Lambda^{h_p} H$ ; in this way we determine a filtration $\{\hat{F}_\infty^p\}$ on $H$. Clearly,

$$\begin{cases}\text{(i)} \quad \lim_{w\to i\infty} F_w^p = \hat{F}_\infty^p \in \bar{D} \subset \check{D}\\ \text{(ii)} \quad T\hat{F}_w^p = \hat{F}_w^p .\end{cases}$$

The second statement implies that $\{\hat{F}_w^p\} \in \partial D = \bar{D} - D$ in case $T \neq I$, and so we have a picture like

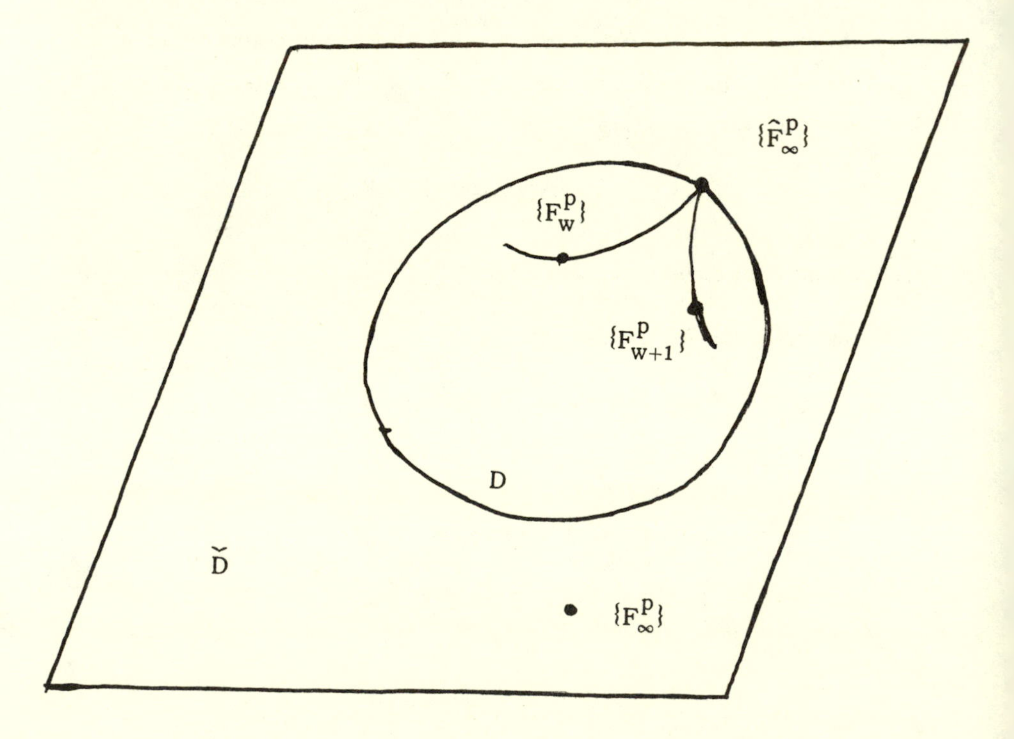

In the naive sense $\{\hat{F}^p_\infty\}$ is the limiting Hodge filtration, so we call it the *naive limiting Hodge filtration.*

The points we wish to make are

a) *The naive limiting Hodge filtration is not the same as the limiting filtration* $\{F^p_\infty\}$ *discussed above*; and

b) *The limiting filtration* $\{F^p_\infty\}$ *is the "correct" object.*

Point a) is clear since $N\hat{F}^p_\infty \subseteq \hat{F}^p_\infty$ while this is certainly false for $F^p_\infty$. Concerning point b), a preliminary remark is that this same property $(N\hat{F}^p_\infty \subseteq \hat{F}^p_\infty)$ makes it unlikely that the limiting Hodge filtration should give a mixed Hodge structure. Before giving a deeper reason for b), we remark that the relation between the two filtrations is obviously

$$\lim_{w\to i\infty} \exp(wN) \cdot F^p_\infty = \hat{F}^p_\infty .$$

Concerning b), let us first agree that the privileged extension $\overline{\mathcal{H}} \to \Delta$ is the "correct" extension for the cohomology bundle $\mathcal{H} \to \Delta^*$. (This claim will be justified algebro-geometrically in Chapter VII below.) Then, since $\mathcal{F}^p_s$ gives the Hodge filtration on $\mathcal{H}_s$ for $s \neq 0$ and $\lim_{s\to 0} \mathcal{F}^p_s =: \overline{\mathcal{F}}^p_0$ exists (by our discussion above), the "correct" limiting Hodge filtration must be given by $\{\overline{\mathcal{F}}^p_0\}$ on $\overline{\mathcal{H}}_0$ ( = the fibre of $\overline{\mathcal{H}} \to \Delta$ over $s = 0$ ).

Now let $e_i \in H$ be any basis, and denote by $e_i(w) \in \mathcal{H}_s$ $(s = e^{2\pi i w})$ the multi-valued horizontal section of $\mathcal{H} \to \Delta$ determined by $e_i$. Setting $f_i(s) = \exp(-wN)\, e_i(w) \in \mathcal{H}_s$ $(s = e^{2\pi i w})$ we obtain a *single-valued* holomorphic framing of $\mathcal{H} \to \Delta^*$. The definition of the privileged extension $\overline{\mathcal{H}} \to \Delta$ is that a holomorphic section $g(s) = \Sigma g^i(s) f_i(s) \in \mathcal{H}_s$ extends across $s = 0$ if, and only if, the holomorphic functions $g^i(s)$ extend across $s = 0$. In other words, via the frame field $\{f_i(s)\}$ we have an isomorphism

$$\overline{\mathcal{H}} \cong \mathcal{O}^{(h)} .$$

Intrinsically this is

$$\overline{\mathcal{H}} \cong \mathcal{O}(H) \tag{9}$$

where $\mathcal{O}(H) \to \Delta$ is the trivial bundle with fibre $H$. Under this isomorphism the subspaces $\mathcal{F}_s^p \subset \mathcal{H}_s$ must go to subspaces of $H$ that are holomorphic *and single-valued* functions of $s \in \Delta^*$. The only such possibility is that $\mathcal{F}_s^p$ maps to $\tilde{F}_s^p$, and therefore $\overline{\mathcal{F}}_0^p$ maps to $F_\infty^p$. More formally we may state this as:

PROPOSITION 10. *Under the isomorphism (9) the subspace* $\mathcal{F}_s^p \subset \mathcal{H}_s$ *maps to* $\tilde{F}_s^p \subset H$. *In particular, at* $s = 0$ *the subspace* $\overline{\mathcal{F}}_0^p$ *maps to* $F_\infty^p$.

The proof consists in unwinding the definitions, and is perhaps therefore best left as a private matter.

Because of the proposition we see that $\{F_\infty^p\}$ is the correct limiting Hodge filtration granted that $\overline{\mathcal{H}} \to \Delta$ is the correct extension of $\mathcal{H} \to \Delta^*$. To make this completely convincing we need to see that $\{F_\infty^p\}$ gives the right answer in the geometric case, and that the limiting mixed Hodge structure has to do with the mixed Hodge structure on the central fibre. This will be done in Chapters VI and VII.

## REFERENCES

[1] M. Cornalba and P. Griffiths, Some transcendental aspects of algebraic geometry, in Proceedings of Symposia in Pure Mathematics, vol. 29 (1975), A.M.S., 3-110.

[2] P. Griffiths and W. Schmid, Recent developments in Hodge theory: a discussion of techniques and results, in *Discrete Subgroups of Lie Groups and Applications to Moduli,* Oxford University Press, 1973.

[3] W. Schmid, Variation of Hodge structure: the singularities of the period mapping, Invent. Math. 22 (1973), 211-319.

PHILLIP GRIFFITHS
MATHEMATICS DEPARTMENT
HARVARD UNIVERSITY
CAMBRIDGE, MA 02138

LORING TU
MATHEMATICS DEPARTMENT
JOHNS HOPKINS UNIVERSITY
BALTIMORE, MD 21218

Chapter V

# MIXED HODGE STRUCTURES, COMPACTIFICATIONS AND MONODROMY WEIGHT FILTRATION

Eduardo H. Cattani

## §0. *Introduction*

In his survey paper [10], Griffiths conjectured the existence of partial compactifications for the arithmetic quotients of classifying spaces for polarized Hodge structures, that would generalize the Satake-Baily-Borel compactification for arithmetic quotients of Hermitian symmetric spaces. The richness of the problem becomes clear in the fundamental work of Schmid [16], on the asymptotic behavior of the period mapping (see Chapter IV); in particular, and as a consequence of his Nilpotent and $SL_2$-orbit theorems, Schmid was able to show—as conjectured by Deligne—the existence of a limiting mixed Hodge structure associated to a one-parameter variation of polarized Hodge structures. (This was also done independently by Steenbrink [18] for the geometric case. His approach will be discussed in Chapter VII.)

Schmid's work is at the core of the topological partial compactification constructed in [5] for the case of Hodge structures of weight two. However, this compactification as well as Satake's in the Hermitian symmetric case contain only part of the "information" in the limiting mixed Hodge structure; namely the Hodge structures on the graded pieces of the weight filtration. On the other hand, Carlson's work [3] has shown that the extension data of the limiting mixed Hodge structure contains significant geometric information. It is then natural to attempt the construction of partial compactifications which incorporate in the boundary the full limiting mixed Hodge structure. When this is done in the Hermitian symmetric case, one recovers the smooth compactifications of Ash, Mumford, Rapoport and Tai [1]; the problem remains open in the arbitrary weight case.

This chapter is divided in two parts: in the first, we will look in some detail at the limiting mixed Hodge structure associated to a variation of Hodge structures of weight one. As seen in Chapter I, the classifying space in this case is a Hermitian symmetric space and thus we shall be able to look at both the Satake and Mumford compactifications from a Hodge-theoretic point of view. We shall mostly limit ourselves to a description of the boundary components; the details of the constructions are beyond the scope of this chapter and we refer the reader to [1], [2], [14] and [15]. Much of the material included here is contained in [4].

The second part of this chapter will be a description of joint work of the author with A. Kaplan ([6], [7]) concerning the monodromy weight filtration in the case of degenerations depending on several parameters. The purpose here is, of course, to obtain a limiting mixed Hodge structure in the several variables case.

§1. *Compactifications of Siegel's upper-half-space*

We recall (cf. Chapter I) that the classifying space for polarized Hodge structures of weight one may be described as follows: Let $H_{\mathbf{Z}}$ be a lattice of rank $2g$, $Q$ a non-degenerate, skew symmetric bilinear form on $H_{\mathbf{Z}}$ and $H = H_{\mathbf{Z}} \otimes \mathbf{C}$. A polarized Hodge structure of weight one on $H$ is a decomposition

$$H = H^{1,0} \oplus H^{0,1} \;; \qquad H^{0,1} = \overline{H^{1,0}}$$

satisfying the bilinear relations

i) $Q(u,u) = 0$ if $u \in H^{1,0}$, and

ii) $iQ(u,\bar{u}) > 0$ if $0 \neq u \in H^{1,0}$

As in Chapter I, it will be convenient to identify a Hodge structure with a filtration $F$ which, in this case, takes the simple form

$$H = F^0 \supset F^1 \supset \{0\} \;; \qquad F^1 = H^{1,0} .$$

The classifying space of all such polarized Hodge structures of weight one on $H$ is then

$$D = \{F^1 \in G(g,H) : Q(F^1,F^1) = 0;\ iQ(F^1,\overline{F^1}) > 0\}$$

while the subvariety of $G(g,H)$ consisting of the maximal $Q$-isotropic subspaces of $H$ is the compact dual $\check{D}$ of $D$.

We also recall that the choice of a symplectic basis; i.e. a rational basis $\mathcal{E} = \{e_1, \cdots, e_g, f_1, \cdots, f_g\}$ relative to which:

$$Q = \begin{bmatrix} & I \\ -I & \end{bmatrix}$$

gives rise to the usual realization of $D$ as Siegel's upper-half space. Let $F^1 \in D$ and $\omega_1, \cdots, \omega_g$ a basis of $F^1$; let

$$\begin{bmatrix} \Omega_1 \\ \\ \Omega_2 \end{bmatrix}$$

be the matrix whose columns give the coefficients of the $\omega_i$'s relative to the basis $\mathcal{E}$. The second bilinear relation guarantees that $\Omega_2$ is non-singular, and therefore $F^1$ has a (unique) basis of the form

$$\omega_i = \sum_j z_{ji} e_j + f_i \,. \tag{1}$$

It is easy to check that the bilinear relations imply that, denoting by $Z$ the matrix $(z_{ij})$,

$$^tZ = Z \quad \text{and} \quad \operatorname{Im}(Z) \quad \text{is positive definite.}$$

We shall refer to $Z$ as the normalized period matrix of $F$. Let $V_{\mathbf{R}}$ (respectively, $V_{\mathbf{C}}$) denote the vector space of symmetric $g \times g$-matrices with real (respectively, complex entries) and $V_{\mathbf{R}}^+$ the cone of positive definite symmetric matrices, then

$$D \cong V_{\mathbf{R}} + iV_{\mathbf{R}}^{+} \subset V_{\mathbf{C}} \subset \check{D} . \tag{2}$$

We consider now a variation of polarized Hodge structures of weight one defined over the punctured disk $\Delta^*$, and its corresponding period mapping:

$$\phi : \Delta^* \to \Gamma \backslash D ; \qquad \Gamma = G_{\mathbf{Z}} = Sp(g,\mathbf{Z}) .$$

Let $\tilde{\phi} : U \to D$ be a global lifting of $\phi$ to the upper half plane $U$ and $\gamma \in \Gamma$ the Picard-Lefschetz transformation, i.e.,

$$\tilde{\phi}(z+1) = \gamma\tilde{\phi}(z) ; \qquad z \in U .$$

We know that $\gamma$ is a quasi-unipotent automorphism of $H$. We will assume, for the sake of simplicity in the exposition, that $\gamma$ is actually unipotent.

Let $N = \log \gamma$ ; $N$ is then a rational element in the Lie algebra

$$\mathfrak{g}_0 = \{X \in \mathrm{Hom}(H_{\mathbf{R}},H_{\mathbf{R}}) : Q(X.,.) + Q(.,X.) = 0\} .$$

By the Monodromy Theorem, $N^2 = 0$ and therefore in this case

$$N = (\gamma - I) .$$

The *monodromy weight filtration* $W(N)$ of $N$ is the filtration

$$\{0\} \subset W_0 \subset W_1 \subset W_2 = H_{\mathbf{C}}$$

given by:

$$W_0 = \mathrm{Im}(N) ; \qquad W_1 = \mathrm{Ker}(N) .$$

We note that $W(N)$ is defined over $\mathbf{Q}$. Moreover, since $N \in \mathfrak{g}_0$ and $N^2 = 0$, it follows that $W_0$ is an isotropic subspace and $W_1$ is the Q-annihilator of $W_0$, that is:

$$W_1 = W_0^{\perp} = \{u \in H : Q(u,v) = 0 \text{ for all } v \in W_0\} .$$

Let $\Omega$ be a maximal totally isotropic subspace of H, and $\mathcal{E}$ a symplectic basis as above and such that $W_0 = \mathrm{Span}\{e_1, \cdots, e_\nu\}$ and $\Omega = \mathrm{Span}\{e_1, \cdots, e_g\}$. Relative to $\mathcal{E}$ we can write

$$N = \begin{bmatrix} 0 & \eta \\ 0 & 0 \end{bmatrix} ; \qquad {}^t\eta = \eta = \begin{bmatrix} \eta_{11} & 0 \\ 0 & 0 \end{bmatrix}$$

where $\eta_{11}$ is a $\nu \times \nu$-symmetric matrix.

With $\phi$, $\tilde{\phi}$ and N as above, let

$$\tilde{\psi} : U \to \check{D}$$

be the map $\tilde{\psi}(z) = \exp(-zN)\tilde{\phi}(z)$. Since $\tilde{\psi}(z+1) = \tilde{\psi}(z)$ it is clear that $\tilde{\psi}$ drops to a map $\psi : \Delta^* \to \check{D}$ given by $\psi(t) = \tilde{\psi}\left(\frac{1}{2\pi i} \log t\right)$.

We can now state and prove (in the weight-one case) Schmid's Nilpotent Orbit Theorem. The proof given here follows the argument of Schmid in [17] (cf. also [10, §13]).

NILPOTENT ORBIT THEOREM.

i) *The map* $\psi : \Delta^* \to \check{D}$ *has a removable singularity at the origin.*

ii) *Let* $F_a = \psi(0) \in \check{D}$, *then for* $\mathrm{Im}(z)$ *sufficiently large* $\exp zN \cdot F_a \in D$.

iii) *Relative to a* $G_{\mathbf{R}}$*-invariant distance in* D

$$d(\exp zN \cdot F_a, \tilde{\phi}(z)) = 0(e^{-2\pi \mathrm{Im}(z)}) \quad \text{as} \quad \mathrm{Im}(z) \to \infty .$$

*Proof.* Let $\tilde{Z}(z)$ denote the normalized period matrix of $\tilde{\phi}(z)$. Since

$$\exp(-zN) = \begin{bmatrix} I & -z\eta \\ & I \end{bmatrix}$$

we have that $\tilde{\psi}(z)$ has as normalized period matrix

$$\tilde{W}(z) = \tilde{Z}(z) - z\eta .$$

Let $\lambda$ be a linear functional on $V_{\mathbf{R}}$, which takes positive values on $V_{\mathbf{R}}^{+}$. We also denote by $\lambda$ its extension to $V_{\mathbf{C}}$ and set:

$$\tilde{f}(z) = \lambda(\tilde{Z}(z)) \quad \text{and} \quad \tilde{g}(z) = \lambda(\tilde{W}(z)) .$$

Since $\operatorname{Im}(\tilde{Z}(z)) \in V_{\mathbf{R}}^{+}$, it follows that $\tilde{f}$ maps the upper half-plane to itself, and thus we can conclude from Schwarz Lemma that

$$|\tilde{f}(z)| \leq K \operatorname{Im}(z)$$

for $|\operatorname{Re}(z)| \leq \alpha$ and $\operatorname{Im}(z) \geq \beta$.

Moreover, since $\tilde{g}(z) = \tilde{f}(z) - \lambda(\eta)z$, the same estimate holds for $\tilde{g}(z)$. But $\tilde{g}$ drops to a function $g : \Delta^{*} \to \mathbf{C}$ for which:

$$|g(t)| = \left|\tilde{g}\left(\frac{1}{2\pi i} \log t\right)\right| \leq K \operatorname{Im}\left(\frac{\log t}{2\pi i}\right) = K' \log\left(\frac{1}{|t|}\right)$$

as $|t| \to 0$. Hence $g$ has a removable singularity at the origin and since the dual space $V_{\mathbf{R}}^{*}$ is spanned by such functionals $\lambda$, we obtain the same conclusion for $\psi$ proving (i).

The nilpotent orbit $\exp zN\cdot F_a$ is given by the normalized period matrix $W(0) + z\eta$, thus (ii) is equivalent to:

$$\operatorname{Im}(W(0) + z\eta) > 0$$

for $\operatorname{Im}(z)$ sufficiently large. But if $\lambda \in V_{\mathbf{R}}^{*}$ is as above, we have

$$\tilde{f}(z) = g(e^{2\pi i z}) + \lambda(\eta)z ,$$

but since $\operatorname{Im}(\tilde{f}(z)) > 0$ for $\operatorname{Im}(z) > 0$, it follows that $\lambda(\eta) \geq 0$, which implies that $\eta$ is positive semidefinite and $\eta_{11}$, being non-singular, is positive definite. Also, for any $\lambda$ such that $\lambda(\eta) = 0$, we must have $\operatorname{Im}(g(0)) > 0$, hence, if we write

$$W(0) = \begin{bmatrix} W_{11} & W_{12} \\ {}^{t}W_{12} & W_{22} \end{bmatrix} . \tag{3}$$

Then the $(g-\nu)\times(g-\nu)$ symmetric matrix $W_{22}$ must have positive definite imaginary part, and (ii) follows. The proof of (iii) follows along similar lines. q.e.d.

We remark then that if we regard the normalized period matrix of $\phi$ as a multivalued map in $\Delta^*$ with monodromy $\gamma$, then the nilpotent orbit theorem gives the expansion

$$Z(t) = W(t) + \frac{1}{2\pi i}(\log t)\eta$$

with $W(t)$ holomorphic on the disk $\Delta$. The nilpotent orbit itself corresponds to the lifting of $W(0) + \frac{1}{2\pi i}(\log t)\eta$.

In this case, it is also very easy to extend the Nilpotent Orbit Theorem to several variables; in fact, let

$$\phi : (\Delta^*)^r \to \Gamma \backslash D$$

be a period map, and let $\gamma_1, \cdots, \gamma_r$ denote the corresponding Picard-Lefschetz transformations, which we will continue to assume unipotent. As before, let $N_i = \log \gamma_i = (\gamma_i - I)$. These are commuting nilpotent elements of $\mathfrak{g}_Q$ and from the Nilpotent Orbit Theorem we have that the symmetric forms $Q(.,N_i .)$ are positive semidefinite.

Let $\sigma \subset \mathfrak{g}_0$ be the *monodromy cone*:

$$\sigma = \left\{ \sum_{i=1}^{r} \lambda_i N_i : \lambda_i \in R;\ \lambda_i > 0 \right\} .$$

For any element $N \in \sigma$ we have

$$\text{(4)} \qquad \mathrm{Ker}(N) = \bigcap_{i=1}^{r} \mathrm{Ker}(N_i)$$

indeed, if $u \in \mathrm{Ker}(N)$, then

$$0 = Q(u,Nu) = \sum_{i=1}^{r} \lambda_i Q(u,N_i u)$$

which implies $Q(u,N_i u) = 0$, $i = 1,\cdots,r$, showing (4). By duality, we also have

$$\text{(5)} \qquad \operatorname{Im}(N) = \sum_{i=1}^{r} \operatorname{Im}(N_i) .$$

Thus: *all elements in the monodromy cone* $\sigma$ *define the same weight filtration.*

In particular, we can choose a maximal totally isotropic real subspace $\Omega$ containing $W_0(N)$ for every $N \epsilon \sigma$, and the proof given above for the Nilpotent Orbit Theorem carries without modifications to the several variables case.

We should note that the limiting Hodge filtration $F_a = \psi(0)$ depends on the choice of parameters in $\Delta^*$. In fact, if $s$ is a new parameter in $\Delta^*$ with $t = h(s)$, and we denote by

$$\psi_1(s) = \exp\left(-\frac{1}{2\pi i}\log s\right)\phi(s) ,$$

then $\psi_1(s) = \exp\left(\frac{1}{2\pi i}\left(\log\frac{h(s)}{s}\right)N\right)\psi(h(s))$ and therefore, $\psi_1(0) = \exp\left(\frac{1}{2\pi i}(\log h'(0))N\right) \cdot \psi(0)$. Similarly, in the several variables case the limiting Hodge filtration is well defined only up to the action of $\exp \sigma_{\mathbf{C}}$, where $\sigma_{\mathbf{C}}$ denotes the $\mathbf{C}$-span of the monodromy cone $\sigma$.

*The limiting mixed Hodge structure*

The limiting Hodge filtration $F_a = \psi(0)$ is an element of the dual space $\check{D}$ and does not, in general, define a Hodge structure. We have seen, however, that the period matrix $W(0)$ is symmetric—reflecting the fact that $F_a$ satisfies the first bilinear relation—and that when written in terms of a symplectic basis adapted to the monodromy weight filtration, as in (3), then its block $W_{22}$ has positive definite imaginary part. These properties may be invariantly stated via the notion of a polarized mixed Hodge structure. We shall give the precise definition in §2. In the present case, it is equivalent to:

(6) The filtration $F_a$ defines a Hodge structure of weight $\ell$ on the graded quotient $Gr_\ell^{W(N)} = W_\ell(N)/W_{\ell-1}(N)$, $N \in \sigma$.

(7) The Hodge structure induced by $F_a$ on $Gr_{\ell+1}^{W(N)}$, $\ell \geq 0$, is polarized by the bilinear form

$$Q_\ell = S(\cdot, N^\ell \cdot)\ ; \quad N \in \sigma .$$

We can deduce (6) and (7) directly from the Nilpotent Orbit Theorem. Indeed, it follows from (4) and (5) that it is enough to consider the case of a single-variable nilpotent orbit:

$$\theta(z) = \exp zN \cdot F_a .$$

Notice also that since $\exp zN$ acts trivially on $Gr^{W(N)}$, we may assume that $F_a \in D$.

For $\ell = 0$, (6) is equivalent to $F_a^1 \cap W_0(N) = 0$, but this is immediate from the bilinear relations and the fact that $W_0(N)$ is a Q-isotropic subspace. By duality, we also obtain

$$H = F_a^1 + W_1(N) = \overline{F_a^1} + W_1(N) ,$$

i.e., $F_a$ defines a Hodge structure of weight two and pure type (1,1) on $Gr_2^{W(N)}$, which is trivially polarized by $Q(\cdot, N\cdot)$.

It remains to show that $F_a$ defines a polarized Hodge structure of weight one on $Gr_1^{W(N)}$; this is equivalent to:

$$W_1(N) = W_0(N) \oplus (F_a^1 \cap W_1(N)) \oplus (\overline{F_a^1} \cap W_1(N)) . \tag{8}$$

If $f_1, f_2 \in F_a^1 \cap W_1(N)$ are such that $f_1 + \overline{f}_2 \in W_0(N)$, then

$$iQ(f_1, \overline{f}_1) = iQ(f_1, \overline{f}_1 + f_2) = 0$$

since $W_1(N) = (W_0(N))^\perp$, therefore, $f_1 = f_2 = 0$. Similar arguments for the remaining cases show that the sum in (8) is direct. On the other hand, we have:

$$g = \dim F_a^1 = \dim(F_a^1 \cap \operatorname{Ker} N) + \dim(N(F_a^1))$$

$$= \dim(F_a^1 \cap \operatorname{Ker} N) + \dim(W_0(N))$$

from which it follows that both sides in (8) have the same dimension. Since the polarization statement is clear (6) and (7) follow.

As remarked above, the Nilpotent Orbit Theorem gives us a nilpotent orbit of Hodge filtrations $\exp zN \cdot F_a$. Notice that the Hodge structures on $\mathrm{Gr}^{W(N)}$ are, however, well defined as is the induced mixed Hodge structure on $W_1(N) = \operatorname{Ker}(N)$.

*Satake compactification of* $\Gamma \backslash D$

We will now sketch a description of the Satake compactification of $\Gamma \backslash D$, which we will denote by $\Gamma \backslash D^*$. The extended space $D^*$ is a union of $D$ and rational boundary components. The arithmetic group $\Gamma$ acts on $D^*$ and there are only finitely many $\Gamma$-equivalence classes of rational boundary components. Each one of these rational boundary components is a Siegel upper-half space and the action of $\Gamma$ restricts to the natural action of the modular group. The rational boundary components are in one-to-one correspondence with the rational totally isotropic subspaces of $H$. Given such a subspace $U$, we set

$$\mathcal{F}_S(U) = \{F^1 \in \check{D} : U = \operatorname{null}_Q(F^1)\}$$

where

$$\operatorname{null}_Q(F^1) = \{f \in F^1 : Q(f, \bar{g}) = 0, \forall g \in F^1\}.$$

It is clear that $\mathcal{F}_S(U)$ may be identified with the classifying space of Hodge structures of weight one on $U^\perp/U$ polarized by $\tilde{Q}$, the form induced by $Q$ on $U^\perp/U$. Thus $\mathcal{F}_S(U)$ is a Siegel upper-half space of rank $g - \nu$, $\nu = \dim U$.

Choosing now a maximal totally isotropic subspace $\Omega$ containing $U$, and a symplectic basis $\mathcal{E}$ as before, we can write the normalized period matrix of $F \in D$ as

$$(9) \qquad Z = \begin{bmatrix} Z_{11} & Z_{12} \\ {}^tZ_{12} & Z_{22} \end{bmatrix}$$

and $Z_{22}$ may be viewed as the normalized period matrix of a point in $\mathfrak{F}_S(U)$. This defines a projection:

$$(10) \qquad \pi : D \to \mathfrak{F}_S(U)$$

by $\pi(Z) = Z_{22}$. We can define this map invariantly as follows: the proof of (6) and (7) shows that any $F^1 \in D$ will define a mixed Hodge structure with the filtration $W$ given by

$$W_0 = U, \qquad W_1 = U^{\perp}$$

and moreover the Hodge structure on $Gr_1^W$ will be polarized by $\tilde{Q}$, we then have

$$\pi(F^1) = F^1(Gr_1^W).$$

In order to describe the fibers of $\pi$, we notice that given $Z$ as in (9), then $Im(Z)$ is positive definite if and only if the matrices $Im(Z_{22})$ and $Im(Z_{11}) - Im(Z_{12})(Im(Z_{22}))^{-1} Im({}^tZ_{12})$ are positive definite. This last condition characterizes $\pi^{-1}(Z_{22})$ for $Z_{22} \in \mathfrak{F}_S(U)$.

The topology of $D^*$, that is the union of $D$ and its rational boundary components may now be roughly described as follows. We begin with the notion of cylindrical neighborhoods; given $\tilde{Z} \in \mathfrak{F}_S(U)$, we consider an open neighborhood $\mathfrak{U}$ of $\tilde{Z}$ and for any $\nu \times \nu$ positive definite symmetric matrix $Y$, we define:

$$\mathcal{C}(\mathfrak{U},Y) = \{Z \in \pi^{-1}(\mathfrak{U}) : Im(Z_{11}) - Im(Z_{12})(Im(Z_{22}))^{-1} Im({}^tZ_{12}) - Y \text{ is positive definite}\}.$$

Saturating the cylindrical sets by the stabilizer $\Gamma(\tilde{Z}) = \{\gamma \in \Gamma : \gamma\tilde{Z} = \tilde{Z}\}$ one obtains a basis of neighborhoods of $\tilde{Z}$ in $D$. A similar construction gives a basis of neighborhoods for $\tilde{Z}$ in $\mathcal{F}_S(V)$ for any rational isotropic subspace $V \subset U$. This cylindrical topology induces the Satake topology on $\Gamma\backslash D^*$ ([11]).

Suppose now that $\phi : (\Delta^*)^r \to \Gamma\backslash D$ is a period mapping, and let $\sigma$ denote its monodromy cone. The boundary component associated to the isotropic subspace $W_0(\sigma)$ may be identified with the classifying space for polarized Hodge structures of weight one on $Gr_1^{W(\sigma)}$. Let $\tilde{F}$ denote the Hodge structure on $Gr_1^{W(\sigma)}$ defined by the limiting Hodge filtration $F_a$. We recall that $\tilde{F}$ is well defined independently of the choice of $F_a$.

PROPOSITION. *The period map* $\phi(t_1, \cdots, t_r)$ *converges, in the cylindrical topology, to* $\tilde{F}$ *as* $t_j \to 0$, $j = 1, \cdots, r$.

*Proof.* By the Nilpotent Orbit Theorem, we can write

$$\phi(t_1, \cdots, t_r) = \exp\left(\sum_{j=1}^{r} \frac{\log t_j}{2\pi i} \cdot N_j\right) \cdot F_a + \psi_1(t_1, \cdots, t_r)$$

where $\psi_1(t_1, \cdots, t_r)$ is holomorphic on $\Delta^r$ and $\psi_1(0, \cdots, 0) = 0$. Relative to a symplectic basis adapted to the filtration $W(\sigma)$, we can write

$$N_j = \begin{bmatrix} 0 & \eta_j \\ 0 & 0 \end{bmatrix} ; \quad \eta_j = \begin{bmatrix} \eta_j^1 & 0 \\ 0 & 0 \end{bmatrix} .$$

The $\nu \times \nu$-matrices $\eta_j^1$ are symmetric and any linear combination $\Sigma \lambda_j \eta_j^1$ with $\lambda_j > 0$ is positive definite. Let

$$W = \begin{bmatrix} W_{11} & W_{12} \\ {}^tW_{12} & W_{22} \end{bmatrix}$$

be the normalized period matrix of $F_a$ relative to the same basis, then the period matrix of $\exp\left(\Sigma \frac{\log t_j}{2\pi i} N_j\right) F_a$ is

$$\begin{bmatrix} W_{11} + \sum \frac{\log t_j}{2\pi i} \eta_j^1 & W_{12} \\ {}^tW_{12} & W_{22} \end{bmatrix}$$

and $W_{22}$ is the period matrix of $\tilde{F}$. It is then clear that given any cylindrical neighborhood $\mathcal{C}(\mathcal{U},Y)$ of $\tilde{F}$, there exists $\varepsilon > 0$ such that

$$\phi(t_1, \cdots, t_r) \in \mathcal{C}(\mathcal{U},Y) \quad \text{for} \quad |t_j| < \varepsilon ,$$

and the proposition follows. q.e.d.

*Extensions of mixed Hodge structures*

We will now discuss briefly a simple example showing the geometric information contained in the extension data of the limiting mixed Hodge structure associated to a degenerating family of curves. We refer to Carlson [3] for the general theory of extensions of mixed Hodge structures.

Let $\{V_t\}$, $t \in (\Delta^*)^r$ be a curve degeneration with central fiber $V_0$ and unipotent monodromy. We have $N_j = \log \gamma_j = (\gamma_j - I)$ and therefore, the monodromy weight filtration of the degeneration is given by

$$W_0 = \sum_j \mathrm{Im}(N_j)$$

$$W_1 = \bigcap_j \mathrm{Ker}(N_j) = \bigcap_j \mathrm{Ker}(\gamma_j - I) .$$

Since $W_1$ consists of invariant cohomology classes on $H^1(V_t)$, it may be identified with $H^1(V_0)$. Moreover, this identification $W_1 \cong H^1(V_0)$ is an isomorphism of mixed Hodge structures, where $H^1(V_0)$ carries Deligne's functorial mixed Hodge structure.

The Hodge structure on $Gr_1^W$ may be identified with $H^1(\tilde{V}_0)$, where $\pi: \tilde{V}_0 \to V_0$ is the normalization of $V_0$. As we have shown, this Hodge structure is the limit point in the Satake compactification of the period map of the degeneration.

We can also regard the mixed Hodge structure on $W_1$ as an extension

$$0 \to W_0 \to W_1 \to Gr_1^W \to 0$$

of a Hodge structure of weight one by a Hodge structure of weight zero. Carlson [3] has shown that these extensions have moduli and has used this to prove the following Torelli-type theorem:

THEOREM ([3], see also [9], [13]). *If* $V_0$ *is an irreducible curve with ordinary singularities, and* $\tilde{V}_0$ *is non-hyperelliptic, then* $V_0$ *is determined by the polarized mixed Hodge structure on* $H^1(V_0)$.

The remaining extension

$$0 \to W_1 \to H \to Gr_2^W \to 0$$

also has geometric significance. To see this, let us consider the extreme case of a degeneration of curves of genus $g$ over a product of $g$ punctured disks, whose monodromy representation, relative to a fixed symplectic basis of $H^1(V_t)$ is given by

$$\gamma_j = \begin{bmatrix} I & \Lambda_j \\ & I \end{bmatrix}$$

where $\Lambda_j$ is the elementary $g \times g$-matrix with a $1$ on the $(j,j)$-position and zeros elsewhere.

The central fiber $V_0$ is then a rational curve with $g$ nodes, $\tilde{V}_0 \cong \mathbf{P}^1$ and we have $2g$-points $p_j, q_j \in \tilde{V}_0$, $j=1,\cdots,g$ which are identified by $\pi$ to give the nodes of $V_0$.

The monodromy weight filtration of the degeneration has $W_0 = W_1$ and, therefore, the period mapping converges to a one-point component in the Satake compactification.

By the Nilpotent Orbit Theorem, we can write the normalized period matrix of the period mapping as:

$$Z(t_1, \cdots, t_g) = W(t_1, \cdots, t_g) + \sum_{j=1}^{g} \frac{\log t_j}{2\pi i} \Lambda_j$$

with $W(t_1, \cdots, t_g)$ holomorphic in $\Delta^g$. The nilpotent orbit is then

$$W(0) + \sum_{j=1}^{g} \frac{\log t_j}{2\pi i} \Lambda_j .$$

Thus, in this case, it is the off-diagonal elements of $W(0)$ that are invariantly defined. In order to see their geometric meaning, we consider as in (1), the elements

$$\omega_j(t) = \sum_k z_{kj}(t) e_k + f_j \qquad j = 1, \cdots, g$$

which form a basis for $H^{1,0}(V_t)$. In the limit the pull-back $\pi^*\omega_j(0)$ is a meromorphic form with poles of opposite residue at $p_j, q_j$. Thus, we can write

$$\pi^*\omega_j(0) = \frac{1}{2\pi i} d\left(\log\left(\frac{s-q_j}{s-p_j}\right)\right)$$

from which it follows that

$$w_{jk} = \frac{1}{2\pi i} \log\left[\left(\frac{q_j-q_k}{q_j-p_k}\right)\left(\frac{p_j-p_k}{p_j-q_k}\right)\right] .$$

We can then conclude that the nilpotent orbit of limiting mixed Hodge structures determines the cross-ratios $(p_j, p_k, q_j, q_k)$.

*The boundary components in Mumford's compactification*

We will now sketch the construction of a compactification for $\Gamma\backslash D$ whose "points at infinity" are limiting mixed Hodge structures for period mappings. As mentioned above, this will lead us naturally to Mumford's smooth compactifications of $\Gamma\backslash D$. Roughly speaking, the idea is to associate to each possible monodromy weight filtration $W$ a boundary component consisting of those filtrations in $D$ which define with $W$ a polarized mixed Hodge structure (modulo the ambiguity in the Nilpotent Orbit Theorem).

As in the construction of the Satake compactification, we begin with a rational totally $Q$-isotropic subspace $W_0$ and denote by $W_1$ the $Q$-annihilator of $W_0$. Let $\mathfrak{n}(W_0)$ denote the Lie algebra of those elements $N \in \mathfrak{g}_0 = \{X \in \mathrm{Hom}(H_{\mathbf{R}}, H_{\mathbf{R}}) : Q(X.,.) + Q(.,X.) = 0\}$ such that

$$\mathrm{Im}(N) \subset W_0 .$$

For $N \in \mathfrak{n}(W_0)$, we have $\mathrm{Ker}(N) \supset W_1$ and thus if $Q_N$ denotes the symmetric form $Q(.,N.)$, $Q_N$ induces a symmetric form $\tilde{Q}_N$ on $H/W_1$. We define

$$\mathfrak{n}^+(W_0) = \{N \in \mathfrak{n}(W_0) : \tilde{Q}_N \text{ is positive definite}\}$$

and denote by $C\ell(\mathfrak{n}^+(W_0))$ its closure in $\mathfrak{n}(W_0)$. The filtration $W$ defined by

$$\{0\} \subset W_0 \subset W_1 \subset H$$

is then the monodromy weight filtration of any element $N \in \mathfrak{n}^+(W_0)$, as well as of any cone

$$\sigma = \left\{ \sum_{i=1}^{r} \lambda_i N_i \,;\, \lambda_i \in \mathbf{R}, \lambda_i > 0 \,,\, N_i \in C\ell(\mathfrak{n}^+(W_0)) \right\}$$

which contains an element in $\mathfrak{n}^+(W_0)$. For any such cone, we define $B(\sigma) \subset \check{D}$ to be the set of all those filtrations in $\check{D}$ which, together with

W, define a mixed Hodge structure polarized by every $N \in \mathrm{Int}(\sigma)$. It is easy to see that the argument used in proving (6) and (7) gives

$$B(\sigma) = \exp(\sigma_{\mathbf{C}}) \cdot D = \exp(\mathfrak{n}_{\mathbf{C}}(W_0)) \cdot D \tag{11}$$

where $\sigma_{\mathbf{C}} = \{ \sum_{i=1}^{r} \lambda_i N_i ; \lambda_i \in \mathbf{C} \}$.

The *boundary component* associated to the cone $\sigma$ is then defined as

$$\mathcal{B}(\sigma) = B(\sigma)/\exp(\sigma_{\mathbf{C}}) .$$

Since $\exp(\sigma_{\mathbf{C}})$ acts trivially on $Gr^W$, $\mathcal{B}(\sigma)$ fibers over the Satake boundary component $\mathcal{F}_S(W_0)$ with $\xi : \mathcal{B}(\sigma) \to \mathcal{F}_S(W_0)$ given by $\xi(F) = F^1(Gr_1^W)$.

We note that if $\tau$ is a face of $\sigma$, then we may regard $B(\tau) \subset B(\sigma)$ by (11) and moreover, we have a natural projection

$$p : \mathcal{B}(\tau) \to \mathcal{B}(\sigma)$$

compatible with the above inclusion. Moreover, if we identify $D \cong B(\{0\}) = \mathcal{B}(\{0\})$, then the composition $\xi \circ p : D \to \mathcal{F}_S(W_0)$ is the map $\pi$ defined in (10).

In order to attach the boundary components to D in such a way that the extended space $D^{**}$ be compatible with the action of the arithmetic group $\Gamma$, it is necessary to introduce the notion of $\Gamma$-admissible rational polyhedral decompositions of a given $\mathfrak{n}^+(W_0)$. This is a collection $\{\sigma_a\}$ of rational polyhedral cones in $C\ell(\eta^+(W_0))$ satisfying:

i) If $\sigma \in \{\sigma_a\}$ and $\tau$ is a face of $\sigma$, then $\tau \in \{\sigma_a\}$.

ii) If $\sigma, \sigma' \in \{\sigma_a\}$ then $\sigma \cap \sigma'$ is a common face of $\sigma$ and $\sigma'$.

iii) Let $\Gamma(W_0) = \exp(\mathfrak{n}(W_0)) \cap \Gamma$, then for $\gamma \in \Gamma(W_0)$ and $\sigma \in \{\sigma_a\}$, $\gamma(\sigma) \in \{\sigma_a\}$.

iv) There are only finitely many $\Gamma(W_0)$-equivalence classes in $\{\sigma_a\}$.

v) $\mathfrak{n}^+(W_0) = \bigcup_a (\sigma_a \cap \mathfrak{n}^+(W_0))$.

Ash [1] has proved the existence of such decompositions. Moreover, they may be constructed in a manner compatible with the natural inclusion

$C\ell(\mathfrak{n}^+(W_0')) \subset C\ell(\mathfrak{n}^+(W_0))$ for $W_0' \subset W_0$ and with the action of $\Gamma$ on rational totally isotropic subspaces of $H$.

The extended space $D^{**}$ will now be the union of the boundary components $\mathcal{B}(\sigma_a)$ associated to a compatible family of admissible decompositions. It is possible to construct suitable fundamental domains for the action of $\Gamma$ on $D^{**}$; this allows us to define a topology in $D^{**}$ as in [15], inducing a Satake-type topology in $\Gamma\backslash D^{**}$. Loosely speaking, given a point $[F] \in \mathcal{B}(\sigma)$, we can regard $[F]$ as a nilpotent orbit and take a neighborhood $\mathcal{U}$ of the orbit in $D$; the cone $\sigma$ may be regarded as a subcone of $C\ell(V^+_{\mathbf{R}})$ relative to a realization of $D$ as a tube domain as in (2) and we define

$$\sigma^* = \{\alpha \in V^*_{\mathbf{R}} : \alpha(u) > 0 \text{ for all } u \in \sigma\} .$$

Given a set of generators $\alpha_1, \cdots, \alpha_r$ of $\sigma^*$ and $\lambda > 0$, we define

$$\mathcal{U}(\lambda) = \{Z \in \mathcal{U} : \alpha_j(Z) > \lambda ;\ j = 1, \cdots, r\} .$$

These are the basic neighborhoods of $[F]$ in $D$.

A similar construction gives the neighborhoods of $[F]$ in the boundary components $\mathcal{B}(\tau)$ where $\tau$ is a face $\sigma$. We refer to [4] for the details.

Although explicit, the above construction does not yield easily the analytic and algebraic properties of $\Gamma\backslash D^{**}$, it is, therefore, more convenient to construct $\Gamma\backslash D^{**}$ using the theory of toroidal embeddings. We shall not discuss this here, but refer instead to [1] and [14] (see also [12] for a very clear sketch of the construction).

## §2. *The monodromy weight filtration in the arbitrary weight case*

We now turn our attention to variations of polarized Hodge structures of arbitrary weight $n$, and in particular to the study of the monodromy weight filtration. As before, we consider a variation defined on the complement of a divisor with normal crossings, and thus we obtain, locally, a period map

$$\phi : (\Delta^*)^r \times \Delta^\ell \to \Gamma \backslash D .$$

Throughout the rest of this chapter we will assume, for simplicity, that $\ell = 0$ and—as we did in §1—that the Picard-Lefschetz transformations $\gamma_i$ are unipotent. We set $N_i = \log \gamma_i \in \mathfrak{g}_0$, and we have

$$[N_i, N_j] = 0 \quad \text{and} \quad N_i^{n+1} = 0 .$$

Schmid's Nilpotent Orbit Theorem says that $\phi$ may be approximated by a nilpotent orbit. We shall only be interested in the existence of a nilpotent orbit supported by the $N_i$'s ; i.e.,

> There exists $F \in \check{D}$ such that the map $\theta : \mathbf{C}^r \to \check{D}$, $\theta(z_1, \cdots, z_r) = \exp(\Sigma z_i N_i) \cdot F$ is a *nilpotent orbit*; that is $\theta$ is horizontal and there exists $\alpha \in \mathbf{R}$, such that $\theta(z_1, \cdots, z_r) \in D$ for $\operatorname{Im}(z_j) > \alpha$.

We note that $\theta$ is horizontal if and only if $N_j(F^p) \subset F^{p-1}$ for $j = 1, \cdots, r$ and $0 \leq p \leq n$.

DEFINITION ([8], [16]). Given a nilpotent element $N \in \mathfrak{g}_0$ such that $N^{n+1} = 0$, its *monodromy weight filtration* $W = W(N)$ is the unique increasing filtration

$$\{0\} \subset W_0 \subset \cdots \subset W_{2n-1} \subset W_{2n} = H$$

such that

(i) $N(W_\ell) \subset W_{\ell-2}$

(ii) $N^j : Gr^W_{n+j} \to Gr^W_{n-j}$ is an isomorphism, $0 \leq j \leq n$.

We recall also the definition of *primitive subspaces* $P_{n+j} \subset Gr^W_{n+j}$ :

$$P_{n+j} = \operatorname{Ker}\{N^{j+1} : Gr^W_{n+j} \to Gr^W_{n-j-2}\} \quad \text{if} \quad j \geq 0$$

$$P_{n+j} = \{0\} \quad \text{if} \quad j < 0 .$$

We have then a *Lefschetz decomposition*:

(1) $$\mathrm{Gr}_\ell^W = \bigoplus_{j \geq 0} N^j(P_{\ell+2j}) .$$

Motivated by the properties of the limiting mixed Hodge structure associated to a period mapping defined over a punctured disk $\Delta^*$, we define:

DEFINITION. A *polarized mixed Hodge structure* is a triple $(W,F,N)$, where $W$ is an increasing filtration of $H$, $F \in \check{D}$ and $N \in \mathfrak{g}_0$ is a nilpotent element such that $N^{n+1} = 0$, satisfying:

(i) $W$ is the monodromy weight filtration of $N$.

(ii) $(W,F)$ is a *mixed Hodge structure*; that is for every $\ell \geq 0$ the filtration induced by $F$ on $\mathrm{Gr}_\ell^W$ is a Hodge structure of weight $\ell$.

(iii) $N(F^p) \subset F^{p-1}$; $0 \leq p \leq n$.

(iv) For $j \geq 0$, the Hodge structure induced by $F$ on the primitive subspace $P_{n+j}$ is polarized by the bilinear form $Q_j = Q(\cdot, N^j \cdot)$.

We thus have:

THEOREM 1 (cf. Theorem (6.16) in [16]). *Let* $\theta(z) = \exp zN \cdot F$ *be a one variable nilpotent orbit, then* $(W(N),F,N)$ *is a polarized mixed Hodge structure.*

REMARKS. a) In the weight one case, Theorem 1 is equivalent to (6) and (7) in §1.

b) One can show that the converse to Theorem 1 holds; i.e., if $(W(N),F,N)$ is a polarized mixed Hodge structure, then $\theta(z) = \exp zN \cdot F$ is a nilpotent orbit. In the case of a polarized mixed Hodge structure split over $\mathbf{R}$, the resulting orbit is an $SL_2$-orbit in the sense of Schmid [16].

Schmid's proof of Theorem 1 uses his $SL_2$-orbit theorem, which does not as yet have a several variables generalization. Steenbrink's approach (cf. Chapter VII) for the geometric case also depends essentially on having a one-parameter degeneration. On the other hand, we saw that in the weight one case, the polarization conditions implied that all elements in the monodromy cone associated to a several-variables degeneration, defined the same weight filtration (cf. (4) and (5) of §1). We will now sketch a proof of this result for variations of arbitrary weight. The statement of the following theorem was conjectured by Deligne based on his analogous result ([8], 1.9.2) for the $\mathbf{Q}$-cone associated to a geometric variation. Details of the proof may be found in [6] and [7].

THEOREM 2. *Let* $\theta(z_1,\cdots,z_n) = \exp \Sigma z_j N_j \cdot F$ *be a nilpotent orbit. Let* $\sigma = \{\Sigma \lambda_i N_i ; \lambda_i \in \mathbf{R}, \lambda_i > 0\}$ *be the corresponding monodromy cone. Then all elements of* $\sigma$ *have the same monodromy weight filtration.*

We will divide the proof into four steps which are of some independent interest.

STEP 1. It is easy to check from the definition of the monodromy weight filtration that if $T_1, T_2 \in \sigma$ are such that $T_i(W_\ell(T_j)) \subset W_{\ell-2}(T_j)$; $i,j = 1,2$, then $W(T_1) = W(T_2)$. We now show that

$$(2) \qquad T_1(W_\ell(T_2)) \subset W_{\ell-1}(T_2); \quad T_1 \in C\ell(\sigma), \quad T_2 \in \sigma .$$

Indeed, for every $z \in \mathbf{C}$, the map

$$w \to \exp wT_2 \cdot (\exp zT_1 \cdot F)$$

is a nilpotent orbit and, therefore, by Theorem 1, $(W(T_2), \exp zT_1 \cdot F, T_2)$ is a polarized mixed Hodge structure for all $z \in \mathbf{C}$. One then obtains for each $j \geq 0$ a horizontal map of $\mathbf{C}$ into the classifying space for Hodge structures of weight $(n+j)$ on $P_{n+j}(T_2)$ polarized by $Q(\cdot, T_2^j \cdot)$. This map has the form

$$z \to \exp z\tilde{T}_1 \cdot \tilde{F}$$

where $\tilde{}$ denotes projection on $P_{n+j}(T_2)$, but as shown in Chapter II, such a classifying space admits an invariant metric whose holomorphic sectional curvatures in horizontal directions are negative and bounded away from zero. This implies that the map (3) must be constant from which it follows that $\tilde{T}_1 = 0$; i.e.,

$$T_1(\text{Ker } T_2^{j+1}) \subset W_{n+j-1}(T_2) .$$

Assertion (2) now follows from the Lefschetz decomposition (1).

STEP 2. Theorem 2 holds in the open and dense subcone:

$$\sigma' = \{T \in \sigma : \dim \text{Ker } T^j \text{ is minimal; } 0 \leq j \leq n\} .$$

This may be regarded as the crucial step in the proof of Theorem 2. The argument is not, however, particularly enlightening. The underlying idea is that since by (1), the dimension of $W_\ell(T)$ is constant in $\sigma'$, one may consider a map from $\sigma'$ into an appropriate Grassmann manifold; the polarization conditions imply that the differential of this map vanishes (cf. (2.11) in [7]).

STEP 3 (Reduction to the "split" case).

We say that a polarized mixed Hodge structure $(W, F, N)$ is *split* over $\mathbf{R}$ if the subspaces

$$V_\ell = \bigoplus_{p+q=\ell} (F^p \cap \bar{F}^q \cap W_\ell); \qquad 0 \leq \ell \leq 2n$$

define a splitting over $\mathbf{R}$ of the weight filtration $W$.

Let $\sigma_0$ be a connected component of $\sigma'$, and let $W(\sigma_0)$ denote the common weight filtration for elements of $\sigma_0$. We then have the following:

PROPOSITION (Deligne). *There exists a splitting* $\{V_\ell\}$ *of* $W(\sigma_0)$ *defined over* $\mathbf{R}$, *compatible with polarizations and such that* $T(V_\ell) \subset V_{\ell-2}$, *for all* $T \in \sigma_0$.

If we denote by $V^{p,q}$, $p+q = \ell$, the image of $H^{p,q}(Gr_\ell^{W(\sigma_0)})$ under the isomorphism $Gr_\ell^{W(\sigma_0)} \cong V_\ell$ and define

$$F_0^p = \bigoplus_{r \geq p} V^{r,s} ,$$

then one can check that:

(a) $F_0 \in \check{D}$.

(b) $(W(\sigma_0), F_0, T)$ is a polarized mixed Hodge structure, split over $\mathbf{R}$, for all $T \in \sigma_0$.

(c) $\exp(i\,\sigma_0) \cdot F_0 \subset D$.

The first two assertions are straightforward; (c) follows from the equivalence between split polarized mixed Hodge structures and $SL_2$-orbits that we have already mentioned.

We notice also that as a consequence of (c), if $T_0 \in \partial\sigma_0 \cap \sigma$, then $(W(T_0), F_0, T_0)$ is also a polarized mixed Hodge structure. Indeed, (c) implies that for $T \in \sigma_0$, the map

$$z \to \exp zT_0(\exp iT \cdot F_0)$$

is a nilpotent orbit and consequently $(W(T_0), \exp iT \cdot F_0)$ is a polarized mixed Hodge structure. But (2) implies that $F_0$ and $\exp iT \cdot F_0$ induce the same filtrations on $Gr^{W(T_0)}$.

STEP 4. We can now complete the proof of Theorem 2. By continuity, we have that:

$$T_0(W_\ell(\sigma_0)) \subset W_{\ell-2}(\sigma_0) .$$

On the other hand, we claim that

$$\operatorname{Ker} \{T_0^j : Gr_{n+j}^{W(\sigma_0)} \to Gr_{n-j}^{W(\sigma_0)}\} = \{0\} .$$

Because of (b), it is enough to check that

$$\operatorname{Ker} T_0^j \cap F_0^p \cap \bar{F}_0^q = \{0\} \quad \text{for} \quad p+q = n+j ,$$

but $\operatorname{Ker} T_0^j \cap F_0^p \cap \bar{F}_0^q \subset W_{n+j-1}(T_0) \cap F_0^p \cap \bar{F}_0^q = \{0\}$ since $(W(T_0), F_0)$ is a mixed Hodge structure. Therefore,

$$T_0^j : Gr_{n+j}^{W(\sigma_0)} \to Gr_{n-j}^{W(\sigma_0)}$$

is an isomorphism for all $j = 0, \cdots, n$, hence, $W(T_0) = W(\sigma_0)$ and consequently $\sigma' = \sigma$. q.e.d.

Similar arguments may be used to study the relation between the monodromy weight filtration of the cone $\sigma$ and those associated to the faces of $\sigma$. If $\tau$ is a face of $\sigma$, then for any $T \in \sigma$ and $0 \leq \ell \leq 2n$ we have a nilpotent endomorphism

$$\tilde{T} : Gr_\ell^{W(\tau)} \to Gr_\ell^{W(\tau)}$$

such that $\tilde{T}^{\ell+1} = 0$. The following theorem was also conjectured by Deligne.

THEOREM 3. *The monodromy weight filtration of* $\tilde{T}$ *is the projection of* $W(\sigma)$ *on* $Gr_\ell^{W(\tau)}$.

We refer to [7] for the proof.

We note that as a consequence of the proof of Theorem 2, given a nilpotent orbit $\theta(z) = \exp(\Sigma z_i N_i) \cdot F$ there exists a "split" nilpotent orbit $\theta_0(z) = \exp(\Sigma z_i N_i) \cdot F_0$, such that $F$ and $F_0$ define the same Hodge structures on the graded quotients $Gr_\ell^{W(\sigma)}$.

PROBLEM. Estimate the asymptotic distance between $\theta(z)$ and $\theta_0(z)$.

Although $F_0$ is not unique, one might hope that with the proper choice of $F_0$, the orbit $\theta_0(z)$ would be the right analog of Schmid's $SL_2$-orbit in the several variables case.

We finish this chapter with a brief, and mostly conjectural, discussion of the compactification question in the arbitrary weight case. Theorem 2, together with the Nilpotent Orbit Theorem imply that we can associate to a period map a nilpotent orbit of polarized mixed Hodge structures. Thus, we would like those orbits to be the "points at infinity" of a mixed Hodge theoretic compactification of $\Gamma\backslash D$. In the weight one case—because of the equivalence between polarized mixed Hodge structures and nilpotent orbits—it is an easy matter to classify nilpotent orbits: modulo $\Gamma$, the monodromy cones correspond to subcones of $\mathfrak{n}^{+}(\Omega)$, where $\Omega$ is a maximal rational isotropic subspace, and the Hodge filtrations $B(\Omega)$ are an open subset of $\check{D}$. In the arbitrary weight case, one should be able to reduce the study of nilpotent orbits to the split ones; however, we do not have a classification of such orbits.

One of the obstacles to producing such a classification is that we only have polarization conditions on the primitive subspaces, and these depend on the element of the monodromy cone. Even very basic questions about the primitive subspaces are open: for example, Deligne—based on his computation of the Goresky-MacPherson cohomology of $\Delta^2$, with values on the local system defined by a variation of polarized Hodge structures of weight $n$ on $(\Delta^*)^2$ —has conjectured that

$$\text{Ker } N_1 N_2 \subset \text{Ker } N_1 + \text{Ker } N_2 + W_{n+1}(\sigma) \,. \tag{4}$$

For $n \leq 3$, one can show that (4) is a consequence of Theorems 2 and 3; however, the general case remains open.

We should also point out that in the arbitrary weight case, the horizontality condition implies that the set of filtrations defining a polarized mixed Hodge structure with a given weight filtration will be of high codimension in $\check{D}$.

## REFERENCES

[1] A. Ash, D. Mumford, M. Rapoport and Y. Tai, "Smooth compactifications of locally symmetric varieties," Math. Sci. Press, Brookline, 1975.

[2] W. L. Baily and A. Borel, "Compactification of arithmetic quotients of bounded symmetric domains," Ann. of Math. 84 (1966), pp. 442-528.

[3] J. Carlson, "Extensions of mixed Hodge structures," Journées de géometrie algébrique d'Angers, 1979, Sijthoff & Noordhoff, 1980, pp. 107-127.

[4] J. Carlson, E. Cattani and A. Kaplan, "Mixed Hodge structures and compactifications of Siegel's space," Journées de géometrie algébrique d'Angers, 1979, Sijthoff & Noordhoff, 1980, pp. 77-105.

[5] E. Cattani and A. Kaplan, "Extension of period mappings for Hodge structures of weight two," Duke Math. J. 44 (1977), pp. 1-43.

[6] ————, "The monodromy weight filtration for a several variables degeneration of Hodge structures of weight two," Inv. math. 52 (1979), pp. 131-142.

[7] ————, "Polarized mixed Hodge structures and the local monodromy of a variation of Hodge structure," Inv. math. 67 (1982), pp. 101-115.

[8] P. Deligne, "La conjecture de Weil, II," Publ. Math. I.H.E.S. 52 (1980).

[9] R. Friedman, "Hodge theory, degenerations and the global Torelli problem," Thesis, Harvard University (1981).

[10] P.A. Griffiths, "Periods of integrals on algebraic manifolds: summary of main results and discussion of open problems," Bull. Amer. Math. Soc. 76 (1970), pp. 228-296.

[11] P. Kiernan, "On the compactifications of arithmetic quotients of symmetric spaces," Bull. Amer. Math. Soc. 80 (1974), pp. 109-110.

[12] D. Mumford, "A new approach to compactifying locally symmetric varieties," Proc. International Colloquium on Discrete Subgroups of Lie Groups and Applications to Moduli, Bombay, 1973, Oxford University Press, 1975, pp. 211-224.

[13] Y. Namikawa, "On the canonical holomorphic map from the moduli space of stable curves to the Igusa monoidal transform," Nagoya J. Math. 52 (1973), pp. 197-259.

[14] ————, "Toroidal compactification of Siegel spaces," Lect. Notes in Math., 812, Springer, Berlin-Heidelberg-New York, 1980.

[15] I. Satake, "On compactification of the quotient spaces for arithmetically defined discontinuous subgroups," Ann. of Math. (2) 72 (1960), pp. 555-580.

[16] W. Schmid, "Variation of Hodge structure: the singularities of the period mapping," Inv. math. 22 (1973), pp. 211-319.

[17] ————, "Abbildungen in arithmetische Quotienten hermitesch symmetrischer Räume," Lect. Notes in Math., 412, Springer, Berlin-Heidelberg-New York, 1974, pp. 243-258.

[18] J. Steenbrink, "Limits of Hodge structures," Inv. math. 31 (1976), pp. 229-257.

EDUARDO H. CATTANI
MATHEMATICS DEPARTMENT
UNIVERSITY OF MASSACHUSETTS
AMHERST, MA 01003

## Chapter VI

# THE CLEMENS-SCHMID EXACT SEQUENCE AND APPLICATIONS

David R. Morrison*

Let $\mathfrak{X}^* \to \Delta^*$ be a smooth family of complex varieties over the punctured disk, and let $\mathfrak{X} \to \Delta$ be a completion to a proper family over the disk, with Kähler total space. The Clemens-Schmid exact sequence relates the topology and Hodge theory of the central fibre of such a map to that of a smooth fibre by means of the monodromy of the family $\mathfrak{X}^* \to \Delta^*$. This sequence yields rather strong restrictions on the monodromy of such a family, and much information about the cohomology and monodromy of the smooth fibre can be derived from the central fibre alone. In this exposition, we have chosen to separate the topological and Hodge theoretic aspects of the sequence, for two reasons. The first is that setting up this exact sequence requires a great deal of linear algebra; by presenting a topological version first (and postponing the discussion of mixed Hodge structures) we sidestep certain technical complications until the reader has been (we hope) sufficiently motivated to wade through them. But a second, more important, reason is that many of the applications of the Clemens-Schmid sequence depend only on this topological version (i.e. depend on the weight filtrations and not the Hodge filtrations), a fact which is often ignored in the literature. It should be pointed out that the proof (which we do not give here) does not appear to separate into topological and Hodge theoretic parts.

Our main references have been [2], [4], [8] and [11]. The citations in the text are intended as guides to the reader, and do not pretend to assign proper credit to the many who have worked in this subject.

---

*Partially supported by the National Science Foundation.

§1. *Semistable degenerations*

Let $\Delta$ denote the unit disk. A *degeneration* is a proper flat holomorphic map $\pi : \mathfrak{X} \to \Delta$ of relative dimension n such that $\mathfrak{X}_t = \pi^{-1}(t)$ is a smooth complex variety for $t \neq 0$, and $\mathfrak{X}$ is a Kähler manifold. A degeneration is *semistable* if the central fibre $\mathfrak{X}_0$ is a divisor with (global) normal crossings; in other words, writing $\mathfrak{X}_0 = \Sigma X_i$ as a sum of irreducible components, each $X_i$ is smooth and the $X_i$'s meet transversally so that locally $\pi$ is defined by

$$t = x_1 x_2 \cdots x_k .$$

The fundamental fact about degenerations is the

SEMISTABLE REDUCTION THEOREM (Mumford [5]). *Given a degeneration* $\pi : \mathfrak{X} \to \Delta$ *there exists a base change* $b : \Delta \to \Delta$ *(defined by* $t \to t^N$ *for some* N *), a semistable degeneration* $\psi : \mathfrak{Y} \to \Delta$ *and a diagram*

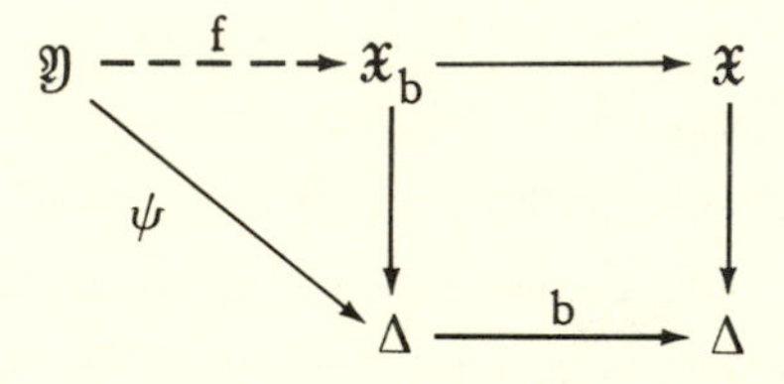

*such that* $f : \mathfrak{Y} \dashrightarrow \mathfrak{X}_b$ *is a bimeromorphic map obtained by blowing up and blowing down subvarieties of the central fibre.*

With this theorem in hand, statements about degenerations which are invariant under blowups, blowdowns, and basechange can of course be proved by considering the special case of semistable degenerations.

Given a semistable degeneration, one can give a fairly precise description of the cohomology of its total space $\mathfrak{X}$. The first step is to construct a retraction $r : \mathfrak{X} \to \mathfrak{X}_0$ which induces isomorphisms $r^* : H^m(\mathfrak{X}_0, \mathbb{Q}) \xrightarrow{\sim} H^m(\mathfrak{X}, \mathbb{Q})$ and $r_* : H_m(\mathfrak{X}, \mathbb{Q}) \to H_m(\mathfrak{X}_0, \mathbb{Q})$. (The details of this construction can be found in Clemens [1] or Persson [8].) We then describe the cohomology of $\mathfrak{X}_0$ by means of a Mayer-Vietoris type spectral sequence, as follows.

Let

$$X_{i_0 \cdots i_p} = X_{i_0} \cap \cdots \cap X_{i_p} ,$$

define the *codimension* p *stratum* of $\mathfrak{X}_0$ as

$$\mathfrak{X}^{[p]} = \bigsqcup_{i_0 < \cdots < i_p} X_{i_0 \cdots i_p}$$

(disjoint union), and let $\iota_p : \mathfrak{X}^{[p]} \to \mathfrak{X}$ be the natural map. Choose an open cover $\mathfrak{U}$ of a neighborhood of $\mathfrak{X}_0$ in $\mathfrak{X}$ such that

(1) for each $U \in \mathfrak{U}$, $\pi|_U$ is defined by

$$t = x_1 \cdots x_k$$

in suitable local coordinates

(2) $\check{H}^*(\mathfrak{U} \cap \mathfrak{X}_0, \mathbb{Q}) \cong \check{H}^*(\mathfrak{X}_0, \mathbb{Q})$

(3) $\check{H}^*(\iota_p^{-1}(\mathfrak{U}), \mathbb{Q}) = \check{H}^*(\mathfrak{X}^{[p]}, \mathbb{Q})$.

Let $E_0^{p,q} = \check{C}^q(\iota_p^{-1}(\mathfrak{U}), \mathbb{Q})$ and define differentials

$d: E_0^{p,q} \to E_0^{p,q+1}$ the Čech coboundary

$\delta : E_0^{p,q} \to E_0^{p-1,q}$ the combinatorial coboundary induced by

$$(*) \qquad \delta\phi(V \cap X_{j_0 \cdots j_{q+1}}) = \sum_{\alpha} (-1)^{\alpha} \phi(V \cap X_{j_0 \cdots \hat{j}_\alpha \cdots j_{q+1}}) .$$

THEOREM. *The spectral sequence with* $E_0$ *term as above and*

$$E_1^{p,q} = \check{H}^q(\mathfrak{X}^{[p]}, \mathbb{Q})$$

*degenerates at* $E_2$, *and converges to* $\check{H}^*(\mathfrak{X}_0, \mathbb{Q})$.

*Proof.* Consider first the opposite spectral sequence

$${}^{op}E_0^{p,q} = E_0^{q,p} .$$

It is easy to check that the complex $0 \to \Gamma(\iota_0^{-1}V, \mathbb{Q}) \xrightarrow{\delta} \Gamma(\iota_1^{-1}V, \mathbb{Q}) \xrightarrow{\delta} \cdots$ has homology $\Gamma(V \cap \mathfrak{X}_0, \mathbb{Q})$ (in degree zero only), for all open sets $V$ satisfying (1). Thus,

$$ {}^{\mathrm{op}}E_1^{p,q} = \begin{cases} \check{C}^p(\mathfrak{U} \cap \mathfrak{X}_0, \mathbb{Q}) & \text{for } q = 0 \\ 0 & \text{for } q \neq 0 \end{cases} $$

and

$$ {}^{\mathrm{op}}E_\infty^{p,q} = {}^{\mathrm{op}}E_2^{p,q} = \begin{cases} \check{H}^p(\mathfrak{X}_0, \mathbb{Q}) & \text{for } q = 0 \\ 0 & \text{for } q \neq 0 \end{cases} $$

which proves the convergence.

To prove the degeneration of the spectral sequence, we introduce its de Rham analogue. Let ${}^{DR}E_0^{p,q} = A^q(\mathfrak{X}^{[p]})$ be the complex $C^\infty$ q-forms on $\mathcal{X}^{[p]}$, with differentials

$$ d : A^q(\mathcal{X}^{[p]}) \to A^{q+1}(\mathcal{X}^{[p]}) \quad \text{the exterior derivative} $$

$$ \delta : A^q(\mathcal{X}^{[p]}) \to A^q(\mathcal{X}^{[p+1]}) \quad \text{induced by } (*). $$

We have ${}^{DR}E_1^{p,q} = H^q_{DR}(\mathcal{X}^{[p]})$, and it is proved in Griffiths and Schmid [4, p. 71] that this spectral sequence degenerates at $E_2$ ; the essential point is the "principle of two types" for forms on a Kähler manifold. This in fact implies that the original $E_1^{p,q}$ degenerates at $E_2$, as we sketch below.

The differentials $d_1$ on $E_1^{p,q}$, $E_1^{p,q} \otimes \mathbb{C}$, and ${}^{DR}E_1^{p,q}$ are all defined by the combinatorial formula (*). We thus get a commutative diagram for $r \geq 1$

$$ \begin{array}{ccc} E_r^{p,q} & \xrightarrow{d_r} & E_r^{p,q+1} \\ \downarrow & & \downarrow \\ E_r^{p,q} \otimes \mathbb{C} & \xrightarrow{d_r} & E_r^{p,q+1} \otimes \mathbb{C} \\ \wr\wr & & \wr\wr \\ {}^{DR}E_r^{p,q} & \xrightarrow{d_r} & {}^{DR}E_r^{p,q+1} \end{array} $$

which implies the degeneration of $E_r^{p,q}$. q.e.d.

We use this spectral sequence to put a filtration, called the *weight filtration,* on $H^m(\mathfrak{X}_0, \mathbf{Q})$ (and hence on $H^m(\mathfrak{X}, \mathbf{Q})$ ) as follows: define

$$W_k = \bigoplus_{q \leq k} E_0^{*,q}$$

and let $W_k(H^m)$ be the induced filtration on cohomology. (This is *not* the usual filtration associated to a spectral sequence, but is more convenient here for technical reasons.) Notice that

$$0 \subset W_0(H^m) \subset W_1(H^m) \subset \cdots \subset W_m(H^m) = H^m .$$

Thus, letting $Gr_k = W_k/W_{k-1}$ denote the graded pieces, we have $Gr_k(H^m) = E_2^{m-k,k}$, and $Gr_k(H^m) = 0$ if $k < 0$ or $k > m$.

We also put a weight filtration on homology $H_m(\mathfrak{X}_0, \mathbf{Q})$ (or $H_m(\mathfrak{X}, \mathbf{Q})$ ) by duality:

$$W_{-k}(H_m) = \mathrm{Ann}(W_{k-1}(H^m)) = \{h \in H_m | (W_{k-1}(H^m), h) = 0\} .$$

With this definition,

$$Gr_k(H_m) \cong (Gr_{-k}(H^m))^*$$

so that $Gr_k(H_m) = 0$ if $k < -m$ or $k > 0$.

We conclude this section with an alternate description of the 0th graded piece of the weight filtration. Define the *dual graph* $\Gamma$ of $\mathfrak{X}_0$ to be a simplicial complex with one vertex $P_i$ for each component $X_i$ of $\mathfrak{X}_0$, such that the simplex $\langle P_{i(0)}, \cdots, P_{i(k)} \rangle$ belongs to $\Gamma$ if and only if $X_{i(0)} \cdots {}_{i(k)} \neq \emptyset$. Then $E_1^{p,0} = H^0(\mathcal{X}^{[p]})$, so that $H^0(\mathcal{X}^{[0]}) \xrightarrow{\delta} H^0(\mathcal{X}^{[1]}) \xrightarrow{\delta} \cdots$ is the Čech complex for $\Gamma$; thus $Gr_0(H^m) \cong E_2^{m,0} \cong H^m(|\Gamma|)$.

## §2. *The monodromy weight filtration*

In this section and the next, all cohomology groups have $\mathbf{Q}$ coefficients unless otherwise specified.

Let $\pi : \mathfrak{X} \to \Delta$ be a degeneration, and $\pi^* : \mathfrak{X}^* \to \Delta^*$ be the restriction to the punctured disk. Fix a smooth fibre $\mathfrak{X}_t$. Since $\pi^*$ is a $C^\infty$ fibration, $\pi_1(\Delta^*)$ acts on the cohomology $H^m(\mathfrak{X}_t)$. The map

$$T : H^m(\mathfrak{X}_t) \to H^m(\mathfrak{X}_t)$$

induced by the canonical generator of $\pi_1(\Delta^*)$ is called the *Picard-Lefshetz transformation.* We have the

MONODROMY THEOREM (Landman [7]).

(1) $T$ *is quasi-unipotent, with index of unipotency at most* $m$. *In other words, there is some* $k$ *such that*

$$(T^k - I)^{m+1} = 0 .$$

(2) *If* $\pi : \mathfrak{X} \to \Delta$ *is semistable, then* $T$ *is unipotent* $(k=1)$.

Thanks to this theorem, we may define the logarithm of $T$ in the semistable case by the finite sum

$$N = \log T = (T-I) - \frac{1}{2}(T-I)^2 + \frac{1}{3}(T-I)^3 - \cdots .$$

$N$ is nilpotent, and the index of unipotency of $T$ coincides with the index of nilpotency of $N$; in particular, $T = I$ if and only if $N = 0$.

Associated to this nilpotent map $N$ with $N^{n+1} = 0$ is an increasing filtration of $\mathbf{Q}$-subspaces

$$0 \subset W_0 \subset W_1 \subset \cdots \subset W_{2m} = H^m(\mathfrak{X}_t)$$

called the *monodromy weight filtration,* which is defined inductively as follows: first let $W_0 = \operatorname{Im} N^m$ and $W_{2m-1} = \operatorname{Ker} N^m$. Now fix some $\ell < m$; if

$$0 \subset W_{\ell-1} \subset W_{2m-\ell} \subset W_{2m} = H^m(\mathfrak{X}_t)$$

have already been defined in such a way that $N^{m-\ell+1}(W_{2m-\ell}) \subset W_{\ell-1}$, then we define

$$W_{\ell}/W_{\ell-1} = \mathrm{Im}(N^{m-2}|_{W_{2m-\ell}/W_{\ell-1}})$$

$$W_{2m-\ell-1}/W_{\ell-1} = \mathrm{Ker}(N^{m-\ell}|_{W_{2m-\ell}/W_{\ell-1}}),$$

and $W_\ell$, $W_{2m-\ell-1}$ to be the corresponding inverse images. Notice that $W_\ell/W_{\ell-1} \subset W_{2m-\ell-1}/W_{\ell-1}$ so that $W_\ell \subset W_{2m-\ell-1}$. Clearly, $N^{m-1}(W_{2m-\ell-1}) \subset W_\ell$, so that the inductive hypotheses are satisfied.

We collect below the important linear algebra facts about this filtration; these follow from the existence of a representation of $S\ell(2;\mathbf{Q})$ on H extending the action of N. (For indications of proof, see Griffiths [3, p. 255].)

PROPOSITION. *Let* $K = \mathrm{Ker}\, N$, $\mathrm{Gr}_k(H) = W_k/W_{k-1}$, $\mathrm{Gr}_k(K) = (W_k \cap K)/(W_{k-1} \cap K)$

(1) $N(W_k) \subset W_{k-2}$

(2) $N(W_k) = (\mathrm{Im} N) \cap W_{k-2}$

(3) $N^k : \mathrm{Gr}_{m+k}(H) \xrightarrow{\sim} \mathrm{Gr}_{m-k}(H)$

(4) *Properties (1) and (3) uniquely determine the filtration.*

(5) *If* $k \leq m$,

$$\mathrm{Gr}_k(H) \cong \bigoplus_{\alpha=0}^{[k/2]} \mathrm{Gr}_{k-2\alpha}(K)$$

(6) *If* $0 < k \leq m$, $N^k : H^m(\mathfrak{X}_t) \to H^m(\mathfrak{X}_t)$ *is the zero map if and only if* $W_{m-k} = 0$; *if and only if* $W_{m-k} \cap K = 0$.

(7) $\mathrm{Gr}_m(H)/_{\mathrm{Im}(N : \mathrm{Gr}_{m+2}(H) \to \mathrm{Gr}_m(H))} \cong \mathrm{Gr}_m(K)$.

§3. *The Clemens-Schmid exact sequence*

Let $\mathfrak{X} \to \Delta$ be a semistable degeneration, $\mathfrak{X}_t$ a fixed smooth fibre, and $i : \mathfrak{X}_t \subset \mathfrak{X}$ the inclusion. We denote $H^m(\mathfrak{X}_t)$ by $H^m_{\lim}$, $H^m(\mathfrak{X}) \cong H^m(\mathfrak{X}_0)$ by $H^m$, and $H_m(\mathfrak{X}) \cong H_m(\mathfrak{X}_0)$ by $H_m$. $H^m_{\lim}$, $H^m$, and $H_m$ all carry filtrations which we have called *weight filtrations*; we define a *weighted vector space* to be a $\mathbf{Q}$-vector space H together with an increasing filtration of $\mathbf{Q}$-subspaces

$$0 \subset \cdots \subset W_k(H) \subset W_{k+1}(H) \subset \cdots \subset H$$

called the *weight filtration.* A *morphism of weighted vector spaces of type* $(r,r)$ is a linear map $\phi : H \to H'$ such that $\phi(W_k(H)) = W_{k+2r}(H') \cap \operatorname{Im} \phi$.

The Clemens-Schmid exact sequence studies the homomorphism $N : H^m_{\lim} \to H^m_{\lim}$ (which is a morphism of weighted vector spaces of type $(-1,1)$ by property (2) of the monodromy weight filtration). The first piece of the sequence is the

LOCAL INVARIANT CYCLE THEOREM. *The sequence*

$$H^m \xrightarrow{i^*} H^m_{\lim} \xrightarrow{N} H^m_{\lim}$$

*is exact. In other words, all cocycles which are invariant under the monodromy action come from cocycles in* $\mathfrak{X}$.

However, the Clemens-Schmid theorem says more. Let

$$p : H_{2n+2-m}(\mathfrak{X}) \to H^m(\mathfrak{X}, \partial\mathfrak{X})$$

and

$$p_t : H^m(\mathfrak{X}_t) \to H_{2n-m}(\mathfrak{X}_t)$$

be the Poincaré duality maps, and define $\alpha : H_{2n+2-m} \to H^m$ as the composite

$$H_{2n+2-m}(\mathfrak{X}) \xrightarrow{p} H^m(\mathfrak{X}, \partial\mathfrak{X}) \longrightarrow H^m(\mathfrak{X})$$

and $\beta : H^m_{\lim} \to H_{2n-m}$ as the composite

$$H^m(\mathfrak{X}_t) \xrightarrow{p_t} H_{2n-m}(\mathfrak{X}_t) \xrightarrow{i_*} H_{2n-m}(\mathfrak{X}) .$$

We can now state

CLEMENS-SCHMID I. *The maps* $\alpha$, $i^*$, $N$, $\beta$ *are morphisms of weighted vector spaces of types* $(n+1, n+1)$, $(0,0)$, $(-1,-1)$ *and* $(-n,-n)$ *respectively, and the sequence*

$$\longrightarrow H_{2n+2-m} \xrightarrow{\alpha} H^m \xrightarrow{i^*} H^m_{\lim} \xrightarrow{N} H^m_{\lim} \xrightarrow{\beta} H_{2n-m} \xrightarrow{\alpha} H^{m+2} \longrightarrow$$

*is exact.*

Notice that our definition of morphism included strictness (with a shift of indices) of the weight filtrations; thus, this exact sequence induces exact sequences of the filtered and graded pieces.

COROLLARY 1. *Let* $K^m$ *denote* $\mathrm{Ker}(N : H^m_{\lim} \to H^m_{\lim})$. *Then* $W_k(H^m) \xrightarrow{\sim} W_k(K^m)$ *for* $k < m$. *In particular,* $W_{m-1}(H^m) \xrightarrow{\sim} W_{m-1}(K^m)$ *as weighted vector spaces.*

*Proof.* We only need to check that $W_{k-2n-2}(H_{2n+2-m}) = 0$ for $k < m$, or equivalently, that $Gr_{k-2n-2}(H_{2n+2-m}) = 0$ for all $k < m$. But in Section 1 we saw that $Gr_j(H_\ell) = 0$ if $j < -\ell$; clearly $k-2n-2 < -(2n+2-m)$. q.e.d.

Properties (3) and (5) of the monodromy weight filtration show that the graded pieces of $H^m_{\lim}$ can be recovered from the graded pieces of $\mathrm{Ker}\, N$; the above corollary allows us to determine all but one of those pieces in terms of the weight filtration on $H^m$.

COROLLARY 2. *For* $k > 0$, $N^k : H^m_{\lim} \to H^m_{\lim}$ *is the zero map if and only if* $W_{m-k}(H^m) = 0$. *In particular,* $N^{m+1}$ *is always zero, and* $N^m = 0$ *if and only if* $H^m(|\Gamma|) = 0$.

*Proof.* Property (6) of the monodromy weight says that $N^k = 0$ if and only if $W_{m-k}(H^m_{\lim}) = 0$. By the previous corollary, this is true if and only if $W_{m-k}(H^m) = 0$. The second statement follows from the isomorphism

$$W_0(H^m) \cong Gr_0(H^m) \cong H^m(|\Gamma|) . \qquad \text{q.e.d.}$$

COROLLARY 3. *The following sequence is exact*

$$0 \to Gr_{m-2}K^{m-2} \to Gr_{m-2n-2}H_{2n+2-m} \to Gr_m H^m \to Gr_m K^m \to 0 .$$

*Proof.* This follows from the strictness of morphisms in the Clemens-Schmid sequence. The only thing that requires checking is

$$\mathrm{Ker}(\mathrm{Gr}_{m-2n-2}H_{2n+2-m} \xrightarrow{\alpha} \mathrm{Gr}_m H^m) \cong \mathrm{Gr}_{m-2}K^{m-2} .$$

But this kernel is isomorphic to

$$\mathrm{Gr}_{m-2}H^{m-2}_{\lim} \big/ \mathrm{Im}(\mathrm{Gr}_m H^{m-2}_{\lim} \xrightarrow{N} \mathrm{Gr}_{m-2}H^{m-2}_{\lim})$$

which, by property (7) of the monodromy weight filtration, is isomorphic to $\mathrm{Gr}_{m-2}K^{m-2}$. q.e.d.

Corollary 2 allows us to compute the index of nilpotency of $N$ from $H^m(\mathfrak{X}_0)$, and Corollary 3 (together with induction) enables us to compute the remaining graded piece of $H^m_{\lim}$. We shall apply these results in some special cases in the next section.

## §4. *Applications*

(a) First cohomology groups

Since $H_{2n+1} = H_{2n+1}(\mathcal{X}_0) = 0$, the Clemens-Schmid sequence becomes

$$0 \longrightarrow H^1 \longrightarrow H^1_{\lim} \xrightarrow{N} H^1_{\lim} .$$

Hence, $\mathrm{Ker}\, N \cong H^1$ as weighted vector spaces. We compute the graded pieces

$$\mathrm{Gr}_2 H^1_{\lim} \cong \mathrm{Gr}_0 H^1_{\lim} \cong \mathrm{Gr}_0 H^1 \cong H^1(|\Gamma|)$$

$$\mathrm{Gr}_1 H^1_{\lim} \cong \mathrm{Gr}_1 H^1 \cong \mathrm{Ker}(H^1(\mathcal{X}^{[0]}) \to H^1(\mathcal{X}^{[1]}) .$$

If we let $\Phi = \dim \mathrm{Gr}_1 H^1$, then

$$b_1(\mathfrak{X}_t) = \Phi + 2h^1(|\Gamma|)$$

$$N = 0 \iff h^1(|\Gamma|) = 0 .$$

(b) Degenerations of curves

The above analysis implies that a semistable degeneration of curves has infinite monodromy if and only if there are cycles in the dual graph of the central fibre (a classical result). The typical picture of such a degeneration is

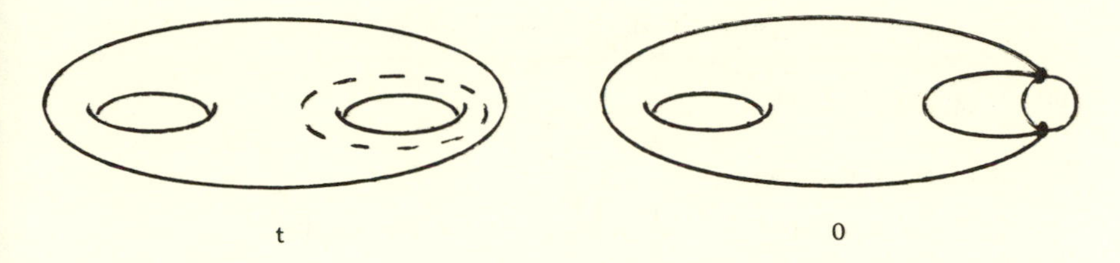


in which the dotted cycle on $\mathfrak{X}_t$ has become a cycle in the dual graph.

In the case of curves, $\Phi = \dim H^1(\mathcal{X}^{[0]}) = 2\,\Sigma\, g(X_i)$, so that an alternate criterion for $N = 0$ is that $g(\mathfrak{X}_t) = \Sigma\, g(X_i)$.

(c) Degenerations of surfaces

Since $N = 0$ on $H^0_{\lim}$, the Clemens-Schmid sequence for $H^2$ breaks into two pieces:

$$0 \longrightarrow H^0 \xrightarrow{i^*} H^0_{\lim} \longrightarrow 0$$

$$0 \longrightarrow H^0_{\lim} \xrightarrow{\beta} H_4 \xrightarrow{\alpha} H^2 \xrightarrow{i^*} H^2_{\lim} \xrightarrow{N} H^2_{\lim}$$

We let

$\Phi = \dim \operatorname{Ker}(H^1(\mathcal{X}^{[0]}) \to H^1(\mathcal{X}^{[1]}))$ as above,

$q = \frac{1}{2} h^1(\mathcal{X}^{[0]})$ the sum of the irregularities of the components

$g = \frac{1}{2} h^1(\mathcal{X}^{[1]})$ the sum of the genera of the double curves.

The 0th and 1st graded pieces of $H^2_{\lim}$ can be computed by Corollary 1 of Section 3:

$$\mathrm{Gr}_0 H^2_{\mathrm{lim}} \cong \mathrm{Gr}_0 H^2 \cong H^2(|\Gamma|)$$

$$\mathrm{Gr}_1 H^2_{\mathrm{lim}} \cong \mathrm{Gr}_1 K^2 \cong \mathrm{Gr}_1 H^2 \cong H^1(\mathcal{X}^{[1]}) \Big/ \mathrm{Im}(H^1(\mathcal{X}^{[0]}) \to H^1(\mathcal{X}^{[1]})) .$$

Thus,

$$\dim \mathrm{Gr}_0 H^2_{\mathrm{lim}} = h^2(|\Gamma|)$$

$$\dim \mathrm{Gr}_1 H^2_{\mathrm{lim}} = \Phi - 2q + 2g$$

and we get the following

MONODROMY CRITERIA.

(1) $N = 0$ *on* $H^1_{\mathrm{lim}} \iff h^1(|\Gamma|) = 0 \iff b_1(\mathfrak{X}_t) = \Phi$.

(2) $N^2 = 0$ *on* $H^2_{\mathrm{lim}} \iff h^2(|\Gamma|) = 0$.

(3) $N = 0$ *on* $H^2_{\mathrm{lim}} \iff h^2(|\Gamma|) = 0$ *and* $\Phi + 2g = 2q$.

Computing $\mathrm{Gr}_2 H^2_{\mathrm{lim}}$ is somewhat more difficult. If we know the Betti numbers of the smooth fibre, the easiest way is to note that

$$b_2(\mathfrak{X}_t) = \dim \mathrm{Gr}_2 H^2_{\mathrm{lim}} + 2 \dim \mathrm{Gr}_1 H^2_{\mathrm{lim}} + 2 \dim \mathrm{Gr}_0 H^2_{\mathrm{lim}}$$

so that

$$\dim \mathrm{Gr}_2 H^2_{\mathrm{lim}} = b_2(\mathfrak{X}_t) - 2\Phi + 4q - 4g - 2h^2(|\Gamma|) .$$

To compute it directly (which will yield an expression for $b_2(\mathfrak{X}_t)$ ) we use

$$\mathrm{Gr}_{-4} H_4 \cong (\mathrm{Gr}_4 H^4)^* = (H^4(\mathcal{X}^{[0]}))^*$$

$$\mathrm{Gr}_2 H^2 = \mathrm{Ker}(H^2(\mathcal{X}^{[0]}) \to H^2(\mathcal{X}^{[1]}))$$

together with Corollary 3 of Section 3 to get

$$\dim \mathrm{Gr}_2 K^2 = h^0(|\Gamma|) - \#\{X_i\} + \dim \mathrm{Ker}(H^2(\mathcal{X}^{[0]}) \to H^2(\mathcal{X}^{[1]}))$$

and hence

$$\dim \mathrm{Gr}_2 H^2_{\lim} = h^2(|\Gamma|) + h^0(|\Gamma|) - \#\{X_i\} + \dim \mathrm{Ker}(H^2(\mathcal{X}^{[0]}) \to H^2(\mathcal{X}^{[1]})) .$$

(d) Degenerations of K3 surfaces

We will illustrate the monodromy criteria for surface degenerations with K3 surfaces. We start with the

THEOREM (Kulikov [6], Persson and Pinkham [9]). *A semistable degeneration of* K3 *surfaces is birational to one for which the central fibre* $\mathfrak{X}_0$ *is one of three types:*

*Type I.* $\mathfrak{X}_0$ *is a smooth* K3 *surface.*

*Type II.* $\mathfrak{X}_0 = X_0 \cup X_1 \cup \cdots \cup X_{k+1}$. $X_\alpha$ *meets only* $X_{\alpha\pm1}$, *and each* $X_\alpha \cap X_{\alpha+1}$ *is an elliptic curve.* $X_0$ and $X_{k+1}$ *are rational surfaces, and for* $1 \leq \alpha \leq k$, $X_\alpha$ *is ruled with* $X_\alpha \cap X_{\alpha-1}$ *and* $X_\alpha \cap X_{\alpha+1}$ *sections of the ruling.*

*Type III. All components of* $\mathfrak{X}_0$ *are rational surfaces,* $X_i \cap (\bigcup_{j\neq i} X_j)$ *is a cycle of rational curves, and* $|\Gamma| = S^2$.

We now apply the monodromy criteria in each case:

*Type I.* $\mathfrak{X}_0 = X$ is regular, so $q = \Phi = 0$. $\mathfrak{X}^{[1]} = \emptyset$ so $g = 0$ as well; $|\Gamma|$ is a point, so $h^2(|\Gamma|) = 0$ and we conclude $N = 0$.

*Type II.* $|\Gamma| = [0,1]$ so that $h^1(|\Gamma|) = h^2(|\Gamma|) = 0$. Since $b_1(\mathfrak{X}_t) = 0$, we conclude $\Phi = 0$.

The components $X_0$ and $X_{k+1}$ are regular, while the $X_\alpha$ for $1 \leq \alpha \leq k$ have irregularity $1$. Thus, $q = k$. The double curves all have genus $1$, so that $g = k+1$. Hence, $\Phi - 2q + 2g = 2 \neq 0$ so that $N^2 = 0$ but $N \neq 0$. The monodromy weight filtration looks like:

$$0 \subset \mathbf{Q}^2 \subset \mathbf{Q}^{20} \subset \mathbf{Q}^{22} \subset \mathbf{Q}^{22} .$$

*Type III.* $h^2(|\Gamma|) = 1 \neq 0$ so that $N^2 \neq 0$. To compute the rest of the weight filtration, note that $\Phi = b_1(\mathfrak{X}_t) - 2h^1(|\Gamma|) = 0$; all the components are rational so that $q = 0$, and all the double curves are rational so that $g = 0$. Thus, $\dim \mathrm{Gr}_1 H^1_{\lim} = 0$ and the monodromy weight filtration looks like

$$\mathbf{Q}^1 \subset \mathbf{Q}^1 \subset \mathbf{Q}^{21} \subset \mathbf{Q}^{21} \subset \mathbf{Q}^{22} .$$

§5. *Mixed Hodge structures*

The Clemens-Schmid sequence contains more information than the topological version presented in Section 3; to explain this, we must introduce mixed Hodge structures.

A *mixed Hodge structure* is a lattice $H_{\mathbf{Z}}$ together with an increasing filtration $W_m = W_m(H)$ (called the *weight filtration*) of $H_{\mathbf{Q}} = H_{\mathbf{Z}} \otimes_{\mathbf{Z}} \mathbf{Q}$, and a decreasing filtration $F^p = F^p(H)$ (called the *Hodge filtration*) of $H = H_{\mathbf{Z}} \otimes_{\mathbf{Z}} \mathbf{C}$, such that the induced Hodge filtrations on $\mathrm{Gr}_m = W_m/W_{m-1}$ define a Hodge structure of weight $m$. More precisely, if we define

$$F^p(\mathrm{Gr}_m) = W_m \cap F^p / W_{m-1} \cap F^p$$

then

$$\mathrm{Gr}_m \cong F^p(\mathrm{Gr}_m) \oplus \overline{F^{m-p+1}(\mathrm{Gr}_m)}$$

for all $p$.

A *morphism of type* $(r,r)$ between two mixed Hodge structures $H_{\mathbf{Z}}$, $W_m(H)$, $F^p(H)$ and $H'_{\mathbf{Z}}, W_m(H')$, $F^p(H')$ is a $\mathbf{Q}$-linear map

$$\phi : H_{\mathbf{Q}} \to H'_{\mathbf{Q}}$$

such that

$$\phi(W_m(H)) \subset W_{m+2r}(H')$$

$$\phi(F^p(H)) \subset F^{p+r}(H') .$$

Notice that such a morphism restricts to a morphism of type $(r,r)$ between the weighted vector spaces $\{H_{\mathbf{Q}}, W_m(H)\}$ and $\{H'_{\mathbf{Q}}, W_m(H')\}$ (using fact (1) below for strictness).

We collect below some linear algebra facts about mixed Hodge structures: proofs can be found in [4].

PROPOSITION.

(1) *A morphism of mixed Hodge structures is strict with respect to both filtrations; in other words,*

$$\phi(W_m(H)) = W_{m+2r}(H') \cap \operatorname{Im} \phi$$

$$\phi(F^p(H)) = F^{p+r}(H') \cap \operatorname{Im} \phi .$$

(2) *If* $\phi : H_Q \to H'_Q$ *is a morphism of mixed Hodge structures, then the induced weight and Hodge filtrations define mixed Hodge structures on* $\operatorname{Ker} \phi$ *and* $\operatorname{Coker} \phi$.

(3) *If* $H_Z$ *carries a mixed Hodge structure, its dual* $H_Z^* = \operatorname{Hom}(H_Z, Z)$ *inherits a mixed Hodge structure with filtrations*

$$W_{-k}(H^*) = \operatorname{Ann}(W_{k-1}(H))$$

$$F^{-p}(H^*) = \operatorname{Ann}(F^{p+1}(H)) .$$

Each of the weighted vector spaces occurring in the Clemens-Schmid sequence actually underlies a mixed Hodge structure. As lattices, we take the integral cohomology modulo torsion; we need only define the Hodge filtrations.

For the cohomology of $\mathfrak{X}_0$, we describe the mixed Hodge structure by means of the spectral sequence

$${}^{DR}E_0^{k,\ell} = A^\ell(\mathcal{X}^{[k]}) \Rightarrow H^*(\mathcal{X}_0, C)$$

introduced in Section 1. Define

$$F^p(A^{k,\ell}) = \bigoplus_{r \geq p} H^{r,\ell-r}(\mathcal{X}^{[k]})$$

as a filtration on the $E_0$ term. This induces filtrations in the $E_n$ terms and on $H^*(\mathfrak{X}_0, C)$; on the $E_1$ term,

$$ {}_{DR}E_1^{k,\ell} = H^\ell_{DR}(\mathcal{X}^{[k]}) $$

we get the usual Hodge filtration. Furthermore, the first differential $d_1$ gives a morphism of Hodge structures; thus,

$$ W_\ell(H^m)/_{W_{\ell-1}(H^m)} \cong {}_{DR}E_2^{m-\ell,\ell} $$

inherits a Hodge structure as well, so that $H^*(\mathfrak{X}_0, C)$ carries a mixed Hodge structure. This is a special case of a theorem of Deligne that every variety carries a canonical functorial mixed Hodge structure.

For the monodromy weight filtration, we let $\pi: \mathfrak{X} \to \Delta$ be a semistable degeneration as usual, and let $f: \mathfrak{h} \to \Delta^*$, $f(z) = e^{2\pi iz}$, be the universal cover. For each $z \in \mathfrak{h}$, there is a canonical isomorphism of $H^m(\pi^{-1}(f(z)))$ with our fixed group $H^m_{\lim} = H^m(\mathcal{X}_t)$. In particular, there are Hodge filtrations $F^p(z)$ on $H^m_{\lim}$ with the property that $TF^p(z) = F^p(z+1)$.

THEOREM (Schmid [10]). *The limit*

$$ F^p_\infty = \lim_{\operatorname{Im}(z)\to\infty} \exp(-zN)\, F^p(z) $$

*exists, and the filtrations* $F^p_\infty$ *and* $W_k(H^m_{\lim})$ *define a mixed Hodge structure on* $H^m_{\lim}$, *called the* limiting mixed Hodge structure. *Furthermore,* $N: H^m_{\lim} \to H^m_{\lim}$ *is a morphism of mixed Hodge structures of type* $(-1,-1)$.

We can now state the Hodge theoretic version of the Clemens-Schmid sequence:

CLEMENS-SCHMID II. *The morphisms in the Clemens-Schmid sequence are morphisms of mixed Hodge structures (of the appropriate types).*

§6. *Further applications*

(a) Degenerations of surfaces

We represent the limiting mixed Hodge structure on $H^2_{\lim}$ pictorially as follows

$$\begin{array}{ccccccc}
 & & H^{2,2} & & & & Gr_4 \\
 & H^{2,1} & & H^{1,2} & & & Gr_3 \\
H^{2,0} & & H^{1,1} & & H^{0,2} & & Gr_2 \\
 & H^{1,0} & & H^{0,1} & & & Gr_1 \\
 & & H^{0,0} & & & & Gr_0
\end{array}$$

so that

$$F^2 \cong H^{2,0} \oplus H^{2,1} \oplus H^{2,2}$$

$$F^1/_{F^2} \cong H^{1,0} \oplus H^{1,1} \oplus H^{1,2}$$

$$F^1/_{F^1} \cong H^{0,0} \oplus H^{0,1} \oplus H^{0,2} .$$

Since $N : H^{2,1} \xrightarrow{\sim} H^{1,0}$ and $N^2 : H^{2,2} \xrightarrow{\sim} H^{0,0}$, we get a formula

$$h^{2,0}(\mathfrak{X}_t) = \dim F^2 \, Gr_2(H^2_{lim}) + \left(\frac{1}{2}\, \Phi - q + g\right) + h^2(|\Gamma|) .$$

We can restrict the Clemens-Schmid sequence to

$$F^{-1}Gr_{-4}H_4 \longrightarrow F^2 Gr_2 H^2 \longrightarrow F^2 Gr_2 H^2_{lim} \xrightarrow{N} F^1 Gr_0 H^2_{lim} .$$

But

$$F^{-1}Gr_{-4}H_4 = \mathrm{Ann}(F^2 Gr_4 H^4) = 0$$

and

$$F^1 Gr_0 H^2_{lim} = 0$$

so that

$$F^2 Gr_2 H^2_{lim} \simeq F^2 Gr_2 H^2 \simeq H^{2,0}(\mathfrak{X}^{[0]}) .$$

Thus,

$$p_g(\mathfrak{X}_t) = \Sigma\, p_g(X_i) + \left(\frac{1}{2}\, \Phi - q + g\right) + h^2(|\Gamma|) .$$

Since $\frac{1}{2}\Phi - q + q \geq 0$, we get $p_g(\mathfrak{X}_t) \geq \Sigma\, p_g(X_i)$, and some

ALTERNATE CRITERIA FOR MONODROMY

(1) $N = 0$ on $H^1_{\lim} \iff h^1(|\Gamma|) = 0$

(2) $N^2 = 0$ on $H^2_{\lim} \iff h^2(|\Gamma|) = 0$

(3) $N = 0$ on $H^2_{\lim} \iff p_g(\mathfrak{X}_t) = \Sigma\, p_g(X_i)$.

A formula for $h^{1,1}(\mathfrak{X}_t)$ can be derived in a similar manner, but it is usually more efficient to simply compute both $h^{2,0}(\mathfrak{X}_t)$ and $b_2(\mathfrak{X}_t)$.

(b) The geometric genus

The analysis of the geometric genus above extends easily to higher dimensions. Let $\pi: \mathfrak{X} \to \Delta$ be a semistable degeneration with fibre dimension $n$. As above, we have

$$F^n \cong H^{n,0}_{\lim} \oplus H^{n,1}_{\lim} \oplus \cdots \oplus H^{n,n}_{\lim}.$$

The relevant part of the Clemens-Schmid sequence is

$$F^{-1}\mathrm{Gr}_{-n-2}H_{n+2} \longrightarrow F^n\mathrm{Gr}_n H^n \longrightarrow F_n\mathrm{Gr}_n H^n_{\lim} \xrightarrow{N} F^{n-1}\mathrm{Gr}_{n-2}H^n_{\lim}$$

while

$$F^{n-1}\mathrm{Gr}_{n-2}H^n_{\lim} = 0$$

and

$$F^{-1}\mathrm{Gr}_{-n-2}H_{n+2} = \mathrm{Ann}(F^2\mathrm{Gr}_{n+2}H^{n+2}) = 0$$

so that

$$F^n\mathrm{Gr}_n H^n_{\lim} \simeq F_n\mathrm{Gr}_n H^n = H^{n,0}(\mathfrak{X}^{[0]}).$$

Thus, $\dim H^{n,0}_{\lim} = \Sigma\, p_g(X_i)$, so that

(**) $$p_g(\mathfrak{X}_t) \geq \Sigma\, p_g(X_i).$$

Notice that $N = 0$ implies equality in (**). The converse is not true; however, if equality holds, then $H_{\lim}^{n,n} = H_{\lim}^{n,n-1} = 0$, which implies $H_{\lim}^{n-1,n} = 0$ and hence $Gr_n = Gr_{n-1} = 0$. Thus, we have the

GEOMETRIC GENUS CRITERION.

(1) $p_g(\mathfrak{X}_t) \geq \Sigma\, p_g(X_i)$.

(2) If $N = 0$, then equality holds in (1).

(3) If equality holds in (1), then $N^{n-1} = 0$.

## REFERENCES

[1] C.H. Clemens, "Degenerations of Kähler manifolds," Duke Math J., 44(1977), pp. 215-290.

[2] M. Cornalba and P. Griffiths, "Some transcendental aspects of algebraic geometry," Proc. Symp. Pur. Math. 29(1974), pp. 3-110.

[3] P. Griffiths, "Periods of integrals on algebraic manifolds: Summary of main results and discussion of open problems," Bull. A.M.S. 76 (1970), pp. 228-296.

[4] P. Griffiths and W. Schmid, "Recent developments in Hodge theory: a discussion of techniques and results," in Discrete Subgroups of Lie Groups, Bombay, Oxford Univ. Press (1973), pp. 31-127.

[5] G. Kempf et al., "Toroidal Embeddings I," Lecture Notes in Math. 339(1973).

[6] V. Kulikov, "Degenerations of K3 surfaces and Enriques surfaces," Math USSR Izvestija 11(1977), pp. 957-989.

[7] A. Landman, "On the Picard-Lefschetz transformations," Trans A.M.S. 181 (1973), pp. 89-126.

[8] U. Persson, "Degenerations of algebraic surfaces," Mem A.M.S. 189 (1977).

[9] U. Persson and H. Pinkham, "Degenerations of surfaces with trivial canonical bundle," Ann. of Math. 113(1981), pp. 45-66.

[10] W. Schmid, "Variation of Hodge structure: the singularities of the period mapping," Inventiones Math. 22(1973), pp. 211-320.

[11] L. Tu, "Variation of Hodge structure and the local Torelli problem," Thesis, Harvard University (1979).

DAVID R. MORRISON
MATHEMATICS DEPARTMENT
PRINCETON UNIVERSITY
PRINCETON, NJ 08544

Chapter VII

# DEGENERATION OF HODGE BUNDLES (AFTER STEENBRINK)

Steven Zucker

Let $(\mathcal{H}, \{\mathcal{F}^p\}, \Delta)$ be the bundle data associated to a polarized variation of Hodge structure on the punctured disc $\Delta^*$. Typically, it will come from the cohomology of a family of projective varieties over the disc $\Delta$, where the variety over the origin may be singular. Schmid has shown that there is a "limit" mixed Hodge structure [10, (6.25)], as described in Chapter IV. The approach is Lie theoretic and differential geometric; the existence of the mixed Hodge structure is deduced directly from the variation of Hodge structure, without regard to where it came from.

If the variation of Hodge structure comes from geometry, there is another way to arrive at the limit mixed Hodge structure, for one can use analytic deRham theory at the central fiber. This is the point of view in [12] (and its sequel [13]) and also [2].[1] Since the existence of the limit mixed Hodge structure has already been discussed in the seminar, the direction of this chapter will be slightly different. We aspire to three goals (in the three sections of this article):

1. An explanation of the language in which [12] is written (namely that of [5, §8]), with the twin hope of making Steenbrink's work more

[1]These two papers, though written in different mathematical languages, are actually quite similar. My understanding of the history is that [2] was conceived well before the appearance of the first version of [12] (Steenbrink's thesis [11]), but the details were worked out independently, though subsequently. Clemens wanted to see whether his approach would yield a fundamentally different construction. He was also interested in another matter, namely a deRham theoretic proof of the Hodge norm asymptotics of [10, (6.6)]; this appears in the last section of his paper.

accessible to the reader, and of presenting Deligne's "recipe" for placing mixed Hodge structures on cohomology groups that arise in algebraic geometry as the natural way of making the construction.

2. Formulas for the "canonical" extension of Hodge bundles in terms of deRham cohomology, that follow from Steenbrink's work. In the process, we will present his construction of the limit mixed Hodge structure.

3. A brief description of some directions of application of the results that are presented in §2.

We will use the following notation and conventions throughout this chapter:

Geometric situation

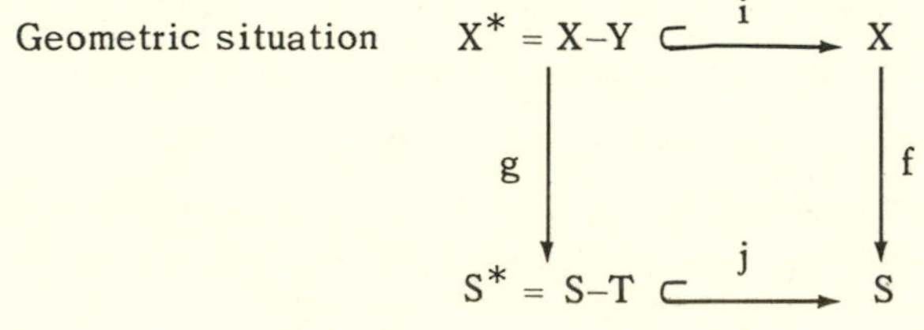


where: X is a smooth projective variety (more generally, a compact Kähler manifold),

S is a smooth complete curve,

T is the image of the singular fibers of f (a finite subset of S ),

$Y = f^{-1}(T)$ is a divisor with normal crossings.

When we wish to work locally near a point of T, we will understand that the global data is replaced by $S^* = \Delta^*$, $S = \Delta$, $T = \{0\}$, and $X, f, Y$ are to be correspondingly restricted. Local and global notation will sometimes be used concurrently; it is an important theme that local calculations imply global formulas.

## §1. *Cohomological mixed Hodge complexes*

In this section, we discuss the construction of mixed Hodge structures from the algebraic point of view of Deligne. We start from the beginning with the following well-known example.

Let X be a compact Kähler manifold of dimension n, $\Omega_X^p$ the sheaf of holomorphic p-forms on X, $\mathcal{E}_X^{p,q}$ the sheaf of $C^\infty$ forms on X of type $(p,q)$. Then one has the following diagram of sheaves:

$$
(1)\qquad
\begin{array}{ccccccccccc}
 & & \vdots & & \vdots & & & & & & \\
 & & \uparrow & & \uparrow & & \Big[\ \uparrow & & \uparrow & & \\
 & & \mathcal{E}_X^{0,1} & \to & \mathcal{E}_X^{1,1} & \to \cdots \to & \Big|\ \mathcal{E}_X^{p,1} & \to \cdots \to & \mathcal{E}_X^{n,1} & \to & 0 \\
 & & \uparrow \bar{\partial} & & \uparrow & & \Big|\ \uparrow & & \uparrow & & \\
 & & \mathcal{E}_X^{0,0} & \xrightarrow{\partial} & \mathcal{E}_X^{1,0} & \to \cdots \to & \Big|\ \mathcal{E}_X^{p,0} & \to \cdots \to & \mathcal{E}_X^{n,0} & \to & 0 \\
 & & \uparrow & & \uparrow & & \Big|\ \uparrow & & \uparrow & & \\
C_X & \to & \Omega_X^0 & \to & \Omega_X^1 & \to \cdots \to & \Big[\ \Omega_X^p & \to \cdots \to & \Omega_X^n & \to & 0\,.
\end{array}
$$

We see three complexes:

$$
(2)\qquad
\begin{array}{ll}
C_X & \text{a single sheaf placed in degree } 0 \\
\big\downarrow \alpha & \\
\Omega_X^{\cdot} & \Omega_X^i \text{ in degree } i \\
\big\downarrow \beta & \\
\mathcal{E}_X^{\cdot} & \text{in degree } i\,,\ \mathcal{E}_X^i = \bigoplus_{r+s=i} \mathcal{E}_X^{r,s}\,.
\end{array}
$$

The mappings $\alpha$ and $\beta$ are *quasi-isomorphisms*: they induce isomorphisms on cohomology sheaves (Poincaré Lemma). As such, the three complexes should be regarded as being equivalent, for they all have for natural reasons the same (hyper-) cohomology, namely the complex cohomology of $X$. (In the derived category, they become isomorphic by decree.)

The filtration $F$ on $\Omega_X^{\cdot}$ and $\mathcal{E}_X^{\cdot}$ is indicated in (1), with $F^p$ being the subcomplex to the right of the vertical line. On $\Omega_X^{\cdot}$ it is determined by truncation from below, "filtration bete." ( N.B.– $\Omega_X^p$ is still in degree $p$ .)

Before going on with the example, we give some general remarks about filtered complexes. Let $K^{\cdot}$ be a complex of (say) sheaves on a space $X$, with decreasing filtration $F$ $(F^{p+1} \subset F^p)$. The successive quotients are denoted

$$\mathrm{Gr}_F^p K^{\cdot} = F^p K^{\cdot} / F^{p+1} K^{\cdot} . \tag{3}$$

Let $\iota^p : F^p K^{\cdot} \to K^{\cdot}$ denote the inclusion. One puts a filtration on $H^i(X, K^{\cdot})$ by

$$F^p H^i(X, K^{\cdot}) = \mathrm{image}\{\iota_*^p : H^i(X, F^p K^{\cdot}) \to H^i(X, K^{\cdot})\} . \tag{4}$$

There is a spectral sequence

$$_F E_1^{p,q} = H^{p+q}(X, \mathrm{Gr}_F^p K^{\cdot}) \tag{5}$$

whose limit is

$$_F E_\infty^{p,q} = \mathrm{Gr}_F^p H^{p+q}(X, K^{\cdot}) , \tag{6}$$

the successive quotients of (4).

We return to the case of $K^{\cdot} = \Omega_X^{\cdot}$. Here, the successive quotients (3) are

$$\mathrm{Gr}_F^p \Omega_X^{\cdot} = \Omega_X^p[-p] , \tag{7}$$

where the $[-p]$ refers to a shift of $p$ degrees to the right (so (7) gives $\Omega_X^p$ in degree $p$ ). Thus, (5) becomes in this case the familiar

$$_F E_1^{p,q} = H^q(X, \Omega_X^p) . \tag{8}$$

The following is a consequence of Hodge theory:

THEOREM 1. i) *The filtration* $F$ *on* $H^i(X, \Omega_X^{\cdot}) = H^i(X, \mathbb{C})$ *is the Hodge filtration of a Hodge structure of weight* $i$.

ii) *The spectral sequence (8) degenerates at* $E_1$. *Thus,* $\iota_*^p$ *is injective for all* $p$, *in all degrees* $i$.

Theorem 1 follows from the existence of a set of cohomology representatives in $H^0(X, \mathcal{E}^i_X)$ that are of pure bidegree, closed under complex conjugation, namely the harmonic forms relative to any Kähler metric on $X$. Though we are using $\mathcal{E}^\cdot_X$ as a filtered resolution of $\Omega^\cdot_X$, it is less cluttered to discuss the Hodge structure in terms of $\Omega^\cdot_X$, and we shall always do so.[2]

One makes the following definition [5, (8.1.2)]:

DEFINITION 1. A *cohomological Hodge complex of weight* $k$ consists of the data $(K^\cdot_{\mathbf{Z}}; K^\cdot, F)$, where

i) $K^\cdot_{\mathbf{Z}}$ is a complex of sheaves of $\mathbf{Z}$-modules,

ii) $K^\cdot$ is a complex of sheaves of $\mathbf{C}$-modules, with decreasing filtration $F$,

iii) $K^\cdot_{\mathbf{Z}} \otimes_{\mathbf{Z}} \mathbf{C}$ is quasi-isomorphic to $K^\cdot$,

iv) the spectral sequence (5) degenerates at $E_1$,

v) $F$ induces on $H^i(X, K^\cdot_{\mathbf{Z}})$ a Hodge structure of weight $i + k$.

In this language, $(\mathbf{Z}_X; \Omega^\cdot_X, F)$ is a cohomological Hodge complex of weight $0$.

For convenience, we will suppress the $K^\cdot_{\mathbf{Z}}$ in Definition 1, and refer to the filtered complex $K^\cdot$ as a cohomological Hodge complex defined over $\mathbf{Z}$. The variants "over $\mathbf{Q}$" or "over $\mathbf{R}$" should require no explanation.

In order to motivate the definition of a cohomological mixed Hodge complex, we recall briefly the construction of the mixed Hodge structure of an open variety $X - Y$ ($Y$ a divisor with normal crossings) [4, §3]. Let $\Omega^\cdot_X(\log Y)$ denote the complex of meromorphic differentials on $X$ with at worst logarithmic poles along $Y$. At a point where $Y$ has local equation $\Pi_{j=1}^{\ell} z_j = 0$, $\Omega^\cdot_X(\log Y)$ is obtained by adjoining $dz_j/z_j$ $(1 \leq j \leq \ell)$

[2] However, it may be convenient to bring back $C^\infty$ forms for other purposes, such as calculating periods.

to $\Omega^{\cdot}_X$. In addition to the filtration F (defined as for $\Omega^{\cdot}_X$), $\Omega^{\cdot}_X(\log Y)$ has a second "weight" filtration W that is increasing: a form lies in $W_k$ at a point if and only if it can be written as a sum of terms having poles along at most k local components of Y. In other words,

$$W_k\Omega^p_X(\log Y) = \Omega^k_X(\log Y)\wedge\Omega^{p-k}_X . \tag{9}$$

Whenever one has two filtrations on a complex $K^{\cdot}$, as above, there are induced filtrations,

$$F^p(Gr^W_k K^{\cdot}) = (F^p\cap W_k)/(F^p\cap W_{k-1}) .$$

Via Poincaré residues, one sees that

$$Gr^W_k\Omega^{\cdot}_X(\log Y) \simeq a^{(k)}_*\Omega^{\cdot}_{\tilde{Y}^{(k)}}[-k] , \tag{10}$$

where $a^{(k)}:\tilde{Y}^{(k)}\to X$ is the obvious mapping of the disjoint union $\tilde{Y}^{(k)}$ of the k-fold intersection of components of Y into X $(\tilde{Y}^{(0)} = X)$. The isomorphism (10) respects F with a shift of k. Since $\tilde{Y}^{(k)}$ is a compact Kähler manifold, $Gr^W_k\Omega^{\cdot}_X(\log Y)$ is a cohomological Hodge complex, whose weight we will carefully determine:[3]

$$\begin{aligned} H^i(X,F^pGr^W_k\Omega^{\cdot}_X(\log Y)) &\simeq F^{p-k}H^{i-k}(\tilde{Y}^{(k)},\mathbb{C}) \\ &\simeq F^p[H^{i-k}(\tilde{Y}^{(k)},\mathbb{C})\langle -k\rangle] , \end{aligned} \tag{11}$$

where $\langle -k\rangle$ indicates a Tate twist: the tensor product with the one-dimensional Hodge structure of type (k,k). We see that F induces a Hodge structure of weight $i+k$ on $H^i(X,Gr^W_k\Omega^{\cdot}_X(\log Y))$.

We now adapt [5, (8.1.6)] to our slightly informal language:

DEFINITION 2. A *cohomological mixed Hodge complex* on X consists of a complex $K^{\cdot}$ of sheaves of $\mathbb{C}$-modules with increasing filtration W

[3] The filtration W is defined over $\mathbb{Z}$, in a sense that the reader may be able to infer (see [4, (3.1.8)] for the precise meaning of this assertion).

and decreasing filtration $F$, such that $W$ is defined over $\mathbb{Q}$, and $(\mathrm{Gr}_k^W K^{\cdot}, F)$ is a cohomological Hodge complex of weight $k$ defined over $\mathbb{Q}$.

Thus, $K^{\cdot} = \Omega_X^{\cdot}(\log Y)$, with its filtrations $W$ and $F$, is a cohomological mixed Hodge complex. One has by [4, (3.2)],

$$H^i(X-Y, \mathbb{C}) \simeq H^i(X, \Omega_X^{\cdot}(\log Y)), \tag{12}$$

and its mixed Hodge structure is induced by $W$ and $F$.

The spectral sequence, analogous to (5), for $W$ begins

$$_W E_1^{p,q} = H^{p+q}(X, \mathrm{Gr}_{-p}^W K^{\cdot}) \tag{13}$$

(as the relabelling $W^k = W_{-k}$ converts $W$ to a decreasing filtration). For purposes of mixed Hodge theory, we violate the analogue of (4) by setting[4]

$$W_{i+k} H^i(X, K^{\cdot}) = \mathrm{image}\{H^i(X, W_k K^{\cdot}) \to H^i(X, K^{\cdot})\}. \tag{14}$$

The significance of Definition 2 lies in the following:

THEOREM 2 ([5, (8.1.9)]). *Let* $(K^{\cdot}, W, F)$ *be a cohomological mixed Hodge complex. Then:*

i) $F$ *and* $W$ *define on* $H^i(X, K^{\cdot})$ *a mixed Hodge structure,*

ii) *The spectral sequence (5) for* $F$ *degenerates at* $E_1$, *and (13) degenerates at* $E_2$.

The key idea in the proof of Theorem 2 is that the differentials $d_r$ of the spectral sequence (13) are compatible with both $F$ and the underlying real structure. In particular, $d_1$ is a morphism of Hodge structures, since it goes between Hodge structures of the same weight. On the other hand, if $r \geq 2$, one sees recursively that $d_r$ goes in the direction of decreasing

---

[4] The definition (4) suggests that the right-hand side of (14) should be called $W_k H^i(X, K^{\cdot})$. Somewhere along the way, the shift by $i$ on $H^i$ must be made. I feel that it is best done *after* taking cohomology, as above; if it were done before, $W$ would fail to give a filtration by subcomplexes.

weight, so is necessarily the zero mapping. This imparts a Hodge structure to the ${}_W E_\infty$ terms. One sees that it is induced directly by $F$ on $\mathrm{Gr}^W_{i+k} H^i(X, K^\cdot)$, because the hypotheses of the lemma on two filtrations [4, (1.3.17)] are satisfied.

All mixed Hodge structures currently used in algebraic geometry are deduced from a cohomological mixed Hodge complex. It should be stressed that one must be scrupulously careful about the numeration in using Theorem 2; then, one has a routine procedure for producing mixed Hodge structures.

## §2. *Extensions of Hodge bundles and the limit mixed Hodge structure*

We recall the bundle data associated to a variation of Hodge structure on $S^*$. By convention, we identify a bundle with its locally free sheaf of sections. A sub-bundle then corresponds to a locally free subsheaf such that the quotient is also locally free. Thus, we have a bundle $\mathcal{H}$ on $S^*$, with connection

$$\nabla : \mathcal{H} \to \Omega^1_{S^*} \otimes \mathcal{H} \tag{15}$$

and flat real structure; and a decreasing filtration $\{\mathcal{F}^p\}$ of $\mathcal{H}$ by sub-bundles such that

$$\nabla(\mathcal{F}^p) \subset \Omega^1_{S^*} \otimes \mathcal{F}^{p-1}, \tag{16}$$

with $\mathcal{F}^p$ inducing a Hodge structure on each fiber of $\mathcal{H}$.

By a general construction [3, p. 95], we select an extension $\mathcal{H}_e$ of $\mathcal{H}$ to all of $S$ as follows. Locally, on a $\Delta^*$, the bundle $\mathcal{H}$ is holomorphically trivial.[5] An extension of $\mathcal{H}$ to $\Delta$ is then specified by a choice of frame, i.e., a locally free subsheaf of $j_*\mathcal{H}$ equaling $\mathcal{H}$ on $\Delta^*$. Then $\mathcal{H}_e$ is uniquely determined by the two properties:

---

[5] In one variable, this follows for topological reasons. In several variables (on $(\Delta^*)^n$, $n > 1$), the same is true, though one must use the integrability of the connection to see it. In fact, a frame for $\mathcal{H}_e$ is constructed from the multi-valued horizontal sections of $\mathcal{H}$ and their monodromy.

(17) i) The connection $\nabla$ extends to have logarithmic poles on T:

$$\nabla_e : \mathcal{H}_e \to \Omega^1_S(\log T) \otimes \mathcal{H}_e .$$

ii) The eigenvalues $\lambda$ of the residues of $\nabla_e$ on T satisfy $0 \leq \operatorname{Re} \lambda < 1$.

REMARKS. i) Condition (17, i) selects the frame up to a change of basis meromorphic at $0 \in \Delta$, whereupon (17, ii) determines it completely.

ii) The construction above has nothing to do with the variation of Hodge structure. In its presence, the $\lambda$'s are real numbers, rational if the variation of Hodge structure is.

It follows from the Nilpotent Orbit Theorem [10, (4.9)] (see Chapter IV) that the Hodge filtration bundles $\mathcal{F}^p$ extend to sub-bundles $\mathcal{F}^p_e$ of $\mathcal{H}_e$ (and then, necessarily, $\mathcal{F}^p_e = \mathcal{H}_e \cap j_* \mathcal{F}^p$). The limit mixed Hodge structure lives naturally on the fiber of $\mathcal{H}_e$ at $0 \in \Delta$; its Hodge filtration is given by the fibers of $\{\mathcal{F}^p_e\}$ at the origin.

Suppose now that $\mathcal{H}$ is the bundle of i-th cohomology of a family of varieties. Then we can write

$$\mathcal{H} = R^i g_* \mathbf{C} \otimes_{\mathbf{C}} \mathcal{O}_{S^*} \simeq R^i g_* \Omega^{\cdot}_{X^*/S^*} . \tag{18}$$

Likewise, we can realize $\mathcal{F}^p$ as the image of the *injection* (compare Theorem 1 (ii))

$$R^i g_* F^p \Omega^{\cdot}_{X^*/S^*} \to \mathcal{H} \tag{19}$$

and also

$$\mathcal{F}^p / \mathcal{F}^{p+1} \simeq R^i g_* \mathrm{Gr}^p_F \Omega^{\cdot}_{X^*/S^*} \simeq R^{i-p} g_* \Omega^p_{X^*/S^*} . \tag{20}$$

We seek extensions of these formulas for $\mathcal{H}_e$ and $\mathcal{F}^p_e$. Recall the relative log complex:

$$\Omega^{\cdot}_{X/S}(\log Y) = \Omega^{\cdot}_X(\log Y) / f^* \Omega^1_S(\log T) \wedge \Omega^{\cdot}_X(\log Y)[-1] . \tag{21}$$

In [8], Katz showed that $R^m f_* \Omega^{\cdot}_{X/S}(\log Y)$, modulo its torsion subsheaf (supported on $T$), satisfies (17). Thus,

$$\mathcal{H}_e \simeq R^i f_* \Omega^{\cdot}_{X/S}(\log Y)/\text{torsion}; \tag{22}$$

this is the regularity theorem for the Gauss-Manin connection.

Consider the locally free sheaves

$$\tilde{\mathcal{F}}^p = R^i f_* F^p \Omega^{\cdot}_{X/S}(\log Y)/\text{torsion}. \tag{23}$$

They, as is easily seen, inject into $\mathcal{H}_e$, so we identify them with their image. One is led to guess that (23) gives $\mathcal{F}^p_e$. A priori, it is clear that $\tilde{\mathcal{F}}^p \subset \mathcal{F}^p_e$; it is possible though, that generators of $\tilde{\mathcal{F}}^p$ vanish at the origin in $\mathcal{H}_e$. In actuality, the best possible result is true:

THEOREM 3 ([13, (2.11)]). *All sheaves* $R^q f_* \Omega^p_{X/S}(\log Y)$ *are locally free.*

From Theorem 3, it follows that the spectral sequence for $F$

$$E_1^{p,q} = R^q f_* \Omega^p_{X/S}(\log Y) \Rightarrow R^{p+q} f_* \Omega^{\cdot}_{X/S}(\log Y)$$

degenerates at $E_1$, the torsion sheaves in (23) are all zero, and $\mathcal{H}_e/\tilde{\mathcal{F}}^p$ is locally free for all $p$. This implies the desired formulas:

COROLLARY (cf. [15, (2.4)]). $\mathcal{F}^p_e = R^i f_* F^p \Omega^{\cdot}_{X/S}(\log Y)$.

We will outline the proof of Theorem 3. The argument uses the de Rham theoretic construction of the limit mixed Hodge structure. One uses the basic criterion:

PROPOSITION 1. *Let* $X_s = f^{-1}(s)$. *Then for a complex* $K^{\cdot}$ *consisting of coherent sheaves,* $R^i f_* K^{\cdot}$ *is locally free if*

$$d(s) = \dim H^i(X_s, K^{\cdot} \otimes \mathcal{O}_{X_s})$$

*is a constant function of* $s$.

Of course, if $s \in S^*$, we have

$$\Omega^{\cdot}_{X/S}(\log Y) \otimes \mathcal{O}_{X_s} \simeq \Omega^{\cdot}_{X_s}.$$

Therefore,

$$d(s) = \dim H^i(X_s, \mathbf{C}) \quad \text{if} \quad s \in S^*$$

(which is independent of $s \in S^*$). Working locally over a disc $\Delta$, we have for $X_0 = Y$:

PROPOSITION 2 ([12, (2.16)]). $H^i(Y, \Omega^{\cdot}_{X/S}(\log Y) \otimes \mathcal{O}_Y)$ *is isomorphic to the* i*-th cohomology of a general fiber. More canonically, it is that of the pull-back of* $X^*$ *to the universal cover of* $\Delta^*$.

COROLLARY. $R^i f_* \Omega^{\cdot}_{X/S}(\log Y)$ *is locally free, and is thus isomorphic to* $\mathcal{H}_e$.

Let $F$ denote, as usual, the filtration bête on $\Omega^{\cdot}_{X/S}(\log Y) \otimes \mathcal{O}_Y$. In case that $Y$ is reduced, it follows directly from [12, (4.15), (4.19)] that $F$ induces the Hodge filtration of a mixed Hodge structure, which, in fact, coincides with the limit mixed Hodge structure [12, (5.10)], [13, Appendix to §2]. However, the weight filtration is not induced by $W$ of $\Omega^{\cdot}_X(\log Y)$, at least not directly. Since we can proceed without it to prove Theorem 3, we postpone the discussion.

We assume that $Y$ is reduced. By Theorems 1 and 2, we have

$$\text{(24)} \quad \text{i)} \quad \dim H^i(Y, \Omega^{\cdot}_{X/S}(\log Y) \otimes \mathcal{O}_Y) = \bigoplus_{p+q=i} \dim H^q(Y, \Omega^p_{X/S}(\log Y) \otimes \mathcal{O}_Y)$$

$$\text{ii)} \quad \dim H^i(X_s, \Omega^{\cdot}_{X/S}(\log Y) \otimes \mathcal{O}_{X_s}) = \bigoplus_{p+q=i} \dim H^q(X_s, \Omega^p_{X/S}(\log Y) \otimes \mathcal{O}_{X_s}).$$

We also know by semi-continuity that for all $p$ and $q$,

$$\text{(25)} \quad \dim H^q(Y, \Omega^p_{X/S}(\log Y) \otimes \mathcal{O}_Y) \geq \dim H^q(X_s, \Omega^p_{X/S}(\log Y) \otimes \mathcal{O}_{X_s}).$$

On the other hand, we have in Proposition 2 that the left-hand sides in (24) are equal. It follows that we actually have equality in (25). By our criterion (Proposition 1), Theorem 3 follows in this case.

In order to obtain the result in general, one first extends the constructions and the above case of Theorem 3 to V-manifolds (spaces with only finite quotient singularities). One can then perform a base-change to replace the given central fiber by a reduced one with V-normal crossings, and argue from there [13, §2]. While it is tempting to replace this line of reasoning by an appeal to semi-stable reduction, it seems that one cannot then recover the result for the original situation if one cares to do so.

We return to the issue of the weight filtration (assuming that $Y$ is reduced). There is no construction of the weight filtration directly on $\Omega^{\bullet}_{X/S}(\log Y) \otimes \mathcal{O}_Y$. Instead, one replaces this complex by $A^{\bullet}$, a bigraded complex with terms

$$A^{p,q} = \Omega_X^{p+q+1}(\log Y)/W_q\Omega_X^{p+q+1}(\log Y) \tag{26}$$

for $p, q \geq 0$, and differentials

(27) i) $d' : A^{p,q} \to A^{p+1,q}$ induced by $d$ in $\Omega^{\bullet}_X(\log Y)$

ii) $d'' : A^{p,q} \to A^{p,q+1}$ induced by $\wedge f^*(dt/t)$.

One has inclusions

$$\Omega^p_{X/S}(\log Y) \otimes \mathcal{O}_Y \to A^{p,0} \text{ induced by } \wedge f^*(dt/t)\ . \tag{28}$$

PROPOSITION 3 ([12, (4.15)]). *$A^{p,\cdot}$ is a resolution of $\Omega^p_{X/S}(\log Y) \otimes \mathcal{O}_Y$.*

One places filtrations $W$ and $F$ on $A^{\bullet}$ as follows [12, (4.17)]:

(29) i) $W_k A^{p,q} = W_{2q+k+1}\Omega_X^{p+q+1}(\log Y)/W_q\Omega_X^{p+q+1}(\log Y)$,

ii) $F^r A^{p,q} = \begin{cases} A^{p,q} & \text{if } p \geq r \\ 0 & \text{if } p < r \end{cases}$ (cf. (1)).

We note that (29, i) is non-zero for some negative $k$ $(k \geq -q)$. It follows from Proposition 3 that

(30) $$\Omega^{\cdot}_{X/S}(\log Y) \otimes \mathcal{O}_Y \to A^{\cdot}$$

is a quasi-isomorphism, and that the respective $F$'s determine the same filtration on cohomology. (The precise wording is that (30) is a *filtered quasi-isomorphism* [with respect to $F$]; see [4, (1.3.6)].)

We now show (modulo the rationality issue, for which see [12, (4.13)])[6] that $A^{\cdot}$ is a cohomological mixed Hodge complex. Observing that passing to $Gr^W$ kills $d''$, we have

$$Gr^W_k A^{\cdot} \simeq \bigoplus_{q \geq 0, -k} Gr^W_{2q+k+1} \Omega^{\cdot}_X(\log Y)[1]$$

$$\simeq \bigoplus_q a_*^{(2q+k+1)} \Omega^{\cdot}_{\tilde{Y}(2q+k+1)}[-2q-k] ,$$

(32) $$F^p Gr^W_k A^{\cdot} \simeq \bigoplus_q F^{p+q+1}(Gr^W_{2q+k+1} \Omega^{\cdot}_X(\log Y)[1]) .$$

Thus,

(33) $$H^i(X, F^p Gr^W_k A^{\cdot}) \simeq \bigoplus_q F^p[H^{i-2q-k}(\tilde{Y}^{(2q+k+1)}, C)\langle -k-q \rangle] ,$$

so $Gr^W_k A^{\cdot}$ is of weight $i+k$, as desired.

REMARK. In the $W$ spectral sequence (13), $d_1$ is given by the numerous Gysin mappings and restriction mappings.

The complex $A^{\cdot}$ has the following additional pleasant property.

PROPOSITION 4 ([12, (4.22)]). *Let*

$$(-1)^{p+q+1} \nu : A^{p,q} \to A^{p-1,q+1}$$

*denote the canonical projection. Then* $\nu$ *defines an endomorphism of the*

---

[6] See also [19, §5].

*complex* $A^{\cdot}$, *which induces the monodromy logarithm* N *on cohomology.*[7] *Moreover,*

$$\nu(W_k A^{\cdot}) \subset W_{k-2}A^{\cdot} ,$$

*and* $\nu^k$ *induces an isomorphism*

$$\mathrm{Gr}_k^W A^{\cdot} \xrightarrow{\sim} \mathrm{Gr}_{-k}^W A^{\cdot} .$$

The last assertion in Proposition 4 looks very much like the assertion that W induces the monodromy weight filtration of the limit mixed Hodge structure. Precisely, it implies that

$$\nu^k : H^i(X, \mathrm{Gr}_k^W A^{\cdot}) \to H^i(X, \mathrm{Gr}_{-k}^W A^{\cdot}) \tag{34}$$

is an isomorphism. But this is only the $E_1$ term of (13); the monodromy weight filtration is characterized by the analogue of (34) for ${}_W E_2 = {}_W E_\infty$ :

$$\nu^k : \mathrm{Gr}_{i+k}^W H^i(X, A^{\cdot}) \xrightarrow{\sim} \mathrm{Gr}_{i-k}^W H^i(X, A^{\cdot}) \tag{35}$$

(recall (14)). That (35) is, in fact, true requires a delicate argument [12, (5.9)].

We summarize:

THEOREM 4. *Assume that* Y *is reduced. Then* $(A^{\cdot}, W, F)$ *is a cohomological mixed Hodge complex defined over* $\mathbb{Q}$, *and the induced mixed Hodge structure on cohomology is isomorphic to the limit mixed Hodge structure.*

Thus, we see that the limit mixed Hodge structure is related to the geometry of the singular fiber.

For the reader's convenience, we include a list of other cohomological mixed Hodge complexes associated to $f: X \to \Delta$, that are defined in [2]. The relations among them and $A^{\cdot}$ give rise to the Clemens-Schmid exact sequence [2, (3.7)] (see Chapter VI). Again, we assume that Y is reduced.

---

[7] By [1, p. 94], [9, (1.7)], or [12, (2.20)], the monodromy is unipotent when Y is reduced.

We define a bigraded complex $L^{\cdot}$ with terms

$$(36)\qquad L^{p,q} = \begin{cases} Gr^{W}_{q+1}\Omega^{p+q+1}_{X}(\log Y) = W_{-q}A^{p,q} & \text{if} \quad q>0\,, \\ \dfrac{\Omega^{p}_{X}(\log Y) \oplus Gr^{W}_{1}\Omega^{p+1}_{X}(\log Y)}{(\text{graph of } -\wedge f^{*}(dt/t))} & \text{if} \quad q=0\,. \end{cases}$$

The differentials in $L^{\cdot}$, as well as the filtrations $W$ and $F$, are the obvious ones which make for morphisms of bifiltered complexes

$$\Omega^{\cdot}_{X}(\log Y) \to L^{\cdot} \to A^{\cdot}\,.^{8}$$

The following is a hybrid of [2, §§7, 9] and [11, (12.3)]:

THEOREM 5. i) $L^{\cdot}$ *is a resolution of* $\Omega^{\cdot}_{X}(\log Y)\otimes\mathcal{O}_{Y}$. *It is a cohomological mixed Hodge complex, giving rise to mixed Hodge structures on*

$$H^{i}(X,L^{\cdot}) \simeq H^{i}(X-Y,\mathbb{C})\,,$$

ii) $W_{0}L^{\cdot}$ *is canonically isomorphic to the usual cohomological mixed Hodge complex which gives* $H^{\cdot}(Y,\mathbb{C})$ [$\simeq H^{\cdot}(X,\mathbb{C})$, *since we are in the local situation*].

iii) $L^{\cdot}/W_{0}L^{\cdot}$ *is canonically isomorphic to the cohomological mixed Hodge complex*

$$\Omega^{\cdot}_{X}(\log Y)/W_{0}\Omega^{\cdot}_{X}(\log Y) \simeq \Omega^{\cdot}_{X}(\log Y)/\Omega^{\cdot}_{X}\,,$$

*which gives* $H^{\cdot}_{Y}(X,\mathbb{C}) \simeq H^{\cdot}(X,X-Y,\mathbb{C})$.

Of the assertions in Theorem 5, perhaps the one that, in the local situation, $\Omega^{\cdot}_{X}(\log Y)\otimes\mathcal{O}_{Y}$ has the same cohomology as $\Omega^{\cdot}_{X}(\log Y)$ seems a bit surprising. It follows, however, from the following result that Steenbrink attributes to Katz:

LEMMA. *Let* $\mathcal{J}_{Y}$ *denote the ideal sheaf of the (reduced) divisor* $Y$. *Then* $\mathcal{J}_{Y}\Omega^{\cdot}_{X}(\log Y)$ *is acyclic along* $Y$.

---

[8] It is not hard to see that $L^{\cdot}$ is quasi-isomorphic to the mapping cone $B^{\cdot}$ of $\nu$, which is used analogously in [12, (4.22)-(4.29)].

One can prove this lemma by induction, along the lines of [11, §6].

§3. *Some applications*

(a) Applications of the construction of the limit mixed Hodge structure from $A^{\bullet}$.

1. *The local invariant cycle theorem.* In the local situation, the following is an exact sequence (of mixed Hodge structures).

$$H^i(Y,C) \xrightarrow{\rho} H^i(X,A^{\bullet}) \xrightarrow{N} H^i(X,A^{\bullet}) . \tag{37}$$

Thus $\rho$ maps onto $\ker N$, the cohomology classes invariant under the local monodromy transformation. A proof of (37) is given in [12, (5.12)], along a line of reasoning proposed by Deligne. This result is also part of the Clemens-Schmid exact sequence [2, (3.7)] (see Chapter VI) that extends (37) to a long exact sequence, providing a very useful description of the kernel of $\rho$. For applications to degenerations of surfaces, see Chapter VI. The first proof of (37) for varieties of low dimension is apparently due to Katz.

2. *Limits of mixed Hodge structures.* In recent work [19],[9] Steenbrink and the author have a construction of a limit mixed Hodge structure for degenerating *mixed* Hodge structures that come from geometry. The theme is to define a complex that plays the role of $A^{\bullet}$ for the cohomology groups underlying the variation of mixed Hodge structure, such that it induces the $A^{\bullet}$ for the successive quotients of the weight filtration (of the variation).

3. *The limit mixed Hodge structure on the vanishing cycles associated to an isolated singularity.* Let $h : Z \to \Delta$ be a flat holomorphic mapping from the contractible Stein space $Z$, such that $h$ is a $C^{\infty}$-fiber bundle over $\Delta^*$. We assume that $h$ is smooth except for an isolated singular point $z_0 \in Z_0 = h^{-1}(0)$.

[9] See also [18].

Let

(38)

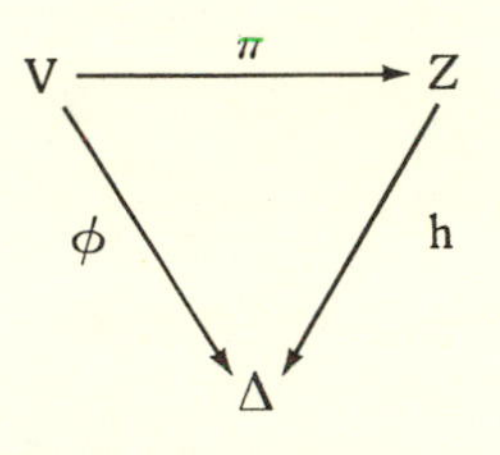


be a resolution of singularities of $h$, so that $V_0 = \phi^{-1}(0)$ is a divisor with normal crossings. We assume that $V_0$ is reduced. We write

$$V_0 = \overline{Z}_0 + E , \tag{39}$$

where $\overline{Z}_0$ is the proper transform of $Z_0$, and $E = \pi^{-1}(z_0)$ is the exceptional divisor of the resolution. Then a complex $K^{\cdot}$, supported on $E$, is defined ([13, (3.4)], [14]) by

$$K^{p,q} = \Omega_V^{p+q+1}(\log V_0)/\Omega_V^1(\log \overline{Z}_0) \wedge W_q\Omega_V^{p+q}(\log E) ; \tag{40}$$

there is an evident surjection of $A^{\cdot}$ (of (26), with $X$ replaced by $V$, $Y$ by $V_0$) onto $K^{\cdot}$, and one takes the induced differentials. Likewise, there are induced filtrations $F$ and $W$ on $K^{\cdot}$. Since

$$W_q\Omega_V^{p+q+1}(\log V_0) \subset \Omega_V^1(\log \overline{Z}_0) \wedge W_q\Omega_V^{p+q}(\log E) \subset W_{q+1}\Omega_V^{p+q+1}(\log V_0) ,$$

one sees that

$$\mathrm{Gr}_k^W K^{\cdot} = \mathrm{Gr}_k^W A^{\cdot} \simeq \bigoplus_{q\geq 0,-k} a_*^{(2q+k+1)}\Omega^{\cdot}_{\tilde{V}_0^{(2q+k+1)}}[-k-2q] , \tag{41}$$

*unless* $k \leq 0$, in which case one just loses the terms for $q = -k$ corresponding to those $(q+1)$-fold intersections of components of $V_0$ for which one component is $\overline{Z}_0$. We note that every non-zero summand of $\mathrm{Gr}_k^W K^{\cdot}$ involves at least one component of $E$, and thus gives the cohomological Hodge complex of a projective variety (up to a shift). It is of the correct weight, because the same was true for $A^{\cdot}$. Thus, $K^{\cdot}$ is a cohomological mixed Hodge complex, and its cohomology is that of the general fiber (Milnor fiber) $V_t$ of (38) [13, (3.8)]. Moreover, if $h$ is the restriction of a projective family $f: X \to \Delta$, then clearly

$$H^i(X, A^{\cdot}) \to H^i(V, K^{\cdot})$$

is a morphism of mixed Hodge structures. If $E$ is not reduced, one can use the same trick as before.

Steenbrink computed in [13, (4.11)] the null-space and signature of cup-product on $H_c^n(V_t, \mathbb{Q})$ ($n = \dim V_t$) in terms of the Hodge numbers of the mixed Hodge structure on its dual space $H^n(V_t, \mathbb{Q})$. (See also [14, (2.23) ff.].) The mixed Hodge structure on the vanishing cycles gives useful invariants of isolated singularities and their smoothings. For more on this, the reader is referred again to [14], and also to [17].

(b) Applications of Theorem 3.

1. *Hodge theory with degenerating coefficients.* For the global situation $f: X \to S$, we consider the Leray spectral sequences

$$\text{(42)} \qquad \text{i)}\quad E_2^{p,q} = H^p(S^*, R^p g_* \mathbb{C}) \Rightarrow H^{p+q}(X^*, \mathbb{C}),$$

$$\text{ii)}\quad E_2^{p,q} = H^p(S, R^q f_* \mathbb{C}) \Rightarrow H^{p+q}(X, \mathbb{C}).$$

We denote by $L$ the associated (Leray) filtration in either case. The spectral sequences (42) degenerate at $E_2$ ([6], [16, (15.15)]). Thus, we may write

$$\text{(43)} \qquad \mathrm{Gr}_L^p H^{i+p}(X^*, \mathbb{C}) \simeq H^p(S^*, R^i g_* \mathbb{C}),$$

$$\text{(44)} \qquad \mathrm{Gr}_L^p H^{i+p}(X, \mathbb{C}) \simeq H^p(S, R^i f_* \mathbb{C}).$$

It is also the case that the Hodge filtration $F$ induces a Hodge structure on (44), and a mixed Hodge structure on (43). Thus, one can use the filtration $L$ to subdivide Hodge theoretic questions on $X$. In order to take advantage of this, we found that it was necessary to reproduce the Hodge theory from cohomological (mixed) Hodge complexes on $S$.

The construction from [16] can be described as follows. Let $\mathcal{H}$ denote, as in (18), the bundle

$$R^i g_* C \otimes_C \mathcal{O}_{S^*}$$

of i-dimensional cohomology along the fibers of $g$. Then for the complex $M^{\cdot}$, defined as

$$\mathcal{H}_e \xrightarrow{\nabla} \Omega^1_S(\log T) \otimes \mathcal{H}_e \,, \tag{45}$$

one has according to [3, p. 105]

$$H^i(S, M^{\cdot}) \simeq H^i(S^*, R^i g_* C) \,. \tag{46}$$

Following Deligne's construction (see [16, (1.12)]) for the case $S = S^*$, we define a filtration $F$ on $M^{\cdot}$ by setting $F^p M^{\cdot}$ to be the complex

$$\mathcal{F}^p_e \xrightarrow{\nabla} \Omega^1_S(\log T) \otimes \mathcal{F}^{p-1}_e \,. \tag{47}$$

There is a filtration $W$ on $M^{\cdot}$ [16, p. 462] such that $(M^{\cdot}, W, F)$ is a cohomological mixed Hodge complex.

We were especially interested in (44) when $p = 1$. We interpret the local invariant cycle theorem as asserting the *surjectivity* of the mapping

$$R^i f_* C \to j_* R^i g_* C \,. \tag{48}$$

One then has

$$H^1(S, R^i f_* C) \simeq H^1(S, j_* R^i g_* C) \,. \tag{49}$$

It is clear that the subcomplex

$$\mathcal{H}_e \xrightarrow{\nabla} \nabla \mathcal{H}_e \tag{50}$$

of $M^{\cdot}$ resolves $j_* R^i g_* C$. With the filtration $F$ induced from (47), (50) becomes a cohomological Hodge complex of weight $i$ [16, (9.1)]; it is, in fact, the lowest non-zero weight filtration level for (45).

Thus, (47) induces Hodge filtrations on the cohomology groups (43), (44). For this construction to be of any use, it must coincide with the induced Hodge filtrations that were mentioned earlier. That this is the case makes use of Theorem 3 and its Corollary (see [16, (15.5)]).

2. *Normal functions*. The bundle $\mathcal{H}_e/\mathcal{F}_e^p$ (for $i = 2p-1$) is the Lie algebra bundle for the family $\bar{J}^p$ of generalized p-th intermediate Jacobians associated to $f: X \to S$. The formula for this bundle is used in [7] to extend the normal functions associated to algebraic cycles across $T$.

The equality of the two Hodge structures on (49) is used to prove the theorem on normal functions, which relates cross sections of $\bar{J}^p$ and Hodge classes on $X$ [16, (9.2)].

These results on normal functions will be discussed in greater detail in Chapter XIV.

## REFERENCES

[1] C. H. Clemens, "Picard-Lefschetz theorem for families of nonsingular algebraic varieties acquiring ordinary singularities," Trans. A.M.S. 136 (1969), pp. 93-108.

[2] ———, "Degeneration of Kähler manifolds," Duke Math. J. 44 (1977), pp. 215-290.

[3] P. Deligne, "Equations différentielles à points singuliers réguliers," Lecture Notes in Math. 163, Springer-Verlag, Berlin-Heidelberg-New York, 1970.

[4] ———, "Théorie de Hodge II," Pub. Math. IHES 40 (1972), pp. 5-57.

[5] ———, "Théorie de Hodge III," Pub. Math. IHES 44 (1973), pp. 5-77.

[6] ———, "Théorème de Lefschetz et critères de dégénérescence de suites spectrales," Pub. Math. IHES 35 (1969), pp. 107-126.

[7] F. El Zein and S. Zucker, "Extendability of normal functions associated to algebraic cycles," this volume, Ch. XV, pp. 269-288.

[8] N. Katz, "The regularity theorem in algebraic geometry," Actes Congrès Intern. Math., Nice, 1970, t.1, pp. 437-443.

[9] A. Landman, "On the Picard-Lefschetz transformation for algebraic manifolds acquiring general singularities," Trans. A.M.S. 181 (1973), pp. 89-123.

[10] W. Schmid, "Variation of Hodge structure: the singularities of the period mapping," Inv. math. 22 (1973), pp. 211-319.

[11] J. Steenbrink, "Limits of Hodge structures and intermediate Jacobians," Thesis, Univ. of Amsterdam, 1974.

[12] ———, "Limits of Hodge structures," Inv. math. 31 (1976), pp. 229-257.

[13] ———, "Mixed Hodge structure on the vanishing cohomology," In: Real and Complex Singularities, Oslo, 1976, pp. 525-563.

[14] ———, "Mixed Hodge structures associated with isolated singularities," In: Singularities, Proc. of Symposia in Pure Math. 40 (1983), Part 2, pp. 513-536.

[15] S. Zucker, "Generalized intermediate Jacobians and the theorem on normal functions," Inv. math. 33 (1976), pp. 185-222.

[16] ———, "Hodge theory with degenerating coefficients: $L_2$ cohomology in the Poincaré metric," Annals of Math. 109 (1979), pp. 415-476.

[17] V. Arnol'd, "Index of a singular point of a vector field, the Petrovskii-Oleinik inequality, and mixed Hodge structures," Functional Analysis and Applications 12 (1978), pp. 1-12.

[18] F. El Zein, "Dégénérescence diagonale, I, II," C. R. Acad. Sci. Paris 296 (1983), pp. 51-54, 199-202.

[19] J. Steenbrink and S. Zucker, "Variation of mixed Hodge structure," 1983.

STEVEN ZUCKER
MATHEMATICS DEPARTMENT
JOHNS HOPKINS UNIVERSITY
BALTIMORE, MD 21218

## Chapter VIII

# INFINITESIMAL TORELLI THEOREMS AND COUNTEREXAMPLES TO TORELLI PROBLEMS

Fabrizio M. E. Catanese*

It seems convenient, to avoid any misunderstanding, to describe the terminology as it seems to be established at the present time.

One has four types of Torelli problems:

a) Infinitesimal Torelli (called Local Torelli in the literature of some years ago)
b) Global Torelli
c) Local Torelli
d) Generic global Torelli (or "Weak global Torelli").

Let me explain the meaning of these problems. Assume for simplicity that $X$ is a complex algebraic variety such that the canonical bundle $K_X$ is ample, so that $K_X$ defines a natural polarization on $X$. From the Kodaira-Spencer-Kuranishi theory ([19], [20]), we know that there exists a semi-universal deformation of $X$, $p : \mathfrak{X} \to (S, s_0)$. This means that $p$ is a smooth proper morphism such that $p^{-1}(s_0) \cong X$; i.e. $p : \mathfrak{X} \to (S, s_0)$ is a deformation of $X$, and moreover any other deformation of $X$, $p' : \mathfrak{Y} \to (B, b_0)$, is obtained as the fibre product over $S$ of $p$ and a suitable morphism $f : B \to S$ (with $f(b_0) = s_0$) whose differential at $b_0$ is uniquely determined by $p'$. In particular the Zariski tangent space of $S$ at $s_0$, $T_{S,s_0}$, is naturally identified with $H^1(X, T_X)$ ( $T_X$ being the tangent

*Ist. Mat. "L. Tonelli" Univ. di Pisa, Institute for Advanced Study, Princeton, a member of G.N.S.A.G.A. of C.N.R., partially supported by NSF grant MCS 81-03365.

bundle to $X$). Moreover, the dimension of $S$ at $s_0$ is at least $\dim H^1(X,T_X) - \dim H^2(X,T_X)$.

Then, for each $1 \leq k \leq \dim X$, we have a corresponding variation of Hodge structure $(H_{\mathbb{Z}} = \mathcal{R}^k p_*(\mathbb{Z})_{prim}, H^{p,q}(s), Q)$, where $p+q=k$, $H_{\mathbb{Z}} \otimes \mathbb{C} = \bigoplus_{p+q=k} H^{p,q}(s)$, and the polarization $Q$ is a quadratic form on $H_{\mathbb{Z}}$ for which the subspaces $H^{p,q}(s)$ are orthogonal to each other. To this variation of Hodge structure is naturally associated a holomorphic map $\phi : S \to D$, where $D$ is the classifying space for polarized Hodge structures of types $(h^{k,0}, h^{k-1,1}, \ldots, h^{0,k})$ and $h^{p,q} = \dim H^{p,q}(s)$. $D$ is usually referred to as the period domain ([14]).

a) asks whether the differential $d\phi$ of $\phi$ is injective on $T_{S,s_0}$: if this is true (notice that this implies the injectivity of $\phi$ in a neighborhood of $s_0$) then one says that the Infinitesimal Torelli Theorem holds for $X$.

Let now $\mathfrak{M}(X)$ be the set of equivalence classes of complex structures on the differentiable manifold underlying $X$, and $[X]$ the point of $\mathfrak{M}(X)$ corresponding to the complex structure $X$. Then we have a global period mapping $\psi : \mathfrak{M}(X) \to D/\Gamma$, where, if $G_{\mathbb{C}}$ is the complex orthogonal group of the quadratic form $Q$, $\Gamma = G_{\mathbb{Z}} = \{g \in G_{\mathbb{C}} | gH_{\mathbb{Z}} \subset H_{\mathbb{Z}}\}$ acts properly discontinuously on the complex manifold $D$. $D/\Gamma$ is called the period space, and

b) asks whether $\psi$ is injective.

If $\mathfrak{M}(X) = \mathfrak{M}$ can be endowed with the structure of a complex analytic space in a functorial way; i.e., there exists a "coarse moduli space" for $X$, we get that $\psi : \mathfrak{M} \to D/\Gamma$ is a holomorphic map. If $X$ is a curve, or a surface of general type, then $\mathfrak{M}(X)$ is a quasi-projective variety, by the results of [11], [24]; for the general case we refer the reader to [29].

Let $[X]$ be the point in $\mathfrak{M}$ corresponding to the complex structure given by $X$: it, therefore, makes sense to ask whether $\psi$ is a local embedding at $[X]$.

c) can be phrased as the problem: is the differential of $\psi$, $d\psi$, injective on the Zariski tangent space of $\mathfrak{M}$ at $[X]$?

d) asks whether there exists a dense open subset $\mathfrak{U}$ of $\mathfrak{M}$ such that $\psi|_{\mathfrak{U}}$ is one-to-one.

§1. *Infinitesimal-Torelli*

Concerning a) the crucial starting point was the result of Griffiths ([13]) giving a cohomological interpretation for $d\phi$

(1) $d\phi$ is induced by cup-product

$$H^1(X,T_X) \otimes H^{p,q}(X) = H^1(X,T_X) \otimes H^q(X,\Omega_X^p) \to H^{q+1}(X,\Omega_X^{p-1}) = H^{p-1,q+1}(X):$$

in other words

$$d\phi : H^1(T_X) \to \bigoplus_{p+q=k} \mathrm{Hom}(H^{p,q}(X), H^{p-1,q+1}(X)) \subset T_{D,\phi([X])}.$$

The first example we want to illustrate is that of a curve $X$ of genus $g \geq 1$: clearly

$$H^1(T_X) \to H^{0,1}(X) \otimes H^{1,0}(X)^{\vee}$$

( $\vee$ denoting the dual of a vector space) is $1-1$ iff dually

$$H^0(K_X)^{\otimes 2} = H^{1,0}(X)^{\otimes 2} \to H^0(2K_X) = H^1(T_X)^{\vee}$$

is surjective. By the classical theorem of M. Noether, this holds iff $g = 1,2$ or iff $g \geq 3$ and $X$ is not hyperelliptic.

So a) does not always have a positive answer for curves, though b) holds by R. Torelli's result ([33], cf. also [1], [36] for modern proofs) and c) holds (in char $0$ always) by work of Oort-Steenbrink ([25]).

To partly explain this phenomenon, observe that if $X$ is of general type (for $n >> 0$, $H^0(nK_X)$ gives a birational map) then $\mathrm{Aut}(X)$ is a finite group (this result, classically due to Schwartz in the case of curves, has been extended to higher dimensional varieties by Painlevé, Matsumura [23], Kobayashi-Ochiai [17]). Then a neighborhood of $[X]$ in $\mathfrak{M}$ is isomorphic to $S/\mathrm{Aut}(X)$, where the action on $S$ is induced by the one of $\mathrm{Aut}(X)$ on $H^1(T_X)$. In the case of hyperelliptic curves, they form a smooth $(2g-1)$ dimensional submanifold $V$ of the smooth manifold $S$ of dimension $(3g-3)$, and the general hyperelliptic curve $X$ has as only nontrivial automorphism the hyperelliptic involution: hence, $\mathfrak{M}$ locally at $[X]$ looks like $\mathbf{C}^{2g-1} \times (\mathbf{C}^{g-2}/\mathbf{Z}/2\mathbf{Z})$, where $\mathbf{Z}/2\mathbf{Z}$ acts by multiplication of a vector by $-1$.

The differential of $\phi$ is of maximal rank on $T_V$ and vanishes in the normal directions: anyhow to check that $d\psi$ is injective, what basically Oort and Steenbrink do is to compute the quadratic behavior of $\phi$ in the directions normal to $V$.

This is somehow a lucky situation as we are going to see. In fact, in a recent preprint, ([35]), S. Usui makes the following remark to explain failures of the infinitesimal Torelli: the action of $\mathrm{Aut}(X)$ extends to $\mathfrak{X} \xrightarrow{p} S$, and actually $\mathrm{Aut}(X) = \mathrm{Aut}(\mathfrak{X}, S, p, s_0)$ (automorphisms of $\mathfrak{X}$ commuting with $p$ and leaving $X = p^{-1}(s_0)$ fixed). Clearly, if $G$ is a subgroup of $\mathrm{Aut}(X)$, $H^1(T_X)$ and the $H^{p,q}(X)$ are representations of $G$. The basic point is that it turns out that $d\phi$ must be a $G$-equivariant homomorphism. But, decomposing $H^1(T_X)$, $H^{p,q}(X)^\vee \otimes H^{p-1,q+1}(X)$ as direct sums of irreducible representations and applying Schur's lemma, one can see a priori if there can be any injective $G$-equivariant homomorphism of $H^1(T_X)$ into $H^{p,q}(X)^\vee \otimes H^{p-1,q+1}(X)$. For instance, in the case of hyperelliptic curves, $(H^0(K_X)^{\otimes 2})^\vee$ is a trivial representation of $\mathbf{Z}/2\mathbf{Z}$, while $H^1(T_X) = T_{S,s_0}$ splits as the $(+1)$ eigenspace $T_V$ plus the $(-1)$ eigenspace normal to $T_V$. On the other hand, if $S^G$ is the variety of fixed points of $G$ in $S$ (above: $V = S^{\mathbf{Z}/2\mathbf{Z}}$), $H^1(T_X)^G = T_{S^G,[x]}$ so $d\phi(T_{S^G})$ must be contained in the $G$-trivial summand of $H^{p,q}(X)^\vee \otimes H^{p-1,q+1}(X)$. It can happen thus that rank $(d\phi|_{T_{S^G}}) < \dim S^G$

so that $\phi$ (hence $\psi$) must have some positive dimensional fibres: this phenomenon actually occurs for some surfaces of general type that we shall discuss later.

In spite of this, infinitesimal Torelli theorems hold for wide class of varieties, and turn out to be useful, e.g., in the classification theory of Kähler varieties ([10]). Let me give a brief and incomplete list of results of type a), and simply point out that, via the cohomological interpretation (1), the proof is reduced often to Koszul-complex-like arguments. Infinitesimal Torelli holds for a Variation of Hodge Structure of weight $n = \dim X$ in the case of

(2) varieties with trivial canonical class $(K \equiv 0)$, given a non-natural polarization (Griffiths [13], G. Tjurina [30]),

(3) hypersurfaces in $\mathbf{P}^n$ with $K_X$ ample (Griffiths [13]),

(4) more generally, complete intersections in $\mathbf{P}^n$, cyclic covers of $\mathbf{P}^n$ or of products of projective spaces, provided $K_X$ is ample (Peters [26], Kiǐ [16]),

(5) some weighted complete intersections (S. Usui [34]): here, though, as we shall see later, the condition that $K_X$ be ample is not sufficient.

More general results are

(6) if $K_X \equiv kL$, $k \geq 1$, $V = H^0(L)$ has no base points, and for each $i = -k, \cdots, -1$ the $H^2$ of the Koszul complex $\mathcal{K}^i = H^0(\Omega_X^{n-1}(L^{i+r})) \otimes \overset{r}{\Lambda} V^\vee$ vanishes (Liebermann, Peters, Wilsker [22], see also [27]).

(7) if $p_g = h^0(K_X) \geq 2$, $h^{n-1,0}(X) = 0$, $K_X = \Sigma\, n_i E_i$ with $E_i$ irreducible, $h^0(E_i) \geq 2$, and $h^0(\Omega_X^{n-1}(E_i)) < h^0(E_i) - 1$ (Kiǐ [16]).

For later use in this series of lectures, we point out a particular case of (7), the following theorem of Kiǐ ([16]):

(8) Let $X \xrightarrow{f} P^1$ be an elliptic surface with nonconstant moduli and without multiple fibres. Then the infinitesimal Torelli theorem holds for $X$.

It is nice to see how the geometrical hypotheses of (8) allow one to apply the more general result (7). In fact, if $F$ is a fibre of $f$, then by the well-known formula for the canonical divisor of an elliptic fibration ([18]) $K_X \equiv rF$, since there are no multiple fibres. Here, if $r < 0$, $X$ would be ruled, hence two general fibres would be isomorphic, contradicting our hypothesis. We can assume $r \geq 1$, the case when $K_X \equiv 0$ being O.K. by (2).

Therefore, $p_g \geq 2$. $h^{1,0}(X) = 0$ follows from a more general fact: if $f : X \to B$ is an elliptic surface without multiple fibres and with non-constant moduli, then there are singular fibres, since otherwise one would get a nonconstant holomorphic map $F : B \to \mathbb{C}$, which is absurd. But the singular fibres consist of rational components, which must be shrunk to a point by the Albanese map $\alpha : X \to \mathrm{Alb}(X)$. Therefore, $\alpha$ factors through $f$; it is then clear that $h^{1,0}(X) = g(B)$.

To apply (7), it suffices thus to prove that $H^0(\Omega^1_X(F)) = 0$. Since the normal bundle $\mathcal{O}_F(F)$ to $F$ is trivial, one has two exact sequences, where we assume $F$ to be a smooth fibre:

$$\begin{array}{ccccccccc}
 & & 0 & & & & & & \\
 & & \| & & & & & & \\
0 & \longrightarrow & H^0(\Omega^1_X) & \longrightarrow & H^0(\Omega^1_X(F)) & \longrightarrow & H^0(\Omega^1_X(F) \otimes \mathcal{O}_F) & \longrightarrow & \\
0 & \longrightarrow & H^0(\mathcal{O}_F) & \longrightarrow & \Omega^0(\Omega^1_X(F) \otimes \mathcal{O}_F) & \xrightarrow{R} & H^0(\Omega^1_F(F)) & \longrightarrow & \\
 & & & & & & \| & & \\
 & & & & & & H^0(\mathcal{O}_F) & &
\end{array}$$

Let $\sigma$ be a section of $H^0(\Omega^1_X(F))$, $z$ a uniformizing parameter on $F$, $t$ a local equation of $F$. If $\sigma \not\equiv 0$, then we can assume $\sigma|_F \not\equiv 0$. We claim that

$$R(\sigma|_F) = 0 .$$

Otherwise, near $F$ we can write

$$\sigma = (a\,dz + b\,dt)t^{-1} .$$

Since $a\,dz|_F$ must be holomorphic on all of $F$, $a|_F$ must be constant. If $a \neq 0$, the one form $(a\,dz + b\,dt)$ would determine a holomorphic foliation transverse to $F$ so that the fibration $f$ would be locally trivial. So $a = 0$ and, since

$$b = \mathrm{Res}_F\, \sigma$$

is constant on the fibres, $\sigma$ would be the pullback of a meromorphic one-form on $\mathbf{P}^1$ with only one simple pole, a contradiction. q.e.d.

§2. *Counterexamples to Torelli problems*

In the rest of this talk, we shall discuss the case when $X$ is a surface of general type: then a), b), c), d) fail for some special classes of surfaces. The new phenomenon that one encounters here is the fact that $X$ has moduli yet has no nontrivial Hodge structure.

In fact, a curve of genus $0$ is $\cong \mathbf{P}^1$, and the hope that a surface with $q = p_g = 0$ would be rational was already broken in 1896, when Castelnuovo ([5]) established the well-known criterion: a surface $X$ is rational iff $P_2 = h^0(2K_X) = 0$, $q = h^0(\Omega^1_X) = 0$. He and Enriques gave examples of nonrational surfaces with $q = p_g = 0$, which, however, were not of general type. In the '30's, Godeaux ([12]) and Campedelli ([4]) constructed surfaces of general type with $q = p_g = 0$, and respectively $K_X^2 = 1$, $\pi_1(X) \cong \mathbf{Z}/5\mathbf{Z}$, $K_X^2 = 2$, $\pi_1(X) = (\mathbf{Z}/2\mathbf{Z})^3$.

Many more examples were later constructed by Godeaux using the technique of taking quotients by group actions, but surfaces of general type with $q = p_g = 0$ have been investigated more thoroughly only after the appearance of Bombieri's paper on canonical models [3] (see [9] for a, by now incomplete, history of the topic).

In many of these cases, one can hope for Torelli type theorems by using the following device: assume that $\Gamma$ is a nontrivial normal subgroup of $\pi_1(X)$, and let $G$ be the quotient group $\pi_1(X)/\Gamma$. Then there exists an unramified covering $\tilde{X} \to X$, Galois with group $G$ (for each deformation of $X$, one can do this), and by using R.R. one sees immediately that $p_g(\tilde{X}) > 0$. Then the Hodge structure on $\tilde{X}$ is a representation of $G$, and one can hope to recover $X$ from the pair ( $H^*(\tilde{X})$, action of $G$ ). This program has been carried out by Horikawa ([15]) for Enriques surfaces (i.e., $q = p_g = 0$, $K \not\equiv 0$, $2K \equiv 0$, so $G = \mathbf{Z}/2\mathbf{Z}$ here) using the global Torelli theorem for K3 surfaces. The classical Godeaux surfaces are quotients of quintic surfaces in $\mathbf{P}^3$ by $\mathbf{Z}/5\mathbf{Z}$ acting on $\mathbf{P}^3$ with the 4 nontrivial characters: it would be interesting to see whether a Torelli theorem as above holds for them.

Unfortunately, the pathologies are not yet over: contrary to a conjecture of Severi, R. Barlow ([2]) has shown the existence of surfaces of general type with $q = p_g = \pi_1 = 0$, and $K^2 = 1$, which (by the rough count $\dim S \geq 10(1-q+p_g) - 2K_X^2 = 8$ ) have at least 8 moduli.

Let's now turn to the case when $X$ has a nontrivial Hodge structure: since by a theorem of Castelnuovo (cf. [3] for a modern proof) $p_g \geq q$, we have $p_g \geq 1$. When $p_g = 1$ let $C$ be the unique effective canonical divisor; here the cohomological interpretation (1) of $d\phi$ takes a very simple form, since $\Omega_X^1(-C) \cong T_X : d\phi$ is multiplication by a nonzero section $s$ of $H^0(K_X)$. Namely, one has the following exact sequence

$$H^0(\Omega_X^1) \longrightarrow H^0(\Omega_X^1 \otimes \mathcal{O}_C) \longrightarrow H^1(T_X) \xrightarrow{\mu} H^1(\Omega_X^1)$$

and we want to see whether $\mu$ is injective.

If $q(X) = 0$, this condition is that $H^0(\Omega_X^1 \otimes \mathcal{O}_C) = 0$. The first counterexample given by Kynef ([21]) for a particular simply-connected surface with $K^2 = p_g = 1$ exploited the existence of an involution $\tau$ on $X$ leaving $C$ pointwise fixed: then one has a splitting in the exact sequence $0 \to \mathcal{O}_C(-C) \to \Omega_X^1 \otimes \mathcal{O}_C \to \omega_C \to 0$ and since $p^{(1)} = K^2+1 = p(C)$ is $2$, $d\phi$ has a 2-dimensional kernel.

The geometry of surfaces with $K^2 = p_g = 1$ has been described by A. Todorov and the present author ([6], [31]). Recall that if you grade the polynomial ring $R = C[x_0, y_1, y_2, z_3, z_4]$ in such a way that $\deg x_0 = 1$, $\deg y_i = 2$, $\deg z_j = 3$, then $\mathrm{Proj}(R) = \mathbf{P}(1,2,2,3,3)$. Then the following holds:

(9) all surfaces with $K^2 = p_g = 1$ are complete intersections of type (6,6) in $\mathbf{P}(1,2,2,3,3)$ and are simply connected. In [6], [31], there appears also a special description of the subvariety $\mathfrak{N} \subset \mathfrak{M}$ representing the surfaces $X$ for which $|2K_X|$ gives a Galois 4-tuple covering of $\mathbf{P}^2$ (in fact, $h^0(2K_X) = 3$, $(2K_X)^2 = 4$ ): just take two cubic curves $F$, $G$ in $\mathbf{P}^2$, and a line $\ell$. If you take the double cover of $\mathbf{P}^2$ branched on $F + G$, you obtain a singular $K-3$ surface $Y$ of which it is possible to take a further double covering ramified at the inverse image of $\ell$ plus some singular points: this is the $X$ which is a $(\mathbf{Z}/2)^2$-Galois cover of $\mathbf{P}^2$, and clearly $X$ admits an involution $\tau$ leaving the canonical curve $C$ (the pull-back of $\ell$ ) pointwise fixed $(X/\tau = Y)$.

Not only the infinitesimal Torelli a), but even the generic global Torelli fails for these surfaces, as we proved in [6], [7].

Namely, we have the following result ([7]).

(10) The moduli space $\mathfrak{M}$ of surfaces with $K^2 = p_g = 1$ is a rational variety and $\psi : \mathfrak{M} \to D/\Gamma$ is a generally finite map of degree $\geq 2$.

Let's sketch very rapidly the arguments used.

(i) set $y_0 = x_0^2$: using basically the normal form for pencils of conics in the plane one can write more explicit equations for $X$ (actually for its canonical model) in $\mathbf{P} = \mathbf{P}(1,2,2,3,3)$, where $y = (y_0, y_1, y_2)$ and $a$, $b$ are linear, $F$, $G$ are cubic forms in $y$.

$$\begin{cases} z_4^2 + x_0 z_3 a(y) + F(y) = 0 \\ z_3^2 + x_0 z_4 b(y) + G(y) = 0 \end{cases}$$

(here one sees easily the Galois covers of $\mathbf{P}^2$ to occur when $a \equiv b \equiv 0$ ).

(ii) using the explicit equations in (i), and the surjection of $H^0(\mathcal{O}_X(6)) \oplus H^0(\mathcal{O}_X(6)) \to H^1(T_X)$ given by the coboundary map associated to the exact sequence $0 \to N_X \to T_{\mathbf{P}}|_X \to T_X \to 0$ ( $N_X$ being the normal bundle of $X$ in $\mathbf{P}$ ) one can determine when the rank of $d\phi$ is not maximal.

(iii) consider the 18-dimensional family $\mathcal{V} \xrightarrow{p} V$ given by

$$\begin{cases} z_3^2 + x_0 z_4 y_1 + F(y) = 0 \\ z_4^2 + x_0 z_3 y_2 + G(y) = 0 \end{cases}$$

where the coefficient of $y_0^2 y_1$ in $F$, and the one of $y_0^2 y_2$ in $G$ are equal to $1$.

Then $\mathcal{V}$ is the Kuranishi family, and has the same dimension as $D$; the local period mapping $\phi$ is ramified on a hypersurface $\Delta \subset V$, and $\phi|_\Delta$ is generally of maximal rank, so that locally around $\Delta$, $\phi$ is a finite covering.

Finally $\mathfrak{M}$ is birational to $V/\sigma$, where $\sigma$ is an involution acting linearly on the vector space $V$ with a 9-dimensional fixed locus: therefore, first $\mathfrak{M}$ is rational; moreover, since $\Delta$ is not contained in the fixed locus of $\sigma$, $\psi$ is of degree at least $2$.

Todorov's counterexamples to the global Torelli problem ([32]) show that $\psi$ can have some positive dimensional fibres and are based on the phenomenon encountered with surfaces with $K^2 = p_g = 1$: constructing double covers $X$ of a $K3$ surface $Y$ in such a way that the unique holomorphic 2-form $\omega$ on $X$ is the pull-back of a 2-form $\omega'$ on $Y$.

If $\tau$ is the involution such that $Y = X/\tau$, $\omega$ is invariant by $\tau$, so the periods of $\omega$ over the cycles antiinvariant by $\tau$ are $0$, and the Hodge structure of $X$ is determined by the one of $Y$.

What happens is that it is possible to find a positive dimensional variety in $\mathfrak{M}$ corresponding to surfaces $X$ with an involution $\tau$ such that $X/\tau$ is fixed: e.g., in the case $K^2 = p_g = 1$, $[X] \in \mathfrak{N}$ if you keep the cubics F, G fixed and move the line $\ell$, you obtain a 2-dimensional fibre of $\psi$.

Todorov ([32]) also constructs a family of surfaces where $\dim \mathfrak{M} \geq 12$, $\dim D/\Gamma = 11$, hence $\psi$ must be a fibration everywhere: let me describe this last example.

Let Y be a Kummer surface in $\mathbf{P}^3$: it has 16 nodes with the property that you can take a double covering ramified exactly at them and a curve $\Gamma$ not passing through them and such that $\Gamma$ is the intersection of Y with a surface of even degree. Take $\Gamma$ a section of $\mathcal{O}_Y(2)$ and let X be the corresponding double cover: X has $K^2 = 8$, $p_g = 1$, $q = 0$, (but a big torsion group).

The family thus constructed depends on 12 parameters: in fact, if $Y'$ is Y reembedded in $\mathbf{P}^9$ by $H^0(\mathcal{O}_Y(2))$, $\Gamma'$ is the image of $\Gamma$, then $Y'$ is the bicanonical image of X, $\Gamma'$ the image of the canonical curve. Since $Y'$ has only a finite number of projective automorphisms, and Y depends on 3 moduli, we obtain the desired assertion: $\dim \mathfrak{M} \geq 12$.

To end this talk, let me mention that it has been conjectured by K. Chakiris ([8]) that when $p_g = 1$ one could get Torelli type results considering the mixed Hodge structure defined on the cohomology of $X - C$.

REMARK ADDED IN PROOF (by P. Griffiths). Recently Catanese has given a proof that, for simply-connected surfaces with $K^2 = \chi = 2$, the differential of the period map is of maximal rank at a general point of the moduli space (cf. F. Catanese, On the period map of surfaces with $K^2 = \chi = 2$, to appear in the Proceedings of the Katata Conference (1982)). In the case of surfaces with the same numerical invariants but with $\pi_1 = \mathbf{Z}/2\mathbf{Z}$, P. Oliverio has shown that the general fibre of the period map has dimension one (cf. P. Oliverio, On the period map for surfaces with $K^2 = 2$, $p_g = 1$, $q = 0$ and torsion $\mathbf{Z}/2\mathbf{Z}$, to appear). Remark that it has been shown by Catanese and Debarre that all surfaces with these numerical characters fall into one of the two cases above (cf. F. Catanese and O. Debarre, Surfaces with $K^2 = 2$, $p_g = 1$, $q = 0$, to appear).

## REFERENCES


[1] Andreotti, A.: On a theorem of Torelli, Am. J. of Math., 80(1958), 801-828.

[2] Barlow, R.: Ph.D. Thesis, Warwick (to appear).

[3] Bombieri, E.: Canonical models of surfaces of general type, I.H.E.S. 42(1973), 171-219.

[4] Campedelli, L.: Sopra alcuni piani doppi notevoli con curva di diramazione del decimo ordine, Atti Accad. Naz. Lincei, 15(1932), 358-362.

[5] Castelnuovo, G.: Sulle superficie di genere zero, Mem. Soc. Ital. Sc. (3), 10(1896), 103-123.

[6] Catanese, F.: Surfaces with $K^2 = p_g = 1$ and their period mapping, Alg. Geom. Proc. Copenhagen, 1978, Springer L.N.M. n.732 (1979), 1-29.

[7] ————.: The moduli and the global period mapping of surfaces with $K^2 = p_g = 1$: a counterexample to the global Torelli problem, Comp. Math. 41, 3(1980), 401-414.

[8] Chakiris, K.: "Counterexamples to global Torelli for certain simply-connected surfaces," Bull, A.M.S., 2(1980), 297-299.

[9] Dolgachev, T.: Algebraic surfaces with $q = p_g = 0$, C.I.M.E. Alg. Surfaces Cortona, 1977, Liguori Napoli (1981), 97-215.

[10] Fujita, T.: On Kähler fibre spaces over curves, J. Math. Soc. Japan, 30(1978), 779-794.

[11] Gieseker, D.: Global moduli for surfaces of general type, Inv. math. 43(1977), 233-282.

[12] Godeaux, L.: Sur une surface algebrique de genere zero et de bigenre deux, Atti Acad. Naz. Lincei, 14(1931), 479-481.

[13] Griffiths, P.: Periods of integrals on algebraic manifolds, I, II, Am. J. of Math., 90(1968), 568-626, 805-865.

[14] Griffiths, P. and Schmid, W.: Recent developments in Hodge theory: a discussion of techniques and results, Proc. Int. Coll. Bombay, (1973), Oxford Univ. Press, 31-127.

[15] Horikawa, E.: On the periods of Enriques surfaces, I, II, Math. Ann. 234(1978), 73-88, 235(1978), 217-246.

[16] Kiǐ, J. T.: The local Torelli theorem for varieties with divisible canonical class, Math. U.S.S.R. Izvestija, 12, 1 (1978), 53-67.

[17] Kobayashi, S. and Ochiai, Z.: Meromorphic mappings onto compact complex spaces of general type, Inv. math. 31 (1975), 7-16.

[18] Kodaira, K.: On compact analytic surfaces, II, Ann. of Math., 77(1963), 563-626.

[19] Kodaira, K. and Morrow, J.: Complex manifolds, New York, Holt-Rinehart-Winston (1971).

[20] Kuranishi, M.: New proof for the existence of locally complete families of complex structures, Proc. Conf. Comp. Anal., Minneapolis: Springer (1965), 142-154.

[21] Kynef, V. T.: An example of a simply-connected surface of general type for which the local Torelli theorem does not hold, C. R. Ac. Bulg. Sc. 30, 3 (1977), 323-325.

[22] Liebermann, D., Peters, C., and Wilsker, R.: A theorem of local Torelli type, Math. Ann. 231 (1977), 39-45.

[23] Matsumura, H.: On algebraic groups of birational transformations, Rend. Acc. Lincei Ser. 8, XXXIV (1963), 151-155.

[24] Mumford, D.: Stability of projective varieties, Ens. Math. XXIII, 1-2, (1977), 39-110.

[25] Oort, F. and Steenbrink, J.: On the local Torelli problem for algebraic curves, Jour. geom. alg. Angers, 1979, Sijhoff and Noordhoff (1980), 157-204.

[26] Peters, C. A. M.: The local Torelli theorem, I, II, Math. Ann. 217 (1975), 1-16, 223 (1976), 191-192, Ann. Sc. Norm. S. Pisa, IV, 3 (1976), 321-340.

[27] Peters, C.: On the local Torelli theorem, Var. Anal. Comp. Nice., 1977, Springer LNM 683 (1978), 62-73.

[28] Piatetski Shapiro, A. and Shafarevitch, T.: A Torelli theorem for algebraic surfaces of type K3, Izv. Akad. Nauk. 35 (1971), 530-572.

[29] Popp, H.: Moduli theory and classification theory of Algebraic Varieties, Springer L.N.M. 620 (1977).

[30] Shafarevitch, et al.: Algebraic Surfaces, Proc. Steklov Inst. Math. 75 (1965).

[31] Todorov, A.: Surfaces of general type with $p_g = 1$ and $K^2 = 1$, Ann. Ec. Norm. Sup. 13, 1 (1980), 1-21.

[32] ———: A construction of surfaces with $p_g = 1$, $q = 0$ and $2 \leq (K^2) \leq 8$, counterexamples of the global Torelli theorem, Inv. math. 63 (1981), 287-304.

[33] Torelli, R.: Sulle varietá di Jacobi, Rend. Acc. Lincei (5) 22 (1914), 98-103.

[34] Usui, S.: Local Torelli theorem for some nonsingular weighted complete intersections, Proc. Int. Symp. Alg. Geom. Kyoto, 1977, Kikokuniya Book Store, Tokyo (1978), 723-734.

[35] ———: Effect of automorphisms on variation of Hodge structure, preprint.

[36] Weil, A.: Zum Beweis des Torellischen Satz, Göttingen Nachrichten (1957), 33-53.

FABRIZIO M. E. CATANESE
UNIVERSITY OF PIDA
via I POSSENTI 37
56100 PISA
ITALY

Chapter IX

# THE TORELLI PROBLEM FOR ELLIPTIC PENCILS

Ken Chakiris

In the theory of algebraic surfaces, the following three problems are of fundamental importance:

(I) Given a degenerating family of surfaces over a disc, then, after a suitable base extension and birational transformations, "standardize" the resulting central fibre.

(II) For surfaces of positive geometric genus, calculate the degree of the period map.

(III) Given two surfaces of the same homotopy type, free from exceptional curves of the first kind, determine whether they are of the same deformation type.

In analogy with the corresponding *theorem* from the theory of algebraic curves, see [3], the first of these will be called the problem of *stable reduction*; in accordance with the terminology established elsewhere in this book, the second will be called the *generic global Torelli* problem. Only for K-3 surfaces have all three problems been successfully treated: (III) was solved by Kodaira; and in the first two problems, the essential breakthroughs are due to, respectively, Kulikov [7], Pjateckii-Sapiro and Safarevic [9]. In accordance with standard usage: The word "surface" will mean a compact complex manifold of dimension two; for any surface $V$, $p_g(V)$ denotes its geometric genus. $V$ will be called an elliptic surface if there is a holomorphic map $\psi : V \to \Delta$ whose generic fibre is a smooth elliptic curve. Note that when $p_g(V) \geq 2$, $\psi$ is uniquely determined up to the action of $\mathrm{Aut}(\Delta)$, and will be called the underlying elliptic fibration on $V$; also, if $V$ is simply connected, then necessarily $\Delta \simeq \mathbf{P}_1$.

Now, among *algebraic* surfaces, elliptic surfaces occupy a kind of middle ground; on the one hand, they form a far more diverse group than algebraic K-3 or Abelian surfaces; but, on the other hand, they do not share the intimidating complexity demonstrated by surfaces of general type. The reasons for this are essentially threefold: First, Kodaira showed that any elliptic surface can be obtained from an elliptic fibration with section by "twisting" and logarithmic transforms; therefore, to a large extent, it suffices to consider only "basic" elliptic surfaces (i.e., those whose fibrations have a section). Second, among such "basic" elliptic fibrations, it was proven by B. Moishezon that those with non-constant functional invariant can all be deformed to an elliptic fibration having only ordinary singular fibres; and, for simple reasons, a locally constant fibration can be constructed as a quotient of a finite cyclic group acting on the product of two curves, one of them elliptic. Third, a "sufficiently generic", nonconstant, elliptic fibration determines, via its functional invariant, a branched covering of $\mathbf{P}_1$; among the branch points of this covering, there are two distinguished ones. When described as a representation of a free group into a permutation group, W. Seiler has shown such a covering can be reduced to a simple normal form. Among other things, this discussion implies that the deformation types of elliptic surfaces can be determined explicitly. For surfaces of general type, we are very far from proving such results.

For the remainder of this chapter, I will discuss, in some detail, my work on elliptic pencils. By definition, an *elliptic pencil* is a pair $(V, \Delta)$: $V$ is a simply-connected elliptic surface free from exceptional curves of the first kind, $p_g(V) \geq 2$, and $\Delta$ is a divisor corresponding to a section of the underlying elliptic fibration on $V$. The concept of an isomorphism between two elliptic pencils is defined in the obvious manner. I was able to solve, in the case of elliptic pencils, both problems (I) and (II). Among other things, I showed that the period mapping for elliptic pencils is generically injective; these results were announced in [2]. Although problems (I) and (II) are interesting for general elliptic surfaces, I restricted my attention to elliptic pencils for two reasons: most (but not all) of the interesting phenomena which can occur already does so in the case I consider; for elliptic fibrations with base $\mathbf{P}_1$, the general case, by "twisting and logarithmic transforms", should be reducible to the case of elliptic pencils.

My approach to the Torelli problem for elliptic pencils has been greatly influenced by the work of Pjateckii-Sapiro and Safarevic. In their paper [9], everything concerning the period mapping for K-3 surfaces is deduced from the properties of special Kummer surfaces. By definition, a special Kummer surface is the minimal resolution of some $E' \times E''/_\iota$; where $E'$ and $E''$ are elliptic curves, $\iota = (\iota', \iota'')$, $\iota^2 = \mathrm{id}$, and $E'/_{\iota'} \simeq E''/_{\iota''} \simeq \mathbf{P}_1$. In analogy to this, I make *special elliptic pencils* the central focus of my investigation of the period mapping for elliptic pencils. By definition, a special elliptic pencil $(V, \vartriangle)$ is obtained from an elliptic curve $E$, a fixed point $p \in E$, and a hyperelliptic curve $\tilde{C}$ in the following manner: $V$ is the minimal resolution of $E \times \tilde{C}/_\iota$ and $\vartriangle$ is the proper transform of $(\{p\} \times \tilde{C})/_\iota$; $\iota^2 = \mathrm{id}$, $\iota = (\iota_E, \iota_{\tilde{C}})$, $\iota_E(p) = p$, and $E/_{\iota_E} \simeq \mathbf{P}_1 \simeq \tilde{C}/_{\iota_{\tilde{C}}}$.

Let $(V, \vartriangle)$ be an elliptic pencil and let $\psi: V \to \mathbf{P}_1$ be the underlying elliptic fibration on $V$; set $C_u = \psi^{-1}(u)$. Let $L_V$ be the subgroup of the Neron-Severi group of $V$ generated by all divisors $\alpha$ satisfying $\alpha \cdot C_u = 0 = \vartriangle \cdot \alpha$ and $\alpha^2 = -2$; given $\alpha \in L_V$, $\alpha^2 = -2$, one of $\alpha$ or $-\alpha$ is effective. Now, if $V$ is a special elliptic pencil, then we have an orthogonal direct sum decomposition $L_V = \bigoplus_{j=1}^{2(n+1)} (G_4)_j$. Where $n = p_g(V)$, and $(G_4)_j$ is a copy of the lattice $\mathbf{Z}\alpha_1 \oplus \mathbf{Z}\alpha_2 \oplus \mathbf{Z}\alpha_3 \oplus \mathbf{Z}\beta$, with the pairing $\alpha_i \cdot \alpha_j = 0$, $\forall i \neq j$; and $\alpha_i^2 = -2$, $\alpha_i \cdot \beta = 1$, $\forall i$. We prove: *$(V, \vartriangle)$ is a special elliptic pencil iff* $L_V \simeq \bigoplus_{j=1}^{2(n+1)} (G_4)_j$; here $p_g = n$. The essential idea of the proof, due to Pjateckii-Shapiro and Safarevic, is to show that each singular fibre of the underlying elliptic fibration on $V$ produces exactly one copy of $G_4$; and, then, from this show that each singular fibre is of type $I_0^*$ in Kodaira's list.

In addition to this, and again in analogy to what occurs in [9], there are a number of purely topological matters to be dealt with; namely, for a special elliptic pencil $V$, we must calculate the position of $L_V$ in $H^2(V;\mathbf{Z})$. This is accomplished by applying Lefschetz theory to the underlying elliptic fibration on $V$; or, rather, we deform $\psi: V \to \mathbf{P}_1$ slightly to a nearby fibration with only ordinary singular fibres, and perform our calculations there. A detailed summary of these calculations would be out of place here; however, constant use is made of the topological information so obtained.

§1. *Let us, once and for all, fix* $n \geq 2$, *and require that all our elliptic pencils* $V$ *satisfy* $p_g(V) = n$. Let $H^2(V;\mathbf{Z})_0$ denote those cohomology classes of $H^2(V;\mathbf{Z})$ orthogonal to both $\Delta$ and $C_u$; in addition, let $L_V^{\perp}$ be the set of all $\beta \in H^2(V;\mathbf{Z})_0$ satisfying $\beta \cdot \alpha = 0$, $\forall \alpha \in L_V$. We know that $H^2(V;\mathbf{Z})_0$ is an even unimodular lattice of rank $10n+8$ with $2n$ positive eigenvalues; there is (up to isomorphism) only one such unimodular lattice, we fix a copy and denote it by $H$. Let $O(H)$ denote the orthogonal group of $H$, this is a linear algebraic group defined over $\mathbf{Q}$. We may fix a sublattice $L$ of $H$, such that for any special elliptic pencil $V$, we can find at least one isomorphism between $H^2(V;\mathbf{Z})_0$ and $H$, sending $L_V$ isomorphically onto $L$. *The pair* $L \subset H$ *will be fixed for the remainder of this chapter.* Let $\alpha \in L$, $\alpha^2 = -2$, we can define an element $s_\alpha$ of $O(H)$ by: $s_\alpha(\beta) = \beta + (\beta\alpha)\alpha$, $\forall \beta \in H$. This is called the reflection in $\alpha$; let $\mathfrak{R}(L)$ denote the subgroup of $O(H)_{\mathbf{Z}}$ generated by the reflections in elements $\alpha \in L$, $\alpha^2 = -2$. We prove: *If* $\sigma \in O(H)_{\mathbf{Z}}$ *satisfies* $\sigma/_{L^{\perp}} = \mathrm{id}_{L^{\perp}}$ *then* $\sigma \in \mathfrak{R}(L)$. Here, as above, $L^{\perp}$ denotes the set of elements in $H$ orthogonal to $L$.

Let $g > 0$, and fix a nondegenerate skew symmetric bilinear form $J$ on $\mathbf{Z}^{2g}$ of degree one; let $Sp(\mathbf{Z}^{2g}, J)$ denote the subgroup of elements in $GL(\mathbf{Z}^{2g})$ which preserve $J$. If $C$ is a smooth curve, and $\phi: H^1(C;\mathbf{Z}) \to \mathbf{Z}^{2g}$ is an isomorphism identifying the cup product on

$H^1(C;Z)$ with the skew pairing $J$, then the pair $(C,\phi)$ is called a marked curve of genus $g$; and $\phi$ is called the marking. Two such pairs $(C_i,\phi_i)$, $i=1,2$, are to be considered equivalent if there is an analytic isomorphism $\psi$ between the curves $C_1$ and $C_2$, satisfying $\phi_1 \circ \psi^* = \phi_2$; where $\psi^*$ is the induced map on cohomology. When $g>1$, the set of marked curves of genus $g$, considered modulo equivalence, is a smooth irreducible analytic space of dimension $3g-3$; this space is acted upon properly discontinuously by $Sp(Z^{2g},J)$; the quotient under $Sp(Z^{2g},J)$ will be the moduli space of curves of genus $g$. When $g=1$, as is well known, we have to modify the concept of a marked curve; namely, a marked elliptic curve is a triple $(E,x,\phi)$, where $E$ is an elliptic curve, $x \in E$, and $\phi$ is a marking. The concept of an isomorphism between such triples is the obvious one.

Now, when $g>1$, the set of marked hyperelliptic curves of genus $g$ is a union of smooth analytic spaces of dimension $2g-1$; fix one of the components and call it $Z_g$; in addition, let $\Gamma_{Z_g}$ be the subgroup of elements in $Sp(Z^{2g},J)$ leaving $Z_g$ invariant. Let $Z_1$ be the space classifying marked elliptic curves. Of course, we know that $Z_1$ is naturally isomorphic to the set of all $\tau \in C^1$ satisfying $\operatorname{Im}\tau > 0$, and, under suitable identifications, $Sp(Z^{2g},J)$ becomes $SL(2,Z)$. Define $Z$ to be $Z_1 \times Z_n$; for each $t \in Z$, $t=(t',t'')$, we have marked curves $(E_{t'},x_{t'},\phi_{t'})$ and $(C_{t''},\phi_{t''})$; using $E_{t'}$, $x_{t'} \in E_{t'}$, and $C_{t''}$ we may construct a special elliptic pencil, call it $V_t$. One sees immediately, that this construction depends holomorphically upon parameters; hence, we may construct a family of special elliptic pencils $\rho_Z : \mathcal{V}_Z \to Z$, such that $\rho_Z^{-1}(t',t'') = V_t$; moreover, setting $\Gamma_Z = \Gamma_{Z_1} \times \Gamma_{Z_n}$, $\Gamma_Z$ will act properly discontinuously on $Z$. The quotient of $Z$ by $\Gamma_Z$ is a (coarse) moduli space for special elliptic pencils.

Considered as lattices, for each $t \in Z$, $t=(t',t'')$, we may identify $H^1(E_{t'},Z) \otimes H^1(C_{t''},Z)$ with $L^{\perp}_{V_t}$; under this identification $H^{1,0}(E_{t'}) \otimes H^{1,0}(C_{t''})$ becomes equal to $H^{2,0}(V_t)$. Here, $H^{1,0}(E_{t'})$ (respectively,

$H^{1,0}(C_{t''})$ ) is the subspace in $H^1(E_{t'},C)$ (respectively, in $H^1(C_{t''},C)$ ) corresponding to the holomorphic one forms on $E_{t'}$ (respectively, on $C_{t''}$ ); $H^{2,0}(V_t) \subset H^2(V_t,C)$ is the subspace corresponding to the holomorphic two forms on $V_t$. We may fix $t_0 \in Z$, and choose an identification between $L_{V_{t_0}} \subset H^2(V_{t_0},Z)_0$ and $L \subset H$. Let $\gamma:[0,1] \to Z$, $\gamma(0)=t_0=\gamma(1)$ be a loop; this determines an isomorphism $\gamma^*: H^2(V_{t_0},Z) \to H^2(V_{t_0},Z)$. Hence, we obtain $\sigma \in O(H)_Z$; moreover, $\sigma/_{L^\perp} = \mathrm{id}_{L^\perp}$, hence $\sigma \in \mathfrak{R}(L)$. Now, $\gamma$ sends the effective divisors in $L_{V_{t_0}}$ to effective divisors, this is more or less clear from our construction of $\mathcal{O}_Z \to Z$; but, then, using some well-known properties of $L$, we conclude that $\sigma = \mathrm{id}$. Therefore, the locally constant sheaf $R^2_{\rho_Z}(Z)$ is, in fact, constant. We can fix an isomorphism between it and the constant sheaf determined by $H \oplus Z\Delta \oplus ZC_u$; sending each $L_{V_t}$ isomorphically onto $L$, $H^2(V_{t_0},Z)_0$ isomorphically onto $H$, the section of $V_t$ to $\Delta$, and $K_{V_t}$ to $(n-1)C_u$. Here, of course, $K_{V_t}$ denotes the canonical divisor on $V_t$; and when we say "the section of $V_t$" we mean the section of the underlying elliptic fibration on $V_t$. This section is obtained from $x_{t'} \times \tilde{C}_{t''}$, where $t = (t', t'')$.

For our fixed point $t_0 \in Z$, let $H'$ and $H''$ denote, respectively, $H^1(E_{t_0'},Z)$ and $H^1(\tilde{C}_{t_0''},Z)$; these are to be considered as lattices with the induced skew pairings. Under the identification of $L_{V_{t_0}}$ with $L$, $H' \otimes H''$ is also identified, as a lattice, with $L^\perp$. Using the markings on each $E_{t'}$ and $\tilde{C}_{t''}$, we may construct identifications between $H^1(E_{t'},Z)$ and $H'$ (respectively, between $H^1(C_{t''},Z)$ and $H''$); let $H^{1,0}(t')$ (respectively, $\tilde{H}^{1,0}(t'')$ ) denote the image of $H^{1,0}(E_{t'})$ in $H'_C$ (respectively of $H^{1,0}(C_{t''})$ in $H''_C$ ); in addition, let $H^{2,0}(t)$ denote the image $H^{2,0}(V_t)$ in $H_C$. Clearly, under all these identifications $H^{1,0}(t') \otimes H^{1,0}(t'') = H^{2,0}(t)$.

§2. Let $\Lambda$ be a lattice, i.e., $\Lambda \simeq \mathbf{Z}^N$; suppose that $\Lambda$ is equipped with a bilinear form defined over $\mathbf{Q}$; the product of $u, v \in \Lambda$ is denoted by $(u,v)_\Lambda$, or, when no confusion will arise, simply by $u \cdot v$. If the pairing on $\Lambda$ is skew symmetric, then $\mathfrak{H}(\Lambda)$ will denote the classifying space for Hodge structures of weight one on $\Lambda_{\mathbf{C}}$ polarized by the skew pairing on $\Lambda$; and, $\mathrm{Sp}(\Lambda)$ will denote symplectic group of the form $(\cdot,\cdot)_\Lambda$. This is a linear algebraic group defined over $\mathbf{Q}$. If the pairing on $\Lambda$ is symmetric, then $D(\Lambda)$ will denote the Hodge structures of weight two on $\Lambda_{\mathbf{C}}$ polarized by the symmetric pairing on $\Lambda$; and $O(\Lambda)$ will denote the orthogonal group of the form $(\cdot,\cdot)_\Lambda$. Also, if $\Lambda_1 \subset \Lambda$ is a sublattice, then $D(\Lambda;\Lambda_1) \subset D(\Lambda)$ will denote those Hodge structures of weight two on $\Lambda_{\mathbf{C}}$ for which each element of $\Lambda_1$ is of type $(1,1)$; $O(\Lambda,\Lambda_1)$ will be the closed subgroup of $O(\Lambda)$ consisting of those elements that act as the identity on $\Lambda \otimes \mathbf{Q}$. Each of these classifying spaces sits naturally inside a suitable Grassmanian. If nonempty, they (respectively, their Zariski closure) are homogenous under the action of the corresponding real (respectively, complex) "orthogonal" groups (see [4]).

Under the identification of $H' \otimes H''$ with $L^\perp$, the symmetric pairing on one becomes a constant multiple of the pairing on the other. Let $\mathfrak{H}'$ and $\mathfrak{H}''$ denote, respectively, $\mathfrak{H}(H')$ and $\mathfrak{H}(H'')$. The point $\xi' \in \mathfrak{H}$ (respectively, $\xi'' \in \mathfrak{H}''$) corresponds to a polarized Hodge structure of weight one on $H'_{\mathbf{C}}$ (respectively on $H''_{\mathbf{C}}$); the tensor product of these Hodge structures defines a polarized Hodge structure of weight two on $(H' \otimes H'')_{\mathbf{C}}$, hence a point $\xi$ of $D(H,L)$; by sending $(\xi', \xi'')$ to $\xi$, we obtain a morphism $F$ from $\mathfrak{H}' \times \mathfrak{H}''$ to $D(H,L)$. It is not difficult to show that $F$ is an analytic embedding; let $\mathfrak{X}$ denote the image of $F$, and let $\check{\mathfrak{X}}$ equal the Zariski closure of $\mathfrak{X}$. If $(\sigma', \sigma'')$ is an element of $\mathrm{Sp}(H')_{\mathbf{R}} \times \mathrm{Sp}(H'')_{\mathbf{R}}$, then define $f(\sigma', \sigma'') = \sigma \in O(H,L)_{\mathbf{R}}$ by: $\sigma(e' \otimes e'') = (\sigma' e') \otimes (\sigma'' e'')$, and $\sigma(v) = v$, $\forall v \in L_{\mathbf{R}}$. The map $F$ is equivariant with respect to the homomorphism $f$; the kernel of $f$ equals $\{\pm \mathrm{id}_{H'}\} \times \{\pm \mathrm{id}_{H''}\}$.

Note that the element $(\sigma', \sigma'')$ of $Sp(H')_{\mathbf{R}} \times Sp(H'')_{\mathbf{R}}$ acts trivially on $\mathfrak{H}' \times \mathfrak{H}''$ iff $\sigma \in \{\pm id_{H'}\}$ and $\sigma \in \{\pm id_{H''}\}$. Using some elementary matrix calculations, together with a result due to Siegel, we prove:

> *The image of* $\check{Sp}(H')_{\mathbf{R}} \times \check{Sp}(H'')_{\mathbf{R}}$ *under* f *equals* $\{\sigma \in O(H,L)_{\mathbf{R}} | \sigma(\mathfrak{X}) = \mathfrak{X}\}$. *Moreover, if* $\sigma \in O(H)_{\mathbf{Z}}$, *and* $\sigma(\mathfrak{X}) = \mathfrak{X}$ *then* $\sigma(L) = L$.

Let $\Gamma^*$ equal $\{\sigma \in O(H,L)_{\mathbf{R}} | \sigma(L^{\perp}) = L^{\perp}\}$; a result due to Borel (see [1]) implies:

$$f(Sp(H')_{\mathbf{Z}} \times Sp(H'')_{\mathbf{Z}}) = \{\sigma \in \Gamma^* | \sigma(\mathfrak{X}) = \mathfrak{X}\}.$$

Let us define $\Phi : Z \to D(H)$ by sending each $t \in Z$ to the polarized Hodge structure on $H_{\mathbf{C}}$ defined by $H^{2,0}(t) \subset H_{\mathbf{C}}$. $\Phi$ is the period mapping for the family $\mho_Z \to Z$; note that $\Phi(Z) \subset D(H,L)$. In addition, for $t = (t', t'') \in Z$, define $\Phi_{1,n}(t) = (\xi', \xi'') \in \mathfrak{H}' \times \mathfrak{H}''$ as follows: $\xi'$ (respectively $\xi''$) corresponds to the polarized Hodge structure on $H'_{\mathbf{C}}$ (respectively on $H''_{\mathbf{C}}$) defined by $H^{1,0}(t')$ (respectively by $H^{1,0}(t'')$). Note that $F(\Phi_{1,n}(t)) = \Phi(t)$. Using the global Torelli theorem for curves, together with the injectivity of $F$, we conclude: *The period mapping for the family* $\mho_Z \to Z$ *is injective*. One can show that the topological closure of $\Phi(Z)$ is a closed analytic subspace of $D(H)$ containing $\Phi(Z)$ as an open subset.

The discrete group $\Gamma_Z$ acts on $\mho_Z \to Z$, hence, also on $R^2_{\rho_Z}(Z)$; and, via our identification of this sheaf with the constant sheaf determined by $H \oplus \mathbf{Z}C_u \oplus \mathbf{Z}\Delta$, we obtain an action of $\Gamma_Z$ on $H$. Let $\Gamma_Z^*$ denote the image of $\Gamma_Z$ in $O(H)_{\mathbf{Z}}$ obtained using this action. Now, $\Gamma_Z$ also sits, more or less by definition, naturally in $Sp(H')_{\mathbf{Z}} \times Sp(H'')_{\mathbf{Z}}$; if $\sigma \in \Gamma_Z$, then the action of $\sigma$ on $L^{\perp}$ agrees with the action of $f(\sigma)$ on $L^{\perp}$; hence, $f(\Gamma_Z)$ commensurable with $\Gamma_Z^* \subset O(H)_{\mathbf{Z}}$. Using the fact that $\Gamma_{Z_n}$ (respectively $\Gamma_{Z_1}$) is Zariski dense in $Sp(H'')_{\mathbf{C}}$ (respectively in $Sp(H')_{\mathbf{C}}$), we prove:

*The Zariski closure of* $\Phi(Z)$ *in* $D(H)$ *equals* $\check{X}$. *In addition, if* $\Gamma'$ *is a subgroup of finite index in* $\Gamma_Z^*$, $F' \subset H_C$ *is a* $\Gamma'$ *invariant subspace, and* $F' \cap (L^{\perp} \otimes C) \neq 0$, *then* $F' \supset (L^{\perp} \otimes C)$.

Let $H^{2,0} \oplus H^{1,1} \oplus H^{0,2} = H_C$ be a polarized Hodge structure of weight two; if $H_Q = F_1 \oplus F_2$, $H^{2,0} = (H^{2,0} \cap (F \otimes C)) \oplus (H^{2,0} \cap (F_2 \otimes C))$, and $H^{2,0} \cap (F_i \otimes C) \neq 0$ for $i = 1,2$, then we say that $H^{2,0} \oplus H^{1,1} \oplus H^{0,2}$ *splits in an essential way.* As a direct consequence of the above statement we have:

*The Hodge structure of the generic special elliptic pencil doesn't split in an essential way.*

We should clarify our use of the qualifier "generic"; by this we mean that the phenomena in question takes place on the complement of a countable union of proper closed analytic subspaces.

Suppose $\sigma \in \Gamma_Z$, $\sigma \neq 1$, and $\sigma$ acts trivially on $Z$; then, writing $\sigma = (\sigma', \sigma'')$ we know that $\sigma' \in \{\pm \mathrm{id}_{H'}\}$, and $\sigma'' \in \{\pm \mathrm{id}_{H''}\}$. By looking at how we constructed $V_{t_0}$ from $E_{t_0'}$, $x_{t_0'} \in E_{t_0'}$, and $C_{t_0''}$, we see that $\sigma/_{L^{\perp}} = \pm \mathrm{id}_{L^{\perp}}$ and $\sigma/_L = \mathrm{id}_L$. Moreover, $\sigma$ acts trivially on $H$ iff $\sigma$ acts trivially on the family $\mathcal{V}_Z \to Z$; in this case, $\sigma = -\mathrm{id}_{H'}$ and $\sigma = -\mathrm{id}_{H''}$.

Let $\sigma \in O(H)_Z$, and assume $\sigma\overline{\Phi(Z)} = \overline{\Phi(Z)}$; then, $\sigma(\check{\mathfrak{X}}) = \check{\mathfrak{X}}$ and hence $\sigma(\mathfrak{X}) = \mathfrak{X}$. There is an element $\sigma^* \in \Gamma^*$ such that $\sigma^*\sigma^{-1}(v) = v$, $\forall v \in L$. Clearly, $\sigma/_{\mathfrak{X}} = \sigma^*/_{\mathfrak{X}}$. Write, using the above, $\sigma^* = f(\sigma_1^*, \sigma_2^*)$, where $\sigma_1^* \in Sp(H')_Z$ and $\sigma_2^* \in Sp(H'')_Z$; we have $(\sigma_1^*, \sigma_2^*)\overline{\Phi_{1,n}(Z)} = \overline{\Phi_{1,n}(Z)}$. Now, $\Phi_{1,n}$ induces a map:

$$Z/\Gamma_Z \to (\mathfrak{H}'/_{Sp(H')_Z}) \times (\mathfrak{H}''/_{Sp(H'')_Z}).$$

From the global Torelli theorem for curves, we know that this map is injective. Hence, there exists $\sigma_1 \in \Gamma_{Z_1}$ and $\sigma_2 \in \Gamma_{Z_n}$ such that

$(\sigma_1^*\sigma_1^{-1}, \sigma_2^*\sigma_2^{-1})$ restricts to the identity on $\Phi_{1,n}(Z)$; but, since $\Phi_{1,n}(Z)$ is Zariski dense in $\mathfrak{H}' \times \mathfrak{H}''$, this implies that $(\sigma_1^*\sigma_1^{-1}, \sigma_2^*\sigma_2^{-1})$ acts trivially on $\mathfrak{H}' \times \mathfrak{H}''$. This only happens when $\sigma_1^*\sigma_1^{-1} \in \{\pm \mathrm{id}_{H'}\}$ and $\sigma_2^*\sigma_2^{-1} \in \{\pm \mathrm{id}_{H''}\}$; therefore, for some $\gamma \in \Gamma_Z$, $\sigma^* = f(\gamma)$. Now, if $\gamma^*$ denotes the image of $\gamma$ in $\Gamma_Z^*$, then $\gamma^*\sigma^{-1}$ restricts to the identity on $\overline{\Phi(Z)}$; but since the Zariski closure of $\Phi(Z)$ is $\check{\mathfrak{X}}$, using previous results, we know that $\gamma^*\sigma^{-1}/_{L^\perp} = \pm \mathrm{id}_{L^\perp}$; hence, again by previous results, $\gamma^*\sigma^{-1}/_L \in \mathfrak{R}(L)$. We have proven:

$$\{\sigma \in O(H)_Z | \sigma\overline{\Phi(Z)} = \overline{\Phi(Z)}\} = \mathfrak{R}(L) \cdot \Gamma_Z^*$$
$$= \Gamma_Z^* \cdot \mathfrak{R}(L)\,.$$

Let us now fix $V_0 = V_{t_0}$, $t_0 \in Z$, with the following properties:

(i) $\{\gamma \in O(H)_Z | \gamma\Phi(t_0) \in \overline{\Phi(Z)}\} = \mathfrak{R}(L) \cdot \Gamma_Z^*$

(ii) The Hodge structure of $V_0$ doesn't split in an essential way.

Set $\xi_0 = \Phi(t_0)$; and let $\Delta_0$ be the fixed section of $V_0$. Suppose $V$ is any elliptic pencil, and that there is an isomorphism from $H^2(V; \mathbf{Z})_0$ to $H^2(V_0; \mathbf{Z})_0$ preserving the cup product and sending $H^{2,0}(V)$ to $H^{2,0}(V_0)$, i.e., there is an isomorphism of polarized Hodge structures. Clearly, since $L_V$ and $L_{V_0}$ are determined, respectively, by the polarized Hodge structures of $V$ and $V_0$; then, $L_V$ will be mapped isomorphically onto $L_{V_0}$. This implies, by the results stated above, that $V$ is also a special elliptic pencil, say, $V \simeq V_{\tilde{t}}$, for some $\tilde{t} \in Z$. Set $\tilde{\xi} = \Phi(\tilde{t})$, then there exists $\sigma \in O(H)_Z$ with $\sigma(\tilde{\xi}) = \xi_0$. By property i) above, $\sigma = \tau \cdot \sigma^*$, for some $\tau \in \mathfrak{R}(L)$, $\sigma^* \in \Gamma_Z^*$; where $\sigma^*$ is the image of $\sigma \in \Gamma_Z$. This implies $\Phi(\sigma\tilde{t}) = \Phi(t_0)$; hence, by the injectivity of $\Phi$, $\sigma(\tilde{t}) = t_0$. *Conclusion*: $V$ *is isomorphic to* $V_0$.

§3. Let $(V, \Delta)$ be an elliptic pencil, and let $\psi: V \to \mathbf{P}_1$ be the underlying elliptic fibration on $V$. For all $m \geq 1$, $R^0_\psi(\mathcal{O}_V(m\Delta))$ is locally free of rank $m$. In addition, if $m \geq 2$, we obtain a regular analytic map $f_m$ from $V$ to the projectivization of $R^0_\psi(\mathcal{O}_V(m\Delta))$. Both of these statements are proven by showing, first, that $\dim H^0(C_u, \mathcal{O}_{C_u}(m\Delta))$ is independent of $u$; and, then by examining the behavior of $H^0(C_u, \mathcal{O}_{C_u}(m\Delta))$, regarded as a linear system on $C_u$. To obtain the latter information, one uses Kodaira's list of singular fibres. One can find an invertible sheaf $\mathcal{L}$ on $P_1$ of degree $-(n+1)$, sections $g_2$ and $g_3$ of, respectively, $\mathcal{L}^{-\otimes 4}$ and $\mathcal{L}^{-\otimes 6}$; such that, $R^0_\psi(\mathcal{O}_V(3\Delta)) = \mathcal{L}^{\otimes 2} \oplus \mathcal{L}^{\otimes 3} \oplus \mathcal{O}_{P_1}$, and the image of $V$ under $f_3$ is defined, with respect to some covering $\{U_j\}_j$ of $P$, by the homogenous equation $y_j^2 z_j = x_j^3 - g_{2j} x_j^2 z_j - g_{3j} z_j^3$. Where $x_j$, $y_j$ and $z_j$ are the fibre coordinates, over $U_j$, for $\mathcal{L}^{\otimes 2} \oplus \mathcal{L}^{\otimes 3} \oplus \mathcal{O}_{P_1}$; $g_{2j}$ and $g_{3j}$ are, respectively, the local representatives, on $U_j$, for $g_2$ and $g_3$. The existence of $g_2$, $g_3$ and $\mathcal{L}$ is proven in a manner entirely similar to what one does when dealing with a single elliptic curve. The map $V \to f_3(V)$ is obtained by collapsing precisely those irreducible curves $a$ on $V$ satisfying $a^2 = -2$, and $a \cdot C_u = 0 = a \cdot \Delta$; hence, $f_3(V)$ is a surface whose only singularities are rational double points. In accordance with the terminology of [8], the pair $(f_3(V), f_3(\Delta))$ will be called a Weierstrass pencil, or, perhaps, the Weierstrass pencil associated to $(V,\Delta)$. Clearly, two elliptic pencils will be isomorphic iff their associated Weierstrass pencils are. The fact that $f_3(V)$ has only rational double points translates immediately into restrictions on $g_2$ and $g_3$. For latter use, we note: $f_2(V)$ is the projectivization of $\mathcal{L}^{\otimes 2} \oplus \mathcal{O}_{P_1}$; the map $f_2 : V \to f_2(V)$ is of degree two; and, over $U_j$, the branch curve of $f_2$ is defined by $z_j(x_j^3 - g_{2j} x_j^2 z_j - g_{3j} z_j^3) = 0$, with $f_2(\Delta)$ being defined by $z_j = 0$. Moreover, $[x_j, y_j, z_j] \mapsto [x_j, -y_j, z_j]$ determines an involution $\iota_V$ on $V$; $\iota_V$ will be called *canonical involution*. Clearly, $\iota_V(\Delta) = \Delta$, and, if $C_u$ is smooth, then $C_u/\iota_V \simeq P_1$. Here, as above, $x_j$, $y_j$ and $z_j$ are the fibre coordinates, over $U_j$, of $\mathcal{L}^2 \oplus \mathcal{L}^3 \oplus \mathcal{O}_{P_1}$.

If we choose an isomorphism between $\mathcal{L}$ and $\mathcal{O}_{P_1}(-n-1)$, then $g_2$ and $g_3$ may be regarded as homogenous polynomials. Using the coefficients of $g_2$ and $g_3$ as coordinates, the set of pairs $(g_2,g_3)$ arising as above, forms an open subset $U$ of $C^{4n+5} \times C^{6n+7}$. For each $t \in U$, $t$ corresponding to $(g_2,g_3)$, let $Y_t$ be the Weierstrass pencil defined by the (weighted) homogenous equation $y^2z = x^3 - g_2xz - g_3z^3$; and let $\Delta_t \subset Y_t$ equal the section determined by $y = 0 = z$. There is a natural action of $C^* \times \mathrm{Aut}(P_1)$ on $U$; two pairs $(Y_{t'}, \Delta_{t'})$, $(Y_{t''}, \Delta_{t''})$ are isomorphic iff $t'$ and $t''$ lie in the same orbit of $C^* \times \mathrm{Aut}(P_1)$. This result and the ones in the preceding paragraph are all due to Kas (see [5]).

Miranda (see [8]) proved that the quotient of $U$ under $C^* \times \mathrm{Aut}(P_1)$ exists as a quasiprojective variety of dimension $10n + 8$; and, forms a coarse moduli space for Weierstrass pencils. Let this moduli space be denoted by $M$; clearly $M$ will also be a coarse moduli space for elliptic pencils. There will be a natural analytic map $h: Z \to M$, obtained by assigning to each $t \in Z$, the isomorphism class of $\rho_Z^{-1}(t)$; where $\mathcal{O}_Z \xrightarrow{\rho_Z} Z$ is the family (marked) special elliptic pencils constructed above. Clearly, $h(t) = h(\sigma t)$, $\forall \sigma \in \Gamma_Z$; and, in fact, $h$ determines an analytic embedding of $Z/\Gamma_Z$ into $M$

Let $U' = \{t \in U | Y_t$ is smooth$\}$; this is an open subset of $U$ invariant under $C^* \times \mathrm{Aut}(P_1)$. Its image in $M$ will be an open set $M'$. For the moment, fix $s_0 \in U'$, and choose an isomorphism, preserving the symmetric pairings, between $H^2(Y_{s_0}, Z)_0$ and $H$. As $t$ varies over $U'$, the $Y_t$ vary in a holomorphic manner. In particular, for any path $\gamma: [0,1] \to U'$, $\gamma(0) = s_0$, and $\gamma(1) = t$, there will be an isomorphism $\gamma^*: H^2(Y_t; Z)_0 \to H^2(Y_{s_0}; Z)_0$; using this, we may construct $\gamma^*(H^{2,0}(Y_t)) \subset H^2(Y_{s_0}, C)_0$, and hence, via our identification, we have a subspace $\gamma^* H^{2,0}(Y_t) \subset H_C$. If $\gamma_1$ is another such path, then, there exists $\sigma \in O(H)_Z$ such that, $\gamma_1^* H^{2,0}(Y_t) = \sigma \gamma^* H^{2,0}(Y_t)$; the collection of such $\sigma$'s forms a subgroup $\Gamma$ of $O(H)_Z$. $\Gamma$ will be called the global monodromy group for the family

$\bigcup_{t \in U'} Y_t$. In this way, sending $t \in U'$ to $\gamma^* H^{2,0}(Y_t) \subset H_{\mathbf{C}}$, we obtain a map $\Psi$ from $U'$ to $D(H)/_{\Gamma}$. $\Psi$ is constant on the orbits of $\mathbf{C}^* \times \mathrm{Aut}(\mathbf{P}_1)$ and descends to a map $\Psi$ of $M'$ to $D(H)/_{\Gamma}$. As is well known, $\Psi$ is analytic and extends across $M - M'$ to a holomorphic map on $M$; denote the resulting extension also by $\Psi$. $\Psi$ is the *period mapping for elliptic pencils*. Let $\pi_{\Gamma}: D(H) \to D(H)/_{\Gamma}$ be the quotient map; clearly, after replacing $\Gamma$ by $\sigma\Gamma\sigma^{-1}$, for some $\sigma \in O(H)_{\mathbf{Z}}$, we may assume $\Psi \circ h = \pi_{\Gamma} \circ \Phi$.

§4. Let $(V, \Delta)$ be any elliptic pencil and let $\psi: V \to \mathbf{P}_1$ be the underlying elliptic fibration on $V$. Let $\Theta_V$ and $\Omega^1_V$ denote, respectively, the sheaf of holomorphic tangent vectors and the sheaf of holomorphic one forms on $V$. One can show that $\dim H^0(V, \Omega^1_V(mC_u)) = m-1$ for all $1 \leq m \leq n-1$; from this we calculate that $\dim H^1(V; \Theta_V) = 11n+8$. Now, $V$ is the minimal resolution of $Y_t$, for some $t \in U$. We may construct a suitable $10n+8$ dimensional polydisc transverse to the orbit of $t$. Using the simultaneous resolution of rational double points, we may therefore construct an effectively parameterized family, with central fibre $V$, of elliptic pencils depending upon $10n+8$ parameters; "twisting" each pencil of that family adds $n$ additional parameters. *Conclusion*: *The Kuranishi family for* $V$ *is smooth of dimension* $11n+8$. In the case where $\psi: V \to \mathbf{P}_1$ is non-constant, this was proven by Kas. Let $a_1, \cdots, a_\ell$ be those irreducible curves on $V$ satisfying $a_i \cdot C_u = 0 = a_i \cdot \Delta$, $\forall i$; of course, each $a_i$ is a smooth rational curve with self intersection number $-2$. Using some general results from the theory of rational double points, we know that we can "vary" the $a_i$ independently; that is to say, in the parameter space for the Kuranishi family, the locus where $a_i$ remains effective is smooth, and, intersects transversely the locus where $a_j$ remains effective, $\forall i \neq j$. By analyzing what happens under "twisting", one can show that, in fact, $a_1, \cdots, a_\ell$, *and* $\Delta$ can all be varied independently. Let $\rho_0: \mathcal{V}_0 \to N_0$ be the subfamily of the Kuranishi family where $\Delta$ remains effective; $\rho_0^{-1}(0) \simeq V$, $0 \in N_0$. $N_0$ is smooth of dimension $10n+8$ and each fibre

$\rho_0^{-1}(t)$ is naturally an elliptic pencil; in fact, $\mathcal{V}_0 \to N_0$ is the *universal deformation* of $(V, \Delta)$.

In the usual way, $\mathfrak{R}(L_V)$ will act upon $N_0$: Let $N' \subset N_0$ be the locus where, say, $\alpha_j$ remains effective. Suppose $\Gamma' \subset \rho_0^{-1}(N')$, $\rho_0/_{\Gamma'} : \Gamma' \to N'$ is a $P_1$-bundle, and $\Gamma' \cap \rho_0^{-1}(0) = \alpha_j$. If we blow up $\Gamma'$, and then blow down the resulting exceptional divisor in the other direction; we obtain a new family, say $\mathcal{Y} \to N_0$ with central fibre $V$. Hence, by the universality of $\mathcal{V}_0 \to N_0$, there exists $\tau : N_0 \to N_0$ such that $\tau^*\mathcal{V}_0 \simeq \mathcal{Y}$ $\tau$ will be a "reflection about $N'$"; we identify $\tau$ with $s_{\alpha_j} \in \mathfrak{R}(L_V)$. In addition, if $\mathrm{Aut}(V, \Delta)$ denotes the automorphisms of $(V, \Delta)$, that is to say, the automorphisms of $V$ which preserve $\Delta$; then, the semidirect product of $\mathrm{Aut}(V, \Delta)$ with $\mathfrak{R}(L_V)$ will act upon $N_0$; temporarily, denote this semidirect product by $\Gamma_{N_0}$.

Let $x \in M$ be the point corresponding to $(V, \Delta)$; then, after shrinking $N_0$, if necessary; $N_0/\Gamma_{N_0}$ will be naturally isomorphic to a small nbd of $x$ in $M$. Let $g_0 : N_0 \to M$ be the natural map. Choose an isomorphism between $H^2(V; Z)_0$ and $H$; and, let $\tilde{\phi}_0 : N_0 \to D(H)$ be the resulting period map. We may arrange matters so that the following diagram will commute:

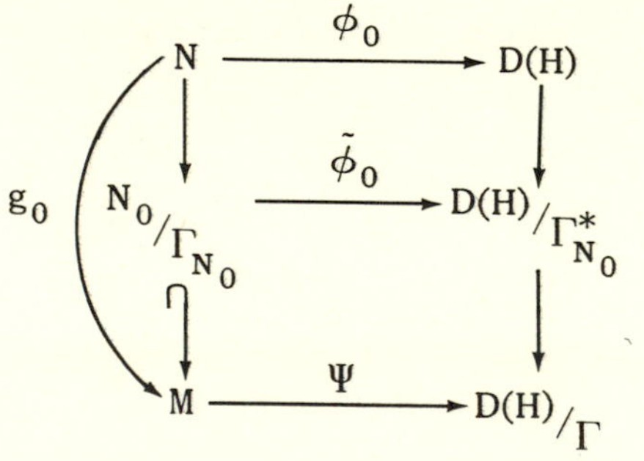


The unlabeled arrows are obtained in the obvious way; and, $\Gamma^*_{N_0}$ denotes the image of $\Gamma_{N_0}$ in $O(H)_Z$. Now, since $H^0(V; \Omega^1_V(C_u)) = 0$, using the results of Kii (see [6]), we know that $\phi_0$ is an analytic embedding (i.e., the local Torelli theorem holds). Clearly, $\tilde{\phi}_0$ will also be an analytic embedding.

Now, let us assume $V = V_0$; where $V_0$ is the special elliptic pencil chosen earlier. In addition to the conditions we required $V_0$ to satisfy; we will now also require:

(iii) $V_0$ has no automorphisms other than the canonical involution $\iota_{V_0}$.

One can show that $(\iota_{V_0})_*/L^{\perp}_{V_0} = -\mathrm{id}_{L^{\perp}_{V_0}}$ and $(\iota_{V_0})_*/L_{V_0} = \mathrm{id}_{V_0}$; also, note that $-\mathrm{id}_{L_{V_0}} \in \mathfrak{R}(L_{V_0})$. From this it follows that $\Gamma^*_{N_0} = \{\pm\mathrm{id}_H\}\cdot\mathfrak{R}(L)$.

Recall that $\xi_0 = \Phi(t_0)$, where $V_0 = V_{t_0}$, $t_0 \in Z$. Let $\sigma \in O(H)_{\mathbf{Z}}$ and suppose $\sigma(\xi_0) = \xi_0$; then, in particular, $\sigma \in \mathfrak{R}(L)\cdot\Gamma^*_Z$. Write $\sigma = \tau\gamma^*$; where $\tau \in \mathfrak{R}(L)$, and $\gamma^* \in \Gamma^*_Z$ comes from $\gamma \in \Gamma_Z$. Clearly, $\gamma(t_0) = t_0$; hence $\gamma$ induces an automorphism of $V_{t_0} = V_0$ which, by assumption, must be the canonical involution of $V_0$. Hence, $\gamma = (\pm\mathrm{id}_{H'}, \pm\mathrm{id}_{H''})$; therefore, $\sigma \in \{-\mathrm{id}_H\}\cdot\mathfrak{R}(L)$. Let $\tilde{U}_{\xi_0}$ be a small nbd of $\xi_0$ in $D(H)$, chosen so that: $\sigma\tilde{U}_{\xi_0} \cap \tilde{U}_{\xi_0} \neq \emptyset$ iff $\sigma\xi_0 = \xi_0$. It follows directly from this discussion that $\tilde{U}_{\xi_0}/\mathfrak{R}(L)$ will be sent isomorphically onto its image in $D(H)/O(H)_{\mathbf{Z}}$. We have thus proven:

*If* $x_0 \in M$ *is the point corresponding to* $(V_0, \vartriangle_0)$; *then* $\Psi$ *maps some* nbd *of* $x_0$ *isomorphically onto its image in* $D(H)/\Gamma$.

Given this, using the results of Griffiths, we know: $\Psi : M \to \Psi(M)$ is of finite degree; $\Psi(M)$ is a constructable analytic subspace of $D(H)/\Gamma$. Actually, this last statement doesn't depend upon whether $\Psi$ is "generically" a local isomorphism.

To finish our calculations, let us assume, for the moment, the truth of the following statement:

(*) Let $(V, \vartriangle)$ be an elliptic pencil, and let $x \in M$ be the point corresponding to $V$. If the polarized Hodge structure of $V$ doesn't split in an essential way; then, there exists a nbd $U_0$ of $\Psi(x)$ such that $\Psi/_{\Psi^{-1}(U_0)}$ is a proper map.

Applying this to the case where $(V, \vartriangle) = (V_0, \vartriangle_0)$; let $U_0$ be a nbd of $\Psi(x_0)$ such that $\Psi/_{\Psi^{-1}(U_0)}$ is a proper map. Clearly, for "most" points $y'$ of $\Psi(M) \cap U_0$, the cardinality of $\Psi^{-1}(y')$ will equal the degree of $\Psi$. From the results given before, we know that $\Psi^{-1}(\Psi(x_0)) = \{x_0\}$. Now, *because* $\Psi/_{\Psi^{-1}(U_0)}$ *is proper*, the above "local calculations" imply: *The period mapping* $\Psi$ *is a degree one map between* $M$ *and* $\Psi(M)$.

Let $Y \subset D(H)$ denote $\pi_\Gamma^{-1}(\Psi(M))$, and suppose we have an element $\sigma$ of $O(H)_Z$ satisfying $\sigma\overline{Y} = \overline{Y}$. Using (*) above, we can see that $\sigma\xi_0 \notin \overline{Y} - Y$. Hence, $\sigma\xi_0 \in \tau\Phi(Z)$, for some $\tau \in \Gamma$. But this implies that $\tau^{-1}\sigma\xi_0 \in \Phi(Z)$, hence, by our assumptions on $V_0$, $\tau^{-1}\sigma \in \mathfrak{R}(L) \cdot \Gamma_Z^* \subset \Gamma$. Hence $\{\sigma \in O(H)_Z | \sigma\overline{Y} = \overline{Y}\}$ equals $\Gamma$. Thus we have proven:

MAIN THEOREM (generic global Torelli). *The generic elliptic pencil is determined by its polarized Hodge structure.*

§5. In this section, and the following one, the notation used will be, for the most part, independent of that used above. Among the other notation to be introduced below, we will consistently use $D_\varepsilon$ to denote $\{\tau \in C | |\tau| < \varepsilon\}$.

Let $\mathcal{V}$ be an analytic space, and let $\rho : \mathcal{V} \to D_\varepsilon$ be a proper, surjective, analytic map; for all $t \in D_\varepsilon$, set $V_t = \rho^{-1}(t)$. If, $\forall t \neq 0$, $V_t$ is a smooth curve (respectively, a surface whose only singularities are rational double points), then $\rho : \mathcal{V} \to D_\varepsilon$ will be called a *degeneration* of curves (respectively, of surfaces). We will call $V_0 = \rho^{-1}(0)$ the *central fibre*, and $\mathcal{V}$ will be called the total space. Note that we consider $\rho^{-1}(0)$ as the analytic subspace of $\mathcal{O}$ defined by the equation $\rho^*\tau = 0$; where $\tau$ is a local parameter at $0 \in D_\varepsilon$. If $\mathcal{O}$ is smooth, and $\rho^{-1}(0)$ has normal crossings, each component appearing with multiplicity one, then $\rho : \mathcal{O} \to D_\varepsilon$ is called a *semistable degeneration.*

Let $\rho_i : \mathcal{O}_i \to D_\varepsilon$ be two degenerations. If $F : \mathcal{O}_1 \to \mathcal{O}_2$ is a proper analytic map satisfying $\rho_2 \circ F = \rho_1$, and $F^{-1}$ exists as a rational map, then we say that $\mathcal{O}_1 \to D_\varepsilon$ dominates $\mathcal{O}_2 \to D_\varepsilon$. If there is birational map F from $\mathcal{O}_1$ to $\mathcal{O}_2$ satisfying $\rho_2 \circ F = \rho_1$ on an open dense subset; then, we say that $\mathcal{O}_1$ is $D_\varepsilon$-birationally equivalent to $\mathcal{O}_2$. Now, according to a theorem of Mumford's, any degeneration $\rho : \mathcal{O} \to D_\varepsilon$ possesses a *semi-stable reduction*; that is to say, for some base extension $D_{\varepsilon'} \to D_\varepsilon$ given by $\tau \mapsto \tau^N$, there will be a semistable degeneration which dominates $\mathcal{O} \times_{D_\varepsilon} D_{\varepsilon'} \to D_{\varepsilon'}$.

Let $\rho : \mathcal{O} \to D_\varepsilon$ be a semistable degeneration of surfaces $\mathcal{S} \subset \mathcal{O}$ an irreducible effective divisor on $\mathcal{O}$ such that, $\forall t \in D_\varepsilon - \{0\}$, the minimal model of $\rho^{-1}(t)$, together with the image of $\mathcal{S} \cap \rho^{-1}(t)$ determines an elliptic pencil. Suppose that $R^2_\rho(\mathbf{Z})$ is isomorphic to the constant sheaf $H \oplus S$; where, under this isomorphism, the image of S in the stalk $R^2_\rho(\mathbf{Z})_t$, $t \neq 0$, corresponds to the subgroup generated by $\mathcal{S} \cap \rho^{-1}(t)$ and the various irreducible components of any effective representative of the canonical divisor on $\rho^{-1}(t)$; H is the unimodular lattice referred to in the previous sections. Under these conditions, we say that $\rho : \mathcal{O} \to D_\varepsilon$ is a (semistable) degeneration of elliptic pencils with trivial monodromy. Given the trivialization of $R^2_\rho(\mathbf{Z})$, we obtain a period mapping $f : D_\varepsilon - \{0\} \to D(H)$; it is well known that f extends to $D_\varepsilon$; f(0) is called the *limiting Hodge structure* of this degeneration. Let us call a component X of $\rho^{-1}(0)$ *distinguished* if $p_g(X) > 0$. If $X_1, \cdots, X_\ell$ are the distinguished components of $\rho^{-1}(0)$, then, as is well known, we must have $\sum_j p_g(X_j) = n$; where n is the geometric genus of any generic fibre. Moreover, again by well-known results, if $\ell > 0$, then the limiting Hodge structure of $\rho : \mathcal{O} \to D_\varepsilon$ must split in an essential way.

Consider the following statement:

(**) If $\rho : \mathcal{Y} \to D_\varepsilon$ is a semistable degeneration of elliptic pencils with trivial monodromy, whose central fibre has a single distinguished component. Then, after a base change $D_{\varepsilon'} \to D_\varepsilon$, $\mathcal{Y} \times_{D_\varepsilon} D_{\varepsilon'} \to D_{\varepsilon'}$ will be $D_{\varepsilon'}$-birationally equivalent to a degeneration whose central fibre is a surface with at most rational double points.

One shows that statement (**) implies statement (*) of the previous section in the usual way; namely, by applying the resolution of singularities to some projective closure $\overline{M}$ of $M$, together with the results of Griffiths [4]; and, in addition, one makes use of the simultaneous resolution of rational double points.

We call a triple $(\mathcal{B}, \mathcal{A}; \mathcal{W} \underset{\vartriangle}{\overset{p_{\mathcal{W}}}{\rightleftarrows}} \mathcal{O} \longrightarrow D_\varepsilon)$ *degeneration data* if:

(i) $\mathcal{O} \to D_\varepsilon$ is a semistable degeneration of rational curves; $\Delta_0 \cup \mathcal{A}$ has normal crossings, and $\mathcal{A}$ is a disjoint union of sections of $p$. Here, $\Delta_t = p(t)$, $\forall t \in D_\varepsilon$.

(ii) $\mathcal{W} \xrightarrow{p_{\mathcal{W}}} \mathcal{O}$ is a $\mathbf{P}_1$-bundle over $\mathcal{O}$ with section $\vartriangle$. Set $W_t = p_{\mathcal{W}}^{-1}(\Delta_t)$.

(iii) $\mathcal{B} \subset \mathcal{W}$ is an effective divisor which is even, that is to say $\mathcal{O}_{\mathcal{W}}(\mathcal{B})$ is the square of some invertible sheaf on $\mathcal{W}$. In addition, $F_u$ intersects $\mathcal{B}$ in four distinct points for all $u \in \mathcal{O}' = \mathcal{O} - \Delta_0 - \mathcal{A}$, where $F_u = p_{\mathcal{W}}^{-1}(u)$.

(iv) Let $\mathcal{V} \xrightarrow{\pi} \mathcal{W}$ be the double covering of $\mathcal{W}$ with branch locus $\mathcal{B}$. Each $\pi^{-1}(W_t)$, $t \neq 0$, when taken together with the reduced preimage of $\vartriangle \cap W_t$, forms a Weierstrass pencil.

For any such degeneration data, define a one complex $\Pi$ as follows: the vertex $v \in \Pi^{(0)}$ corresponds to the component $\Delta(v) \subset \Delta_0$; the edge $e \in \Pi^{(1)}$ corresponds to the point $q(e) \in \Delta(v') \cap \Delta(v'')$ iff $\partial e = \{v', v''\}$.

In addition, define $\Sigma(v)$ to be $p_{\mathcal{W}}^{-1}(\Delta(v))$, and $X(v)$ to be $\pi^{-1}(\Sigma(v))$. For each vertex v in $\Pi$ : $\Delta(v)$ equals the self-intersection of the divisor $\Delta(\mathcal{O}) \cap \Sigma(v)$ in $\Sigma(v)$; and, $p_g(v)$ is the geometric genus of the minimal resolution of $X(v)$.

It is not difficult to see that the only degenerations $\rho : \mathcal{Y} \to D_\varepsilon$ we need consider in statement (**) all arise as semistable reductions of some $\rho_{\mathcal{O}} : \mathcal{O} \to D_\varepsilon$; where, $\mathcal{O}$ is constructed from degeneration data as in (iv) above, and $\rho_{\mathcal{O}} = p \circ p_{\mathcal{W}} \circ \pi$.

Let $f'$, $f''$ be two compactly supported (i.e., $f'$ and $f''$ act as the identity outside a compact set) homeomorphisms of $\mathbf{C}^1$ which leave $\{0,1,2\}$ invariant. Let us say that these two are to be considered equivalent iff there is a one-parameter family of compactly supported homeomorphisms of $\mathbf{C}^1$, $g_t, 0 \leq t \leq 1$, $g_0 = \mathrm{id}_{\mathbf{C}^1}$, such that $g_t(\{0,1,2\}) = \{0,1,2\}$ $\forall t$, and $f' \circ (f'')^{-1} = g_1$. The group of compactly supported homeomorphisms of $\mathbf{C}^1$ preserving $\{0,1,2\}$, modulo this equivalence, defines the braid group (on three strands) $B_3$. $B_3$ is generated by elements $x_1$ and $x_2$ with the relation $x_1x_2x_1 = x_2x_1x_2$; the center of $B_3$ is generated by $(x_1x_2)^6$. As is well known, $B_3/_{\{(x_1x_2)^{6n} | n \in \mathbf{Z}\}}$ is isomorphic to $SL(2,\mathbf{Z})$; let x and y denote, respectively, the images of $x_1$ and $x_2$ in $SL(2,\mathbf{Z})$. For any group G and $g \in G$, we let $[g]$ denote the conjugacy class of g. We define a numerical function $\mu(\cdot)$ as follows:

| w | $\mu([w])$ |
|---|---|
| $x^k$ | $k$ |
| $(xy)^3x^k$ | $6+k$ |
| $(xy)^r$ | $2r$, $\forall\ 0 \leq r < 6$ |
| $(xyx)^r$ | $3r$, $\forall\ 0 \leq r < 4$ . |

One can check that $\mu(\cdot)$ is well defined.

Let $B \subset \mathbf{P}_1 \times D_\varepsilon$ be an analytic subspace such that $\{\infty\} \times D_\varepsilon \subset B$ and $(\mathbf{P}_1 \times \{t\}) \cap B$ consists of 4 distinct points $\forall t \neq 0$. To the pair

$(B, \{\infty\} \times D_\varepsilon)$ of subspaces in $\mathbf{P} \times D_\varepsilon$, one can associate a conjugacy class $[w] \subset B_3$. Namely, we can find compactly supported homeomorphisms $f_t : \mathbf{C}^1 \to \mathbf{C}^1$, $f_t(\{0,1,2\}) = B \cap (\mathbf{C}^1 \times \exp\{\varepsilon' \pi r t\})$, $\forall \leq t \leq 1$; set $[w] = [f_1 f_0^{-1}]$. Assume $\{\infty\} \times D_\varepsilon$ is an isolated component of $B$; then one can show that there are the following possibilities for $[w] : [x_1^k (x_1 x_2)^{3m} (x_1 x_2)^{6s}]$, $k \geq 0, 0 \leq m \leq 1$; $[x_1^k (x_1 x_2)^{6s}]$, $k \geq 0$; $[(x_1 x_2)^{r+6s}]$, $0 \leq r < 6$; and $[(x_1 x_2 x_1)^r (x_1 x_2)^{6s}]$, $0 \leq r < 4$; $s > 0$. Let us say $B \subset \mathbf{P}_1 \times D_\varepsilon$ is in *standard position* if, in the above expression for $[w]$, $s = 0$; we call such a word $w$ *reduced.*

Let $(\mathcal{B}, \mathcal{A}, \mathcal{W} \underset{\Delta}{\overset{p_{\mathcal{W}}}{\rightleftarrows}} \mathcal{O} \xrightarrow{p} D_\varepsilon)$ be degeneration data. For each $v \in \Pi^{(0)}$ (respectively, component $A \subset \mathcal{A}$) define $\theta(v) = [w] \subset B_3$ (respectively, $\theta(A) = [w] \subset B_3$). Here, the conjugacy class $[w]$ is associated to the pair $(\phi^* \mathcal{B}, \phi^* \Delta)$ of subspaces in $\phi^* \mathcal{W}$; and, $\phi : D_{\varepsilon'} \to \mathcal{O}$ is transverse to some "generic" point of $\Delta(v)$ (respectively, of $A$). Let $T(v)$ (respec- $T(A)$) denote the image of $\theta(v)$ (respectively of $\theta(A)$) in $SL(2, \mathbf{Z})$. Now, let us orient each edge of $\Pi$. Let $e \in \Pi^{(1)}$, and write $\partial e = v_1 - v_2$. If $U$ is a suitable small nbd of $q(e)$ in $\mathcal{O}$, then the fundamental group of $U - \Delta$ is abelian. Hence, by choosing suitable representatives $w_i \in T(v_i)$, obtained by using, say, discs lying totally within $U$, we may assign to $e$ the conjugacy class $T(e) = [w_1 w_2^{-1}]$. One can check that this assignment is well defined.

Let us say that the degeneration data $(\mathcal{B}, \mathcal{A}, \mathcal{W} \underset{\Delta}{\rightleftarrows} \mathcal{O} \longrightarrow D_\varepsilon)$ is properly prepared if:

(a) $\Delta(\mathcal{O})$ is an isolated component of $\mathcal{B}$.

(b) For all components $A \subset \mathcal{A}$ and $v \in \Pi^{(0)}$, $\theta(A)$ and $\theta(v)$ are reduced (i.e., for all $\phi : D_\varepsilon \to \mathcal{O}$, $\phi(D_\varepsilon)$ transverse to a simple point of $\mathcal{A} \cup \Delta_0$, $\phi^* \mathcal{B} \subset \phi^* \mathcal{W}$ is in standard position).

(c) If $T(e) = [x^k]$, and $\partial e = v' - v''$; then $T(v'') = [x^{k''}]$ and $T(v') = [x^{k'}]$.

(d) Let $T(v) = [su]$ be the "Jordan decomposition"; where $s$ is semisimple and $u$ unipotent. If $s \neq 1$ then $v$ lies in precisely two edges of $\Pi$. Note that if $u \neq 1$, then the only possibility for $[su]$ is $[(xy)^3 x^k]$.

For a properly prepared degeneration, we prove the following simple, but important, fact:

*Let* $\mathcal{V} \xrightarrow{\pi} \mathcal{W}$ *be the double covering of* $\mathcal{W}$ *branched along* $\mathcal{B}$; $\mathcal{B} \subset \mathcal{W}$ *as above (i.e. properly prepared). If* $\mathcal{Y} \to D_\varepsilon$ *is any semistable reduction of* $\mathcal{V} \to D_\varepsilon$, *and* $X$ *a distinguished component of the central fibre of* $\mathcal{Y}$; *then, the map* $\mathcal{Y} \to \mathcal{V}$, *restricted to* $X$, *will induce a birational isomorphism between* $X$ *and* $X(v)$, *for some* $v \in \Pi^{(0)}$.

§6. Our basic tool, for modifying a given degeneration, is the *elementary transformation*. Namely, if $D \subset \Sigma(v)$ (or $D \subset p_{\mathcal{W}}^{-1}(A)$ ) is a section of $\Sigma(v) \to \Delta(v)$ (respectively of $p_{\mathcal{W}}^{-1}(A) \to A$ ). We blow up $D$, and then blow down the proper transform of $\Sigma(v)$ (respectively, the proper transform of $p_{\mathcal{W}}^{-1}(A)$ ); thus constructing a new $P_1$-bundle $\tilde{\mathcal{W}}$ over $\mathcal{O}$ and new degeneration data: $(\tilde{\mathcal{B}}, \mathcal{A}, \tilde{\mathcal{W}} \underset{\Delta}{\overset{p_{\mathcal{W}}}{\rightleftarrows}} \mathcal{O} \xrightarrow{p} D_\varepsilon)$. Where, if $\mathcal{B}$ contains $D$ with odd multiplicity, then $\tilde{\mathcal{B}}$ consists of the proper transform of $\mathcal{B}$ together with $p_{\tilde{\mathcal{W}}}^{-1}(\Delta(v))$; otherwise, $\tilde{\mathcal{B}}$ is simply the proper transform of $\mathcal{B}$. This process of going from $\mathcal{W}$ to $\tilde{\mathcal{W}}$ is called the *elementary transformation with center* $D$; note that, if $D = \Delta(\mathcal{O}) \cap \Sigma(v)$ then $\Delta(v)$ changes to $\Delta(v) - (\Delta(v))^2$; where $(\Delta(v))^2$ is the self-intersection number of $\Delta(v)$ as a divisor in $\mathcal{O}$. Moreover, if $e', \cdots, e^{(\ell)}$ are the edges meeting $v$, $\partial e^{(j)} = \{v_j, v\}$, then $\Delta(v_j)$ changes to $\Delta(v_j) - 1$. *After a suitable base extension and elementary transforms, any degeneration can be made properly prepared.*

Let $(\mathcal{B}, \mathcal{A}, \mathcal{W} \underset{\Delta}{\overset{p_{\mathcal{W}}}{\rightleftarrows}} \mathcal{O} \xrightarrow{p} D_\varepsilon)$ be properly prepared degeneration data. Let us define the following numerical functions. $\mu(v)$ equals $\Sigma\, \mu(T(A))$, where this sum is taken over all distinct components A of $\mathcal{A}$ which meet $\Delta(v)$. Let $\mu = \sum_v \mu(v)$. Suppose the vertex v is such that $T(v) = [x^k]$, $k > 0$; write $\mathcal{B} \cap \Sigma(v)$ as $\vartriangle + \vartriangle_0 + 2\vartriangle_1$, $\vartriangle = \vartriangle(\mathcal{O}) \cap \Sigma(v)$. Then, due to our assumptions regarding $\mathcal{B}$; $\vartriangle_0 + \vartriangle_1$ will have only ordinary nodes; let their number be $\tilde{\eta}(v)$. Clearly, $\tilde{\eta}(v) \equiv 0 \mod 2$. For any vertex v, satisfying $T(v) \neq [x^k]$, for all $k > 0$, define $\tilde{\eta}(v) = 0$.

Define a subcomplex $L \subset \Pi$ as follows: $v \notin L^{(0)}$ iff $\Sigma(v) \cap \mathcal{B}$, when reduced, consists of either three or four disjoint sections of $\Sigma(v) \to \Delta(v)$; $e \in L^{(1)}$ iff $T(e) \neq [x^k]$, $\forall k$. If $e \in L^{(1)}$, then $\partial e$ is contained $L^{(0)}$; so that L is a full subcomplex of $\Pi$. Let e(L) denote its (topological) Euler characteristic.

We prove that the following *basic formula* holds:

$$\frac{\mu}{12} - \sum_{r} p_g(v) = e(L) + \sum_{\tilde{\eta}(v)\neq 0} \left\{\frac{\tilde{\eta}(v)}{2} - 1\right\}.$$

For the purposes of proving statement (**) above (i.e., to finish our proof of the Torelli theorem) it suffices to consider only the case where $\frac{\mu}{12} = \sum p_g(v) + 1$. *Hence*:

($\alpha$) $e(L) = 1$

($\beta$) $\tilde{\eta}(v) = 0$ or $1$ for all vertices v of $\Pi$.

We now deal only with cases where these hold; in fact, to prove statement (**) above, we need only consider the case where there is a single vertex v satisfying $p_g(v) = \frac{\mu}{12} - 1$.

Let $I \subset \Pi$ be a subcomplex. Assume $I^{(0)} = \{v_1, \cdots, v_n\}$ and $I^{(1)} = \{e_1, \cdots, e_{n-1}\}$; and if $n = 1$, assume that $I^{(1)} = \phi$. The subcomplex I will be called a *standard chain* if: $\forall\ 1 \leq i \leq n$ $T(v_i)$ is not the conjugacy class of a unipotent element. In addition, there must exist edges $e_0$, $e_1$, $\partial e_j = v_j - v_{j+1}$, $\forall\, 0 \leq j \leq n$, such that $T(v_0)$ and $T(v_{n+1})$ are the conjugacy classes of unipotent elements. For such a standard chain, a simple braid calculation establishes:

$$\delta(v_i) = \begin{cases} 2, & j = n \\ \\ 0, & 0 \leq j < n \end{cases}$$

or, this holds with the numbering on the $v_j$ reversed.

Now, because of assumptions ($\alpha$) and ($\beta$) above; we can write:

$$\Pi - L = \{e_1, \cdots, e_n\} \cup J_1 \cup \cdots \cup J_n$$

where $e_j \in \Pi^{(1)}$ and $J_j$ equals one of the two contractable components of $\Pi - \{e_j\}$. Define,

$$\Delta(J_i) = \bigcup_{v \in J_i^{(0)}} \Delta(v) .$$

Clearly, $\forall i = j$ $\Delta(J_i) \cap \Delta(J_j) = \emptyset$; and, in fact, $p_{\mathfrak{W}}^{-1}(\Delta(J_i))$ will be isomorphic to $\Delta(J_i) \times \mathbf{P}_1$. One can show without difficulty that $\Delta(J_i)$ is a *generalized exceptional* curve of the first kind on $\mathcal{O}$. We may contract the $\Delta(J_i)$'s and $p_{\mathfrak{W}}^{-1}(\Delta(J_i))$'s simultaneously. Thus, after this has been done, we have a degeneration for which the subcomplex $L$ equals $\Pi$. $\Delta_0$ of the resulting degeneration still has normal crossings but $\mathfrak{A} \cup \Delta_0$ may not; this, however, is unimportant.

A vertex $v$ of $\Pi$ will be called *extremal* if $v$ meets exactly one edge. Our next task will be to eliminate all extremal vertices which satisfy $p_g(v) = 0$. Let $v_0$ be such an extremal vertex; let $e_0$ be the unique edge meeting $v_0$. There is a standard chain $I$; with vertices $\{v_1, \cdots, v_n\}$, and edges $\{e_1, \cdots, e_{n-1}\}$, satisfying $\partial e_j = v_j - v_{j+1}$ for all $j$ between $0$ and $n$. One can see quite readily that the vector $(\delta(v_0), \cdots, \delta(v_n))$ equals either $(-2, 2, 0 \cdots 0)$, or, $(-2, 0, \cdots, 0, 2)$. Let us perform a "double" elementary transform to $\mathfrak{W}$ with center $\delta(\mathcal{O}) \cap \Sigma(v_0)$ (i.e., do it twice, taking as the second center the proper transform of $\delta(\mathcal{O})$ intersected with the component over $\Delta(v_0)$ ). We thus change $(\delta(v_0), \cdots, \delta(v_n))$ to either $(0, 0, \cdots, 0)$, or, to $(0, -2, 0, \cdots, 0, 2)$. In the first case, we may eliminate $I$ (i.e., blow down $p_{\mathfrak{W}}^{-1}(\Delta(I))$ and

$\Delta(I)$ simultaneously); in the second case, we blow down $\Delta(v_0)$ and $p_{\widetilde{W}}^{-1}(\Delta(v_0))$ simultaneously. By repeating this as many times as required, we eventually eliminate $\{v_0\} \cup \{e_0\} \cup I$ from $\Pi$.

There are some details to be checked; but, it is not difficult to see that, by repeated applications of this process, we may eliminate all extremal vertices $v$ with $p_g(v) = 0$. Now, if a semistable reduction of our original degeneration satisfied the hypothesis of statement (**) (i.e., a single distinguished component in the central fibre); then, at this stage, $\Pi$ *will consist of only a single vertex* $v$, and *the corresponding component* $X(v)$ *will have only rational double points as singularities.* This proves statement (**).

In any event, after some further elementary transformations and blowing down, we may "standardize" the resulting central fibre; for a description of the final configuration, see my research announcement [2].

CONCLUDING REMARKS. As one can see above, my techniques for proving the Torelli theorem for elliptic pencils are mostly elementary. For this reason, I expect the ideas presented above to prove quite useful in other cases. In particular, using methods similar to the above, one should, be able to compute the degree of the period mapping for various surfaces whose canonical series are far from being very ample, e.g., surfaces of geometric genus one with $K^2 > 0$.

## REFERENCES

[1] A. Borel, *Density and maximality of arithmetic groups,* J. Reine Angew. Math. 224 (1966),

[2] K. Chakiris, *A Torelli theorem for simply connected elliptic surfaces with a section and* $p_g \geq 2$, Bull. A.M.S. 7 (1982), No. 1.

[3] P. Deligne and D. Mumford, *The irreducibility of the space of curves of given genus,* Pub. IHES, No. 36 (1969).

[4] P. Griffiths, *Periods of integrals on algebraic manifolds III (some global differential-geometric properties of the period mapping),* Pub. IHES, No. 38 (1970).

[5] A. Kas, *Weierstrass normal forms and invariants of elliptic surfaces,* Trans. Am. Math. Soc. 225 (1977).

[6] K. I. Kii, *The local Torelli problem for varieties with divisible canonical class*, Math. USSR Izvestija, vol. 12 (1978), No. 1.

[7] V. S. Kulikov, *Degenerations of* K-3 *surfaces and Enriques surfaces*, Math. USSR Izvestija, vol. 11 (1977), No. 5.

[8] R. Miranda, *The Moduli of Weierstrass fibrations over* $\mathbf{P}_1$, Math. Ann. 255 (1981), No. 3.

[9] I. I. Pyateckii-Sapiro and I. R. Safarevic, *A Torelli theorem for algebraic surfaces of type* K-3, Math. USSR Izvestija, vol. 5 (1971), No. 3.

KEN CHAKIRIS
MATHEMATICS DEPARTMENT
UNIVERSITY OF IOWA
IOWA CITY, IA 52242

# Chapter X
# THE PERIOD MAP AT THE BOUNDARY OF MODULI

Robert Friedman

## §1. *General philosophy*

In its crudest form, the global Torelli problem is a question about the degree of a map, namely the period map

$$\rho : M \to \Gamma \backslash D ,$$

where $M$ is an appropriate moduli space of polarized varieties and $\Gamma \backslash D$ the corresponding classifying space for Hodge structures. To calculate the degree of $\rho$, a natural approach is to degenerate the varieties in question, in the hope that the moduli and periods will simplify in the limit. Globally, this means that we seek extensions of spaces and maps

$$\begin{array}{ccc} \rho : M & \longrightarrow & \operatorname{Im} \rho \subseteq \Gamma \backslash D \\ \cap| & & \cap| \\ \bar{\rho} : \bar{M} & \longrightarrow & \overline{\operatorname{Im} \rho} \end{array} ,$$

where $\bar{M}$ and $\overline{\operatorname{Im} \rho}$ are (not necessarily compact) spaces containing $M$ and $\operatorname{Im} \rho$ as dense open sets. To prove that $\rho$ has degree one usually involves the following four steps:

a) Construct $\bar{M}$, $\overline{\operatorname{Im} \rho}$, and the extension $\bar{\rho}$;

b) Prove that $\bar{\rho}$ is proper;

c) Find $x \in \bar{M}$ with $\bar{\rho}^{-1}\bar{\rho}(x) = \{x\}$;

d) Show that the differential $(d\bar{\rho})_x$ is injective (in practice, this often amounts to calculating a Picard-Lefschetz transformation).

It follows from the properties of the degree of a map that a)-d) imply that $\rho$ has degree one onto its image.

Consider first the problem of finding a suitable candidate for $\overline{\operatorname{Im}\rho}$. Little is known in this direction unless D is Hermitian symmetric. In the Hermitian symmetric case, we have the choice

$$\Gamma\backslash D \subseteq (\Gamma\backslash D)^* = \text{Satake-Baily-Borel compactification}$$

or

$\Gamma\backslash D \subseteq \overline{\Gamma\backslash D}$, one of Mumford's (infinitely many) toroidal compactifications.

$(\Gamma\backslash D)^*$ has many nice properties. It is unique, and, by Borel's Extension Theorem, there is usually no problem in extending maps $\rho$ to $(\Gamma\backslash D)^*$. Unfortunately, such extensions tend to have positive dimensional fibers over $(\Gamma\backslash D)^* - (\Gamma\backslash D)$. Moreover, $(\Gamma\backslash D)^*$ is very singular at infinity, so that it is unsuitable for making differential computations.

Both of these drawbacks suggest using some kind of blow up of $(\Gamma\backslash D)^*$, such as Mumford's compactifications $\overline{(\Gamma\backslash D)}$.

$\overline{(\Gamma\backslash D)}$ has in its turn many nice properties. It is smooth, aside from very mild singularities in high codimension and quotient singularities of finite group actions, and $\overline{(\Gamma\backslash D)} - (\Gamma\backslash D)$ is, with these same reservations, a divisor with normal crossings. The price that must be paid for these properties is that $\overline{\Gamma\backslash D}$ is not unique and that it is usually highly nontrivial to check if maps $\rho$ extend to $\overline{\rho}$.

On the moduli side, there are various candidates for the varieties corresponding to points of $\overline{M} - M$. However, except in the case of stable curves, where all reasonable notions coincide, the relationship between the various points of view does not seem to be well understood. From the point of view of Hodge theory, however, there are two properties that we would like $\overline{M} - M$ to have.

1) We would like $\overline{M} - M$ to be a divisor with normal crossings in an essentially smooth variety $\overline{M}$. This is a reasonable requirement if we want to extend $\rho$ to $\overline{\rho}: \overline{M} \to$ a toroidal compactification. An additional

motivation is the theory of regular singular points (and nilpotent orbit theorem in several variables), which works well in studying the period map on the complement of a divisor with normal crossings.

2) Let $f:\Delta \to \overline{M}$ with $f(0) \in \overline{M}-M$ be a generic arc in moduli. We would like this to correspond to

$$\begin{array}{ccc} \mathfrak{X} & \supseteq & X_t \\ \pi\downarrow & & \downarrow \\ \Delta & \ni & t \end{array}$$

a semi-stable degeneration, with $\mathfrak{X}$ smooth, $X_t \in M$ for $t \neq 0$, and $X_0 = \pi^{-1}(0)$ a reduced divisor with normal crossings. This condition may seem artificial, but enables one to keep track of the monodromy and to relate the limiting mixed Hodge structure (which will be a point in $\overline{\Gamma\backslash D}$) to something geometric, via Clemens-Schmid. Moreover, even if one wants to allow other singularities in varieties in $\overline{M}-M$, blowing up $\overline{M}$ along $\overline{M}-M$ will often lead to the geometric situation described above.

It turns out that 1) and 2) are closely related. To explain this, we need some definitions.

DEFINITION 1. Let $X$ be a reduced variety with normal crossings of dimension $n$, and $D = X_{sing}$. There is an intrinsically defined line bundle $\mathcal{O}_D(X)$ on $D$, the *infinitesimal normal bundle*, which may be described as follows. If $V$ is a smooth $(n+1)$-fold, and $X \subseteq V$ as a divisor, then

$$\mathcal{O}_D(X) = \mathcal{O}_V(X)|_D \,;$$

this description of $\mathcal{O}_D(X)$ is actually intrinsic.

Hence, if $X \subseteq \mathfrak{X}$ is the central fiber in a semi-stable degeneration, we may take $V = \mathfrak{X}$, so that

$$\mathcal{O}_D(X) = \mathcal{O}_{\mathfrak{X}}(X)|_D = \mathcal{O}_{\mathfrak{X}}|_D = \mathcal{O}_D \,.$$

DEFINITION 2. If X is a reduced variety with normal crossings and $\mathcal{O}_D(X) = \mathcal{O}_D$, X is d-*semi-stable.*

Infinitesimally, then, d-semi-stable varieties look like fibers in semi-stable degenerations. Their deformation theory is quite manageable, and, if smoothable, the total space of the smoothing $\mathfrak{X}$ will typically be smooth itself. Moreover, at least infinitesimally, the set of singular deformations of X looks like a smooth divisor in the versal deformation if D is connected, and like a divisor with normal crossings in general, so that requiring 2) also gives 1) (morally). Another property of d-semi-stable varieties, which is important for their deformation theory, is that they carry an intrinsic "limiting" mixed Hodge structure (suitably interpreted), independent of the choice or even existence of a one variable smoothing. The deformation theory of varieties with normal crossings and d-semi-stable varieties in particular is described in [6].

§2. *The case of curves*

Let $\mathfrak{M}_g$ = the coarse moduli space for smooth curves of genus g ;
$\mathfrak{A}_g$ = the coarse moduli space for principally polarized abelian varieties of dimension g
$= \Gamma\backslash D$, where $D = \mathfrak{H}_g$ is the Siegel upper half plane,
= classifying space for Hodge structures of weight one, and
$\Gamma$ is the Siegel modular group.

For compactifications, there are

$\overline{\mathfrak{M}}_g$ = the coarse moduli space of stable curves, constructed by Mayer-Mumford-Deligne [10], [5];

$\overline{\mathfrak{A}}_g$ = the *Voronoi compactification,* a toroidal compactification described in detail in [12], [13].

THEOREM 1 (Mumford, Namikawa [12]). *The period map extends*:

$$\begin{array}{ccc} \rho : \mathfrak{M}_g & \longrightarrow & \mathfrak{A}_g \\ \cap| & & \cap| \\ \overline{\rho} : \overline{\mathfrak{M}}_g & \longrightarrow & \overline{\mathfrak{A}}_g \end{array}.$$

Without going into the subtleties of the toroidal construction or the combinatorics involved in extending $\rho$, we want to describe $\overline{\mathfrak{A}}_g$ near the image of an irreducible stable curve and the corresponding behavior of the extension $\overline{\rho}$ at such a stable curve. In particular, we show that $d\overline{\rho}$ is injective for generic irreducible stable curves and that $\overline{\rho}(C)$ contains enough geometric information to recover C if C is irreducible.

I) *Description of* $\overline{\mathfrak{A}}_g$

Let

$$(U_{g-1})_{\mathbf{Z}} = \left\{ \left( \begin{array}{c|c} I & \begin{matrix} 0 & & & \\ & 0 & & \\ & & \ddots & \\ & & & 0 \\ & & & & n \end{matrix} \\ \hline 0 & I \end{array} \right) : n \in \mathbf{Z} \right\} \subseteq \Gamma = \mathrm{Sp}(g, \mathbf{Z}) \ .$$

Similarly, define, for each $g' < g$, with $g = g' + g''$,

$$(U_{g'})_{\mathbf{Z}} = \left\{ \left( \begin{array}{c|c} I & \begin{matrix} 0 & \\ & B \end{matrix} \\ \hline 0 & I \end{array} \right) : B \in M_{g''}(\mathbf{Z}),\ {}^tB = B \right\} \subseteq \Gamma \ .$$

As a partial step toward $\Gamma \backslash \mathfrak{H}_g$, we might consider $D_{g'} = (U_{g'})_{\mathbf{Z}} \backslash \mathfrak{H}_g$. Since, for $Z \in \mathfrak{H}_g$, $Z = (z_{ij})$, ${}^tZ = Z$, $\operatorname{Im} Z > 0$,

$$\left( \begin{array}{c|c} I & \begin{matrix} 0 & & & \\ & \ddots & & \\ & & & B \end{matrix} \\ \hline 0 & I \end{array} \right) \cdot Z = Z + \begin{pmatrix} 0 & & \\ & \ddots & \\ & & B \end{pmatrix} ,$$

$D_{g'}$ has the natural choice of coordinates $(w_{ij}) = {}^t(w_{ij})$, with

$$w_{ij} = z_{ij}, \quad i \text{ or } j \leq g',$$

$$w_{ij} = \exp(2\pi\sqrt{-1}\, z_{ij}), \quad i,j > g'.$$

Partially enlarge $D_{g'}$ by allowing the values

$$w_{ii} = 0, \quad i > g';$$

this corresponds to letting $z_{ii} \to i\infty$ and also to the simplest form of a torus embedding, namely $(\mathbf{C}^*)^{g''} \subseteq \mathbf{C}^{g''}$. (By itself, this will not compactify $\mathfrak{A}_g$, but we do get enough points in this manner to extend $\bar{\rho}$ to all *irreducible* stable curves.)

II) *Behavior of the period map*

Let $C$ be an irreducible stable curve with $g''$ nodes, and let $\Delta^{3g-3}$ be a small polydisk which is the base of the universal deformation of $C$. There exists a choice of coordinates $\{t_1, \cdots, t_{3g-3}\}$ on $\Delta$ so that the discriminant locus $D$ = set of curves in the versal deformation which are singular is the hypersurface

$$D = \{t_1 \cdots t_{g''} = 0\} \supseteq W = \{t_1 = \cdots = t_{g''} = 0\}$$

and $\{t_i = 0\}$, $1 \leq i \leq g''$, is a smooth divisor corresponding to the locus where we "keep the $i^{th}$ node singular." If $C_{t_0}$ is a general curve in the universal family over $\Delta$, for an appropriate choice of symplectic basis of $H_1(C_{t_0}; \mathbf{Z})$, the period matrix of a curve $C_t$, $t \notin D$, will be of the form

$$Z(t) = \tilde{Z}(t) + \begin{pmatrix} 0 & & & \\ & \ddots & & \\ & & \frac{\log t_1}{2\pi\sqrt{-1}} & & \\ & & & \ddots & \\ & & & & \frac{\log t_{g''}}{2\pi\sqrt{-1}} \end{pmatrix},$$

where $\tilde{Z}(t)$ is holomorphic and single valued for all $t \in \Delta^{3g-3}$. This is a consequence of the Picard-Lefschetz formula, and is worked out for an arbitrary stable C in the Clemens-Griffiths IAS Seminar on degenerations [4].

Compose $Z(t)$ with the projection map $\mathfrak{H}_g \to D_{g'}$. Then $Z(t)$ becomes single valued. Moreover,

$$w_{i+g',i+g'} \circ Z(t) = t_i\,, \quad i = 1, \cdots, g'' \,.$$

Near $Z(0) = \overline{\rho}(C)$, $\overline{\mathfrak{A}}_g$ contains the divisor

$$\overline{\mathfrak{A}}_g - \mathfrak{A}_g = \left\{ \prod_{i=1}^{g''} w_{i+g',i+g'} = 0 \right\}$$

and the subvariety $V = \{w_{1+g',1+g'} = \cdots = w_{g,g} = 0\}$. Thus, visibly, $d\overline{\rho}$ induces an isomorphism

$$d\overline{\rho} : N_{W,\Delta} \xrightarrow{\sim} N_{V,\overline{\mathfrak{A}}_g}$$

at C. Hence $d\overline{\rho}$ is injective in the normal directions to W in $\Delta$; we will consider $d\overline{\rho} : T_{W,C} \to T_{V,C}$ shortly.

III) *Value of* $\overline{\rho}$ *at* C

Again assume C is irreducible, and let

$$a : \tilde{C} \to C$$

be the normalization map, where $g(\tilde{C}) = g'$. Let

$$\omega_C = \text{dualizing sheaf of } C$$

= sheaf of meromorphic 1-forms $\phi$ on $\tilde{C}$ of the third kind, with poles only at $a^{-1}(C_{sing})$ and satisfying, for each $x \in C_{sing}$,

$$\sum_{p \in a^{-1}(x)} \operatorname{res}_p \phi = 0 \,.$$

If $C_{sing} = \{x_1, \cdots, x_{g''}\}$, let $a^{-1}(x_i) = \{p_i, q_i\}$. Draw a dissection of $\tilde{C}$ into a canonical polygon as follows:

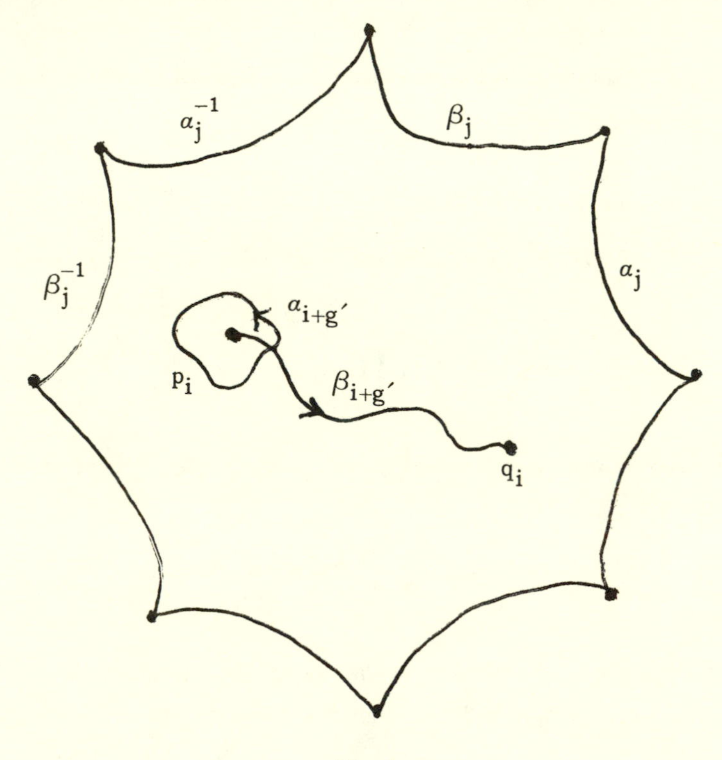


Fig. 1

The boundary curves are thus A and B cycles for $\tilde{C}$.*) The $\alpha_{i+g'}$, $\beta_{i+g'}$ are additional "cycles" which satisfy the following conditions:

1) $\alpha_{i+g'}$ is a loop around $p_i$, homologous to 0 in $\tilde{C}$, and enclosing $p_i$ simply but not enclosing any other $p_j$, $q_j$ or $q_i$, or passing through any singular point.

2) The $\beta_{i+g'}$ are paths in the interior of the polygon joining $p_i$ to $q_i$ and not passing through any other singular point or meeting $\beta_{j+g'}$, $j \neq i$.

---

*) i.e., the $\alpha_i$, $\beta_i$ are a canonical basis in the sense of [4], p. 16.

In a formal sense, then, $\{a_i, \beta_i : i = 1, \cdots, g\}$ is a symplectic basis for $C$, i.e.,

$$a_i \cdot \beta_j = \delta_{ij}, a_i \cdot a_j = \beta_i \cdot \beta_j = 0 .$$

There is a unique basis of $H^0(C, \omega_C)$, $\{\omega_1, \cdots, \omega_g\}$, satisfying

1) $\omega_1, \cdots, \omega_{g'}$ are holomorphic on $\tilde{C}$ and

$$\int_{a_i} \omega_j = \delta_{ij}, \quad 1 \leq i,j \leq g'.$$

2) $\omega_{1+g'}, \cdots, \omega_g$ have no A periods and

$$\int_{a_i} \omega_j = 2\pi\sqrt{-1}\ \mathrm{res}_{p_i}\omega_j = \delta_{ij}, \quad i,j > g' .$$

For $\lambda_1, \cdots, \lambda_{g''}$ arbitrary complex numbers and $\lambda = (\lambda_1, \cdots, \lambda_{g''})$, define the *period matrix* $Z_\lambda$ of $C$ by

$$Z_\lambda = \left(\int_{\beta_i} \omega_j\right),$$

where, for $\int_{\beta_i} \omega_i$, $i > g'$, we simply assign the value $\lambda_i$. Hence

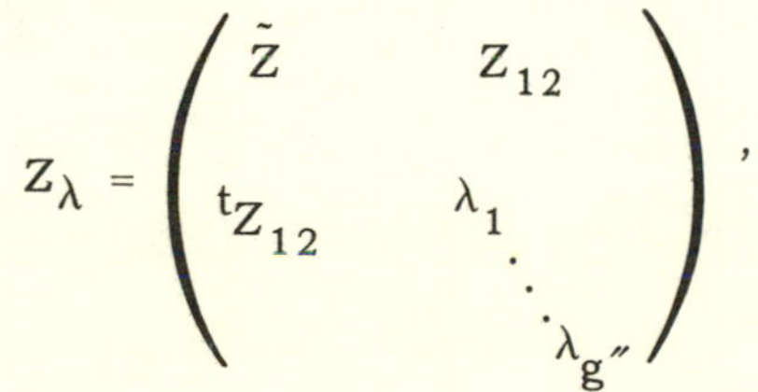

$$Z_\lambda = \begin{pmatrix} \tilde{Z} & Z_{12} \\ {}^tZ_{12} & \begin{matrix} \lambda_1 & & \\ & \ddots & \\ & & \lambda_{g''} \end{matrix} \end{pmatrix},$$

a symmetric $g \times g$ matrix (which does not, however, satisfy $\mathrm{Im}\, Z_\lambda > 0$). $\{Z_\lambda : \lambda \in \mathbf{C}^{g''}\}$ is in fact a full nilpotent orbit under the action of the unipotent group

$$\left\{ \begin{pmatrix} I & \begin{matrix} 0_{\lambda_1} & & \\ & \ddots & \\ & & \lambda_{g''} \end{matrix} \\ \hline 0 & I \end{pmatrix} : \lambda_i \in \mathbf{C} \right\}$$

$$= \left\{ \exp\left(\sum_1^{g''} \mu_i N_i\right), \mu_i \in \mathbf{C}, N_i = \begin{pmatrix} I & \begin{matrix} 0 & & & \\ & \ddots & & \\ & & 1_0 & \\ & & & \ddots_0 \end{matrix} \\ \hline & I \end{pmatrix} \text{ with a 1 in the } i+g' \text{ place} \right\}$$

$Z_\lambda$ is in fact well defined up to a certain subgroup of $\Gamma$. The intrinsic information that one can extract from $Z_\lambda$ includes

1) the period matrix $\tilde{Z}$ mod $Sp(g',\mathbf{Z})$

2) the vectors $\int_{\beta_i} \omega_j$, mod periods of $\tilde{C}$, where $1 \leq j \leq g'$, $g'+1 \leq i \leq g$; in other words, we obtain

$$\left(\int_{p_i}^{q_i} \omega_1, \cdots, \int_{p_i}^{q_i} \omega_{g'}\right) \in J(\tilde{C}),$$

the Abel-Jacobi image of $p_i - q_i$ (up to sign change and permutation).

Now if $W \subseteq \Delta$ is the singular locus of $g''$-nodal curves, we can fiber $W$ via

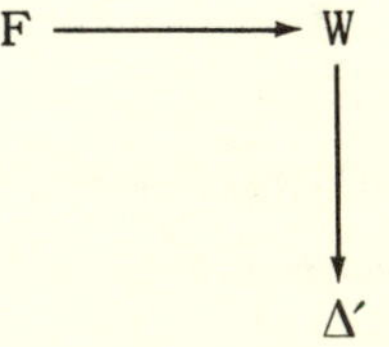

where $\Delta'$ = the universal deformation of curves of genus $g'$ and F is essentially $\mathrm{Sym}^{g''}(\mathrm{Sym}^2 \tilde{C})$, a collection of $g''$ pairs of 2 marked points on $\tilde{C}$.

Similarly, we can fiber V.

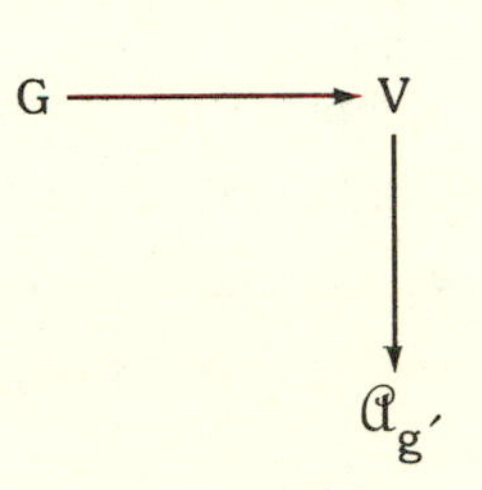


where G is essentially $\mathrm{Sym}^{g''} A$, where A is the abelian variety of dimension $g'$ corresponding to $\tilde{Z} \in \mathfrak{A}_{g'}$. Hence $d\bar{\rho}$ fits into an exact sequence (where we write C and $\tilde{C}$ for the images $\bar{\rho}(C)$, $\rho(\tilde{C})$, by abuse of notation):

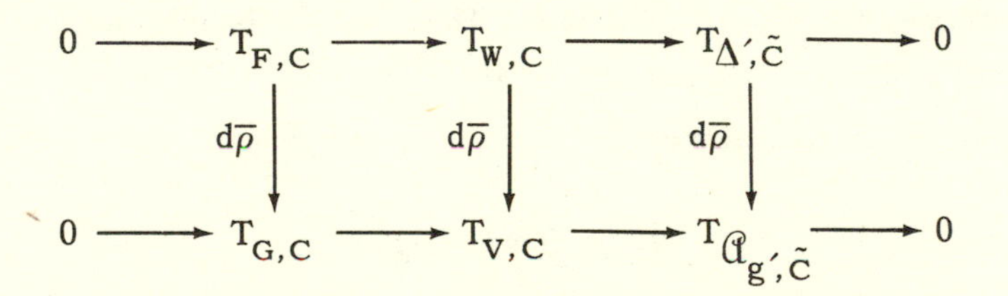


and the injectivity of $d\bar{\rho} \,|\, T_{W,C}$ is implied by the injectivity of $d\bar{\rho}: T_{\Delta',\tilde{C}} \to T_{\mathfrak{A}_{g'},\tilde{C}}$ (local Torelli for $\tilde{C}$) and of $d\bar{\rho}: T_{F,C} \to T_{G,C}$ (computation of the derivative of the Abel-Jacobi map).

IV) *Cohomological formalism*

The above remarks show that $d\bar{\rho}$ is injective in most cases, but they do not yield the best result possible—we must assume that $\tilde{C}$ is non-hyperelliptic. Now, for a smooth curve, the codifferential of the period map is essentially given by cup product

$$H^0(C, \Omega^1_C) \otimes H^0(C, \Omega^1_C) \to H^0(C, \Omega^{1\otimes 2}_C),$$

where $H^0(\Omega_C^{1\otimes 2}) = H^1(\theta_C)^*$ is the cotangent space to moduli. If C is a stable curve, the cotangent space to moduli at C is

$$H^0(\Omega_C^1 \otimes \omega_C) ,$$

where $\Omega_C^1$ is the sheaf of Kähler differentials (which is not locally free) and $\omega_C$ is the sheaf of dualizing differentials described in III). It is therefore natural to ask if we can check the surjectivity of the codifferential of $\overline{\rho}$ from the analogous cup product map

$$H^0(\Omega_C^1) \otimes H^0(\omega_C) \to H^0(\Omega^1 \otimes \omega_C) .$$

It turns out that this condition is sufficient for the surjectivity of $(d\overline{\rho})^*$, but not necessary, and does not in general hold, even if C is irreducible, if C has more than one singular point. To explain why this is so, assume that C is irreducible. Then the line bundle $\omega_C$ has no base locus and either embeds C in $\mathbf{P}^{g-1}$ or maps C 2-1 onto a rational normal curve. Hence, just as for smooth curves, we can think of

$$H^0(\omega_C) \otimes H^0(\omega_C)$$

as quadrics in canonical space. There is a natural map $\Omega_C^1 \to \omega_C$, hence a map

$$H^0(\Omega_C^1) \otimes H^0(\omega_C) \to H^0(\omega_C) \otimes H^0(\omega_C)$$

whose image is contained in the set of quadrics containing the linear span of the double points. A more natural geometric object to consider (and one which is more closely connected with the surjectivity of the codifferential) is the larger set of quadrics containing the double points of C. One can then prove:

PROPOSITION 2. *If* C *is irreducible, the differential* $d\overline{\rho}$ *is injective (at the level of the universal deformation)* $\Longleftrightarrow$ $g = 2$ *or* $g > 2$ *and* C *is non-hyperelliptic.*

Here $C$ is *hyperelliptic* $\Longleftrightarrow$ the canonical series $|\omega_C| : C \to P^{g-1}$ is 2-1 onto a rational normal curve $\Longleftrightarrow$ $\omega_C$ is not very ample $\Longleftrightarrow$ $C$ has a $g_2^1$; if $g(\tilde{C}) > 1$, these conditions are equivalent to the assertion that $\tilde{C}$ is hyperelliptic and, for all $i$, $p_i + q_i$ is in the $g_2^1$ of $\tilde{C}$.

The proof of Proposition 2 involves the following steps:

a) We must identify the tangent bundle to $\overline{\mathfrak{A}}_g$ at $\rho(C)$. This is an easy consequence of the theory of regular singular points.

b) Breaking $d\overline{\rho}$ up into a component normal to $V$ and a component tangent to $V$, we must check injectivity in directions normal to $V$ separately; this was described in II).

c) Via explicit analysis of the extensions of the Hodge bundles, one can reduce to the surjectivity of $H^0(\omega_C) \otimes H^0(\omega_C) \to H^0(2\omega_C)$.

d) The analogue of Noether's lemma for irreducible stable curves holds, so that $H^0(\omega_C) \otimes H^0(\omega_C) \to H^0(2\omega_C)$ is surjective $\Longleftrightarrow$ $g = 2$ or $g > 2$ and $C$ is non-hyperelliptic.

V) *Torelli at the boundary*

We consider the geometric information in $\overline{\rho}(C)$.

Going in one direction, using the Torelli theorem for $\tilde{C}$, we can deduce a precise Torelli type theorem for $\overline{\rho}(C)$. Moreover, simply knowing a degree one Torelli theorem for $\tilde{C}$, we can deduce a degree one Torelli theorem for curves of genus $g$, and so for all $g$ by induction.

Let us first then assume the strong global Torelli theorem for curves $\tilde{C}$ of genus $g' \geq 1$:

THEOREM 3. 1) *The map* $C \to J(C)$ *is injective from* $\mathfrak{M}_{g'} \to \mathfrak{A}_{g'}$.

2) *If* $\tilde{C}$ *is non-hyperelliptic, the map*

$$\mathrm{Aut}(\tilde{C}) \to \mathrm{Aut}(J(\tilde{C})/\{\pm 1\}$$

*is bijective, if* $C$ *is hyperelliptic, the map*

$$\mathrm{Aut}(\tilde{C}) \to \mathrm{Aut}(J(\tilde{C}))$$

*is bijective.*

THEOREM 4 (Namikawa [11]). *If* $C$ *is an irreducible stable curve,* $\overline{\rho}^{-1}(\overline{\rho}(C)) = \{C\}$.

*Proof.* Let $C, C' \in \overline{\rho}^{-1}(\overline{\rho}(C))$, where $C'$ is any stable curve. A straightforward argument shows that we may assume that $C'$ is irreducible with $g''$ nodes (in the degree one case, for $g'' = 1$, one can simply use a parameter count). First assume $g'' = 1$, so that $Z_\lambda$ has the form

$$(*) \qquad \left(\begin{array}{ccc|c} & \tilde{Z} & & \\ \hline \int_{p_1}^{q_1} \omega_1 & \cdots & \int_{p_1}^{q_1} \omega_{g'} & \lambda_1 \end{array}\right), \quad \lambda_1 \in C,$$

where $\tilde{Z}$ is the period matrix of $\tilde{C}$, $\{p_1, q_1\} = a^{-1}(C_{sing})$, and $\omega_1, \cdots, \omega_{g'}$ is a normalized basis for $H^0(C, \Omega^1_{\tilde{C}})$.

By the strong form of the Torelli theorem, we may assume that the limiting period matrix of $C'$ is of the form (*), and that there is an isomorphism $\tilde{C} \to \tilde{C}'$ (unique up to a possible hyperelliptic involution) taking the period matrix of $C$ into that of $C'$. Thus, we may assume $\tilde{C} = \tilde{C}'$ and are reduced to: given $p_1, q_1, p_1', q_1' \in \tilde{C}$, with

$$p_1 - q_1 = \pm(p_1' - q_1') \quad \text{in} \quad J(\tilde{C}),$$

is it true that $\{p_1, q_1\} = \{p_1', q_1'\}$?

Now either $\tilde{C}$ is hyperelliptic or it is non-hyperelliptic. If $\tilde{C}$ is non-hyperelliptic, the difference map

$$\tilde{C} \times \tilde{C} - \Delta(\tilde{C}) \to J(\tilde{C})$$

is injective, where $\Delta(\tilde{C})$ is the diagonal. If $\tilde{C}$ is hyperelliptic,

$$p_1 - q_1 = \pm(p_1' - q_1')$$

$\Longleftrightarrow$ $p_1' = \iota(q_1)$, $p_1 = \iota(q_1')$ or $p_1' = \iota(p_1)$, $q_1' = \iota(q_1)$, where $\iota$ is the hyperelliptic involution, in which case the pair

$$\tilde{C}, \{p_1, q_1\} \xrightarrow{\approx} \tilde{C}, \{p_1', q_1'\}$$

via $\iota$. Hence $C \cong C'$.

In the general case, use induction on the number $g''$ of nodes, normalize $C$ and $C'$ at the $(g'')^{th}$ node, and call this partial normalization $C_1$ (resp. $C_1'$). By induction, $C_1 \cong C_1'$, and the bottom row of $Z_\lambda = Z_\lambda'$ gives the Abel-Jacobi image

$$p_{g''} - q_{g''} \in J(C_1) = \text{generalized Jacobian of } C_1 .$$

Using Abel's theorem for the singular curve $C_1$, we can simply repeat the argument for one node, q.e.d. for Theorem 4.

Notice that the argument of Theorem 4 gives us the step 3) of §1 to complete the proof of

THEOREM 5. *The period map* $\mathfrak{M}_g \to \mathfrak{A}_g$ *has degree one.*

VI) *Reducible curves* [7]

The main tool in the above degree one Torelli theorem was the degeneration of smooth curves to irreducible curves with one node. If one considers *reducible* curves with one node, $C = C_1 \cup C_2$, one need never compactify $\mathfrak{A}_g$; the limiting value of the Jacobian is the reducible principally polarized abelian variety of dimension $g$

$$J(C) = J(C_1) \oplus J(C_2) .$$

Unfortunately, for $g > 2$, the period map has positive dimensional fibers at $\bar{p}(C)$; this is because $J(C)$ does not record the moduli of $p = C_1 \cap C_2 = C_{sing}$. This situation may be remedied by blowing up $\mathfrak{A}_g$ along the locus of reducible abelian varieties

$$T = \{A_1 \oplus A_2 : A_1 \in \mathfrak{A}_{g_1}, A_2 \in \mathfrak{A}_{g_2}, g_1 + g_2 = g\}$$

and calculating an appropriate normal derivative. Briefly, the normal space to $T$ in $\mathfrak{A}_g$ at $A_1 \oplus A_2 = A$ is

$$N_{T|\mathfrak{A}_{g,A}} = \mathrm{Hom}(H^0(\Omega^1_{A_1}), H^1(\mathcal{O}_{A_2}))$$

and the result is

THEOREM 6. *The normal vector to the locus* $\{C_1 \cup_p C_2 : C_1 \in \mathfrak{M}_{g_1}, C_2 \in \mathfrak{M}_{g_2}\}$ *maps via* $d\bar{p}$ *to the rank one transformation* $\psi \in \mathrm{Hom}(H^0(\Omega^1_{C_1}), H^1(\mathcal{O}_{C_2}))$ *whose kernel = all differentials vanishing on* $p \in C_1$ *and whose image is the line corresponding to the canonical image of* $p \in C_2$ *in* $PH^1(\mathcal{O}_{C_2})) = P^{g_2-1}$

It is not hard to deduce the degree one Torelli theorem from Theorem 6.

§3. *The global Torelli theorem for* K3 *surfaces*

The global Torelli theorem for algebraic K3 surfaces was first announced by Piatetskii-Shapiro and Shafarevitch in [15], with additional contributions by Shioda-Mitani. In [1], Burns-Rapoport deduced a strong Torelli theorem for Kähler K3 surfaces, using the ideas of Piatetskii-Shapiro and Shafarevitch. A very nice account of this work is given by Looijenga-Peters in [9]. The rough idea common to these works is as follows.

Let $A$ be an abelian surface, or, more generally, a complex torus of dimension 2, and $\iota : A \to A$ the involution $\iota(a) = -a$, with 16 fixed points (the points of order 2). Setting

$$\bar{S} = A/\iota ,$$

$\bar{S}$ is singular at the images of the 16 fixed points, which are ordinary double points. Let

$$\tilde{S} \to \bar{S}$$

be the minimal desingularization of $\bar{S}$; $\tilde{S}$ is then a K3 surface, the *Kummer surface* associated to $A$. The proof of global Torelli proceeds in the following steps:

*Step I*: If $S$ is a K3 surface and $\{H^2(S;Z), F^{\cdot}\}$ is isomorphic to $\{H^2(\tilde{S};Z); F^{\cdot}\}$ as polarized Hodge structures, then $S$ is a Kummer surface (this follows by studying $\operatorname{Pic} S \cong \operatorname{Pic} \tilde{S}$).

*Step II*: With $S$ as in I), $S \cong \tilde{S}$ (and even stronger statements relating $\operatorname{Aut}(S)$ to the automorphisms of the Hodge structure of $S$ are true)–to prove this, write $S$ as the resolution of $B/\iota$, for some complex torus $B$, produce isomorphisms $\{H^2(A;Z),F^{\cdot}\} \cong \{H^2(B;Z),F^{\cdot}\}$ and then show that, in this case, one can deduce

$$\{H^1(A;Z),F^{\cdot}\} \cong \{H^1(B;Z);F^{\cdot}\}.$$

From this it follows trivially that $A \cong B$, hence $S \cong \tilde{S}$.

*Step III*: Kummer surfaces are dense, in an appropriate sense, in any Kuranishi family of K3 surfaces (i.e., any locally universal family of deformations of a K3 surface $T$). Using III) and II), one deduces the Torelli theorem, roughly, by considering two K3 surfaces $T$ and $T'$ with the same Hodge structure, looking at their universal deformations, and finding a sequence of Kummer surfaces converging to $T$ and $T'$ with isomorphic Hodge structures. The sequence of isomorphisms between the Kummer surfaces then specializes to an isomorphism between $T$ and $T'$.

We sketch here a different proof of the global Torelli theorem for K3 surfaces based on the point of view described in §1. The departure for our study of K3 surfaces is the following theorem of Kulikov-Persson-Pinkham:

THEOREM 7 ([8], [14]). *Let $\pi: \mathfrak{X} \to \Delta$ be a semistable degeneration of K3 surfaces with all components of $X = \pi^{-1}(0)$ algebraic. Then, after birational modifications affecting only $\pi^{-1}(0)$, we may assume further that $K_{\mathfrak{X}} = \mathcal{O}_{\mathfrak{X}}$. It follows that $X$ is one of the following*:

*Type I*: $X$ *is a smooth* K3 *surface.*

*Type II*: $X = X_1 \cup \cdots \cup X_n$ *is a chain of surfaces,* $X_1, X_n$ *are rational, and* $X_2, \cdots, X_{n-1}$ *are elliptic ruled. All double curves are elliptic curves on* $X_1$, $X_n$ *or* 2 *disjoint elliptic curves on* $X_2, \cdots, X_{n-1}$ *and each is a section of the ruling on* $X_2, \cdots, X_{n-1}$, *hence all isomorphic; all double curves are sections of* $-K_{X_i}$. *The picture of* X *is as follows*:

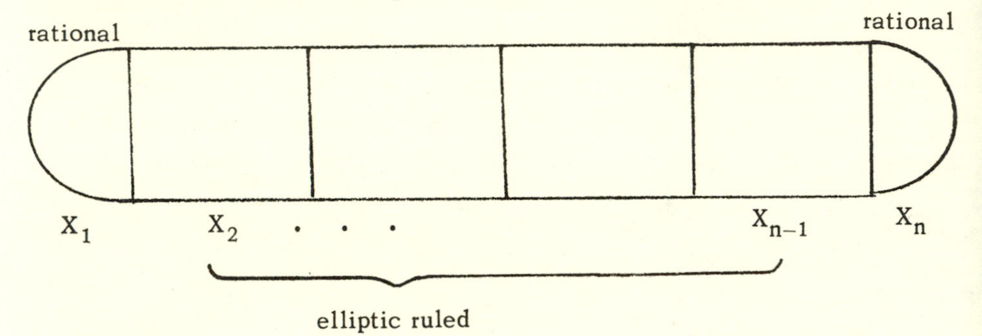


Fig. 2

*Type III*: X *is a union of rational surfaces* $X_i$. *All double curves are cycles of rational curves in each* $X_i$ *and are sections of* $-K_{X_i}$. *The dual graph of* X, $\Gamma$, *is a triangulation of* $S^2$. *(More precise results have been obtained by Miranda-Morrison* [16].*)*

*Moreover, the three types may be distinguished by monodromy, and therefore Hodge-theoretically*: *if* $N = \log T$, *where* T *is the monodromy transformation, then*

$$N = 0 \text{ for Type I degenerations,}$$
$$N^2 = 0,\ N \neq 0 \text{ for Type II degenerations,}$$
$$\text{and } N^3 = 0,\ N^2 \neq 0 \text{ for Type III degenerations.}$$

It is this last statement which will enable us to ignore Type III degenerations systematically in what follows.

To use the theorem of Kulikov-Persson-Pinkham to compactify moduli spaces, it is necessary to add a polarization statement. The following result was obtained independently by Nick Shepherd-Barron (who also had results in the Type III case):

THEOREM 8 ([16]). *Let* $\pi : \mathfrak{X} \to \Delta$ *be a Type II degeneration and* $\mathfrak{L}$ *a line bundle on* $\mathfrak{X}$ *such that* $\mathfrak{L}|X_t$ *is ample,* $t \neq 0$. *Then*

1) *After further birational modifications preserving the hypothesis that* $\mathfrak{X} \to \Delta$ *is a Type II degeneration, we may assume that* $\mathfrak{L}|X$ *is numerically effective.*

2) *Replacing* $\mathfrak{L}$ *by* $N \cdot \mathfrak{L}$, $N >> 0$ *but independent of* $\mathfrak{X}$ *and* $\deg \mathfrak{L}|X_t$, *we may further assume:*

(i) *all elliptic ruled components are minimal and* $X_1 \cup X_n$, *glued along the double curves in the obvious way, is again d-semi-stable;*

(ii) *the restriction of* $\mathfrak{L}$ *to any elliptic ruled component is a sum of fibers;*

(iii) $\mathfrak{L}$ *defines a Cartier divisor on* $X_1 \cup X_n$ *which is very ample, except possibly for contracting some rational double point configurations of smooth rational curves of self-intersection* $-2$, *which therefore do not meet the singular locus of* $X_1 \cup X_n$.

The point of (iii) is to be able to control the embedded deformation theory of $\phi_{\mathfrak{L}}(X_1 \cup X_n)$; $\phi_{\mathfrak{L}}$ is an embedding of some neighborhood of the double curve of $X_1 \cup X_n$, so that the embedded deformation theory of $\phi_{\mathfrak{L}}(X_1 \cup X_n)$ is almost like that of the abstract d-semi-stable variety $X_1 \cup X_n$. Moreover, we have blown down all elliptic ruled components, which did not play an essential role. Passing from $X_1 \cup X_2 \cup \cdots \cup X_n$ to $X_1 \cup X_n$ is, therefore, very much like passing from a semi-stable curve to a stable curve. Thus,

DEFINITION 3. A *polarized stable Type II* K3 *surface* (X,L) is a pair consisting of a d-semi-stable K3 surface of Type II, $X = X_1 \cup X_2$ (without elliptic ruled components) and a Cartier divisor L on X such that, for some $N >> 0$, NL is very ample except possibly for contracting some smooth rational curves on X of self-intersection $-2$ (and hence not meeting $X_{sing} = X_1 \cap X_2$).

Let $\mathcal{F}_{2k}$ = the coarse moduli space of polarized K3 surfaces (X,h), where $h^2 = 2k$ and h is primitive, i.e., not the multiple of any class in Pic(X). We will allow h to be only numerically effective, so that h may fail to meet rational double point configurations in X. Embed X by Nh, N >> 0 but fixed as in Theorem 8, 2), we get

$$U \subseteq \mathrm{Hilb}(P^m),$$

$U$ = the open subset of a component of Hilb consisting of subschemes $\phi_{Nh}(X)$

$\cap |$

$\bar{U}$ : add to U those subschemes $\phi_{NL}(X_1 \cup X_2)$ described above.

THEOREM 9. $\bar{U}$ *is smooth and* $\bar{U} - U$ *is a smooth divisor in* $\bar{U}$.

THEOREM 10. *There exists a partial compactification* $\bar{\bar{\mathcal{F}}}_{2k}$ *of* $\mathcal{F}_{2k}$, *where* $\bar{\bar{\mathcal{F}}}_{2k}$ *is separated and quasi-projective. Up to finite group actions,* $\bar{\bar{\mathcal{F}}}_{2k}$ *is smooth and* $\bar{\bar{\mathcal{F}}}_{2k} - \mathcal{F}_{2k}$ *is a smooth divisor. The points of* $\bar{\bar{\mathcal{F}}}_{2k} - \mathcal{F}_{2k}$ *correspond to polarized stable* K3 *surfaces of Type II, modulo the closure of* PGL(m+1) *equivalence (which can be described more explicitly).*

We can now state

THEOREM 11 (global Torelli theorem for algebraic K3 surfaces).

1) $\mathcal{F}_{2k}$ *is irreducible for all* $k \geq 1$.

2) *The period map* $\rho: \mathcal{F}_{2k} \to \Gamma \backslash D$ *is a bijection.*

There is also a comparison between Aut(X,h) and the automorphisms of the corresponding Hodge structure which gives the strong form of the Torelli theorem. We give an outline of the proof of Theorem 11.

*Step I*: *Some easy reductions*

1) $\rho$ has degree one $\Rightarrow \rho$ is a bijection.

The surjectivity part follows from Theorem 7. As for the injectivity, this can be deduced from Zariski's Main Theorem via local Torelli, which implies that $\rho$ has no positive dimensional fibers.

2) It suffices to prove Theorem 11 for one value of $k$, say $k = 1$.
Indeed, one can use induction on $k$, using the surjectivity of $\rho$ for all $k$ and the classical device (dating back at least to Enriques) of "projecting from a double point."

3) Proving the theorem for one value of $k$ also leads to the Burns-Rapoport strong Torelli theorem for Kählerian K3 surfaces, e.g., by the arguments of [9, §9].

*Step II*: *Extend the period map*

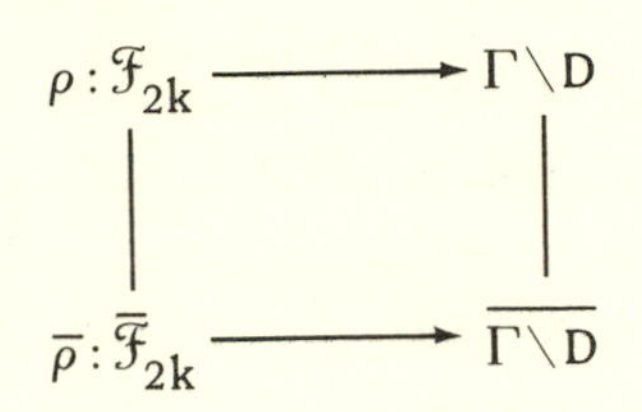

*and prove that* $\overline{\rho}$ *is proper.*

The extension $\overline{\mathcal{F}}_{2k}$ has already been discussed. As for $\overline{\Gamma\backslash D}$, one uses a partial Mumford toroidal compactification which, essentially, only involves those limiting mixed Hodge structures with $N^2 = 0$. In this case, $\overline{\Gamma\backslash D}$ is *unique* and does not involve any subtle combinatorics of toroidal embeddings. The extension $\overline{\rho}$ exists via the nilpotent orbit theorem (which is particularly simple in the Hermitian symmetric case), and the properness is a straightforward consequence of the theorem of Kulikov-Persson-Pinkham.

*Step III*: *Calculate the differential of* $\overline{\rho}$ *at the boundary.*

The essential point in checking that $\rho$ does not ramify along the divisor $\overline{\mathcal{F}}_{2k} - \mathcal{F}_{2k}$ is the generalized Picard-Lefschetz formula of Clemens [3]. With this formula, the argument is much the same as that for curves given in §2, II. The full differential can be handled via cohomological formalism, as in §2, IV, or we can consider $d\bar{\rho}$ in those directions normal to $\bar{\mathcal{F}}_{2k} - \mathcal{F}_{2k}$ and tangent to $\bar{\mathcal{F}}_{2k} - \mathcal{F}_{2k}$ as in §2, III, and show that the resulting variation of mixed Hodge structures has injective differential. In any case, it turns out that $d\bar{\rho}$ is everywhere an isomorphism.

*Step IV: Calculate the fiber* $\bar{\rho}^{-1}(\bar{\rho}(X))$.

This is the most interesting step, and the part of the proof where we must assume $k = 1$. If $X \in \bar{\mathcal{F}}_{2k} - \mathcal{F}_{2k}$, $\bar{\rho}(X)$ computes the data of the limiting mixed Hodge structure on $X$ together with the polarizing class. Since $N^2 = 0$, the weight filtration on $H^2_{\lim}(X)$ looks like

$$W_1 \subseteq W_2 \subseteq W_3$$

where $W_1$ has the Hodge structure of the elliptic curve $D = X_1 \cap X_2$ and $W_2$ is essentially the cohomology $H^2(X)$; as

$$W_2H^2(X)/W_1H^2(X) = \ker\{H^2(\tilde{X}) \to H^2(D)\},$$

$W_2/W_1$ is a pure weight 2 Hodge structure of type $(1,1)$. The extension of Hodge structures

$$0 \to W_1 \to W_2 \to W_2/W_1 \to 0$$

has been studied and classified by Jim Carlson [2]. It is determined by a homomorphism of abelian groups

$$(*) \qquad (W_2/W_1)_Z \to J(D) = \text{the Jacobian of } D \cong D\,.$$

Moreover, this map is the obvious one: a class in $W_2H^2(X)/W_1H^2(X)$ can be represented by a pair of line bundles $(L_1, L_2)$, where $L_i \in NS(X_i)$ and $L_1 \cdot D = L_2 \cdot D$. The line bundle on $D$ defined by

$$L = (L_1|D) \otimes (L_2|D)^{-1}$$

has degree zero, but it is not necessarily linearly equivalent to zero, and this construction induces the map (*).

This information is too flabby for our purposes— $W_2/W_1$ has an intersection form, but it is indefinite. One must consider the *polarized* limiting mixed Hodge structure as well. Denoting by $(W_2/W_1)_0$ the primitive integral vectors in $(W_2/W_1)$, $(W_2/W_1)_0$ is a negative definite lattice (non-unimodular) of dimension 17.

Returning to polarizations of degree 2, let $\pi : \mathfrak{X} \to \Delta$ be a semi-stable degeneration of Type II with $X = \pi^{-1}(0)$ a stable type II K3 surface and $\mathfrak{L}$ a line bundle in $\mathfrak{X}$ such that $\mathfrak{L}^2 \cdot X_t = 2$ and $\mathfrak{L}|X$ numerically effective. There are essentially just four possibilities for L (where we have drawn the components $X_1$, $X_2$ of X and indicated the genera of all components of $L|X_i$; all intersections are supposed to be transverse, so that the generic member of L depicted below has ordinary double points only).

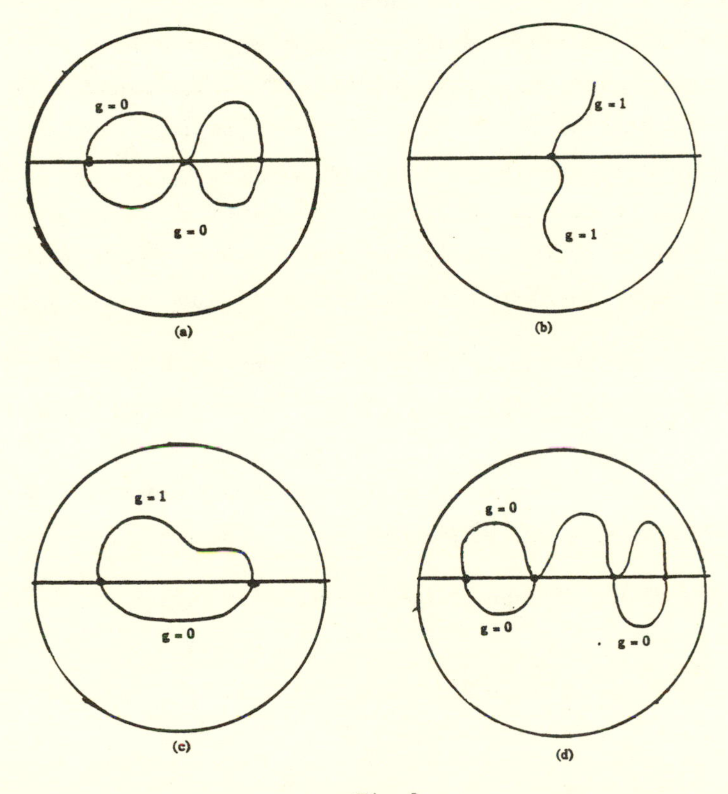


Fig. 3

Examples of (a) - (d) arise as follows:

(a) Take 2 copies of $\mathbf{P}^2$ with an isomorphic cubic D in each, blow up 18 points on the cubic (in one or the other of the copies) and glue back along the proper transforms of D. Let L = inverse image of $\mathcal{O}_{\mathbf{P}^2}(1)$ in each component.

(b) Start as in (a), and in each component take the linear series of cubics passing through 8 points on D, so that the ninth point of intersection (the unique base point of the series) matches up under the isomorphism $D \subseteq X_1 \xrightarrow{\sim} D \subseteq X_2$. Blow up the 16 points in their appropriate components, plus 2 more, and let $L|X_i$ = the proper transform of the cubics with the 8 base conditions.

(c) As in (b), but consider cubics in one component passing through 7 points on D and lines in the other component passing through one point of D. This gives 8 points to blow up, and blow up 10 more.

(d) Choose conics in one component passing through 2 points of D and twice the linear system of lines in the other passing through a point on D; blow up these three points plus 15 more.

Of course, we must choose our points so that the divisors L constructed above are Cartier and the modified variety is d-semi-stable. The reader can check that imposing these conditions always gives 18 moduli.

*Fact 1.* The descriptions given above are the *only* possibilities for the pictures (a) - (d) in Figure 3, and so all (X,L) arise this way. Moreover, in case (a), the linear series $L|X_i$ maps each component $X_i$ birationally to $\mathbf{P}^2$, without fixed components or base locus, and is precisely the blowing down of the points we blew up.

*Fact 2.* (a) - (d) look different Hodge theoretically. In fact, if

$$V = \{\gamma \in (W_2/W_1)_0 : \gamma^2 = -2\}$$

then V is a root system of dimension 17 in all cases:

(a) $V = A_{17}$

(b) $V = E_8 + E_8 + A_1$

(c) $V = E_7 + D_{10}$

(d) $V = D_{16} + A_1$ .

Focusing attention on case (a), for instance, we can pick a set of simple roots (unique up to sign change and permutation), and use these plus the mixed Hodge structure to reconstruct the j-invariant of $D$ and the moduli of the 18 points on $D$ modulo projective embeddings. This concludes the final step in the proof of the global Torelli theorem for K3 surfaces.

## REFERENCES

[1] D. Burns and M. Rapoport, "On the Torelli problem for Kählerian K3 surfaces," Ann. E.N.S. 8 (1975), pp. 235-274.

[2] J. Carlson, "The obstruction to splitting a mixed Hodge structure over the integers I," University of Utah preprint (1979).

[3] C.H. Clemens, "Picard-Lefschetz theorem for families of non-singular algebraic varieties acquiring ordinary singularities," Trans. Amer. Math. Soc. 136 (1969), pp. 93-108.

[4] Clemens, Griffiths, et al., IAS Seminar on degenerations of algebraic varieties, mimeographed notes, 1968.

[5] P. Deligne and D. Mumford, "The irreducibility of the space of curves of a given genus," Publ. Math. I.H.E.S. 36 (1969), pp. 75-109.

[6] R. Friedman, "Global smoothings of varieties with normal crossings," to appear.

[7] R. Friedman and R. Smith, "The generic Torelli theorem for the Prym map," to appear in Inv. Math.

[8] V. Kulikov, "Degenerations of K3 surfaces and Enriques surfaces," Math. USSR Izvestiya 11 (1977), pp. 957-989.

[9] E. Looijenga and C. Peters, "Torelli theorems for K3 surfaces," Comp. Math. 42 (1980), pp. 145-186.

[10] D. Mumford, "Further comments on boundary points," AMS Summer Institute in Algebraic Geometry (Woods Hole), 1964.

[11] Y. Namikawa, "On the canonical holomorphic map from the moduli space of stable curves to the Igusa monoidal transform," Nagoya J. Math. 52 (1973), pp. 197-259.

[12] Y. Namikawa, "A new compactification of the Siegel space and degeneration of Abelian varieties I, II," Math. Ann. 221 (1976), pp. 97-142 and pp. 201-242.

[13] ———, *Toroidal Compactifications of Siegel Space*, Springer Lecture Notes 812 (1980).

[14] U. Persson and H. Pinkham, "Degenerations of surfaces with trivial canonical bundle," Annals of Math. 113 (1981), pp. 45-66.

[15] I. Piatetskii-Shapiro and I. Shafarevitch, "A Torelli theorem for algebraic surfaces of type K3," Math. USSR Izvestiya 35 (1971), pp. 530-572.

[16] *The Birational Geometry of Degenerations*: Proceedings of the Harvard Summer Algebraic Geometry Seminar, to appear in Progress in Math.

ROBERT FRIEDMAN
MATHEMATICS DEPARTMENT
COLUMBIA UNIVERSITY
NEW YORK, NY 10027

## Chapter XI

# THE GENERIC TORELLI PROBLEM FOR PRYM VARIETIES AND INTERSECTIONS OF THREE QUADRICS

Roy Smith

In this talk we give a sketch of recent joint work with Ron Donagi and Robert Friedman, and we wish to thank them for giving us the pleasure of making this presentation. Earlier contributions by Beauville, Mumford, Namikawa, Recillas, and Tjurin figure importantly as well, and as usual in work on these questions, the influence of Clemens and Griffiths is fundamental. Conversations with Ken Chakiris and David Mumford at Harvard and a talk by Alberto Collino in Varenna were particularly helpful at crucial points. We also thank Phillip Griffiths for the invitation to speak, and Ron Donagi, Robert Friedman, Ted Shifrin, Robert Varley, and the referee, for pointing out mistakes and points needing more explanation in an earlier version of these notes.

## §1. *The Torelli problem for an intersection of 2 or 3 quadrics*

The Torelli problem asks whether an algebraic variety $X$ is determined by the information in its middle dimensional cohomology space $H^n(X;\mathbf{Z})\otimes\mathbf{C}$; which means, when $n$ is odd, by the (polarized) intermediate Jacobian of $X$:

$$J(X) = \Big( \bigoplus_{0\leq p\leq \frac{n}{2}} H^{p,n-p}/H^n(X;\mathbf{Z}); <,>_{\mathbf{Z}}\Big) .$$

Since Torelli's problem has been solved for curves, it is natural to try to reduce the problem for higher dimensional varieties to that for curves. We present two illustrations of this approach taken from the very instructive paper of Tjurin [22].

EXAMPLE 1. The intersection of two general quadrics

$$X = Q_0 \cap Q_1 \subseteq \mathbf{P}^{2k+1}, \qquad k \geq 2 .$$

Since each $Q_i$ is $2k$ dimensional, a quadric in the general pencil $\{t_0 Q_0 + t_1 Q_1\}$ has two rulings (by k-planes) if non-singular and one if singular. Thus the singular quadrics given by $\{[t_0, t_1] : \det |t_0 Q_0 + t_1 Q_1| = 0\} \subseteq \mathbf{P}^1$, are the branch points of a natural double cover of $\mathbf{P}^1$ associated to the pencil:

$$\{\text{set of rulings}\} = C \xrightarrow{2-1} \mathbf{P}^1 = \{\text{set of quadrics}\} .$$

Since $Q_0, Q_1$ are general, these $2k+2$ branch points are distinct and therefore define a smooth hyperelliptic curve $C$ of genus $k$.

If $J(V)$ denotes the intermediate Jacobian of the variety $V$, then one can prove $J(X) \cong$ the variety of $(k-1)$ planes lying in $X$, and:

THEOREM.

1) $J(X) \cong J(C)$,
2) $J(C)$ *determines* $C$,
3) $C$ *determines* $X$.

*Proof of 3).* (For 1) and 2), see [7] and [1], respectively.) If $[\lambda_0, -1], \cdots, [\lambda_{2k+1}, -1] \in \mathbf{P}^1$ are coordinates for the branch points of the canonical map $\phi : C \to \mathbf{P}^1$, then the two quadrics $Q_0 : \{\sum_0^{2k+1} z_i^2 = 0\}$, $\{Q_1 : \sum_0^{2k+1} \lambda_i z_i^2 = 0\}$, define a pencil with singular quadrics at the given branch points. Since any two quadrics defining a pencil containing just $2k+2$ distinct singular elements can be diagonalized simultaneously in this way, this determines $X = Q_0 \cap Q_1$ uniquely. q.e.d.

COROLLARY. *Assuming 1), the Torelli theorem for* $X$ *is implied by 2), the classical Torelli theorem for the curve* $C$.

REMARK. Parts 1) and 3) of the theorem, and the corollary, were proved by Miles Reid in his (unpublished) Ph.D. dissertation, Cambridge, 1972.

EXAMPLE 2. The intersection of three general quadrics

$$X = Q_0 \cap Q_1 \cap Q_2 \subseteq \mathbf{P}^{2k}, \qquad k \geq 3.$$

This time, since the quadrics in the general net $\{\sum_0^2 t_i Q_i\}$ are $2k-1$ dimensional, the non-singular ones have one ruling (by $(k-1)$ planes) and the singular ones have two, and so we have a natural double cover of a plane curve $\Delta$ of degree $2k+1$:

$$\eta: \{\text{rulings on singular quadrics}\} = \tilde{\Delta} \xrightarrow{2-1} \Delta = \{\text{singular quadrics}\}$$
$$= \{[t_0, t_1, t_2] : \det |\sum_0^2 t_i Q_i| = 0\} \subseteq \mathbf{P}^2.$$

If $\eta_* : J\tilde{\Delta} \to J\Delta$ is the natural map of Jacobians induced by $\eta$, then the results in example one generalize as follows:

THEOREM.

1) $J(X) \cong P(\eta) \overset{\text{def}}{=} (\operatorname{Ker}(\eta_*))^0$
$= (\textit{connected component of zero in } \operatorname{Ker}(\eta_*))$.

2) $\eta : \tilde{\Delta} \to \Delta$ *determines* $X$.

REMARKS. This theorem is due independently to Beauville [5], and to Reid and Tjurin [22]. Beauville also credits a result of A. C. Dixon with inspiring his proof of 2).

*Proof of 2).* We summarize the main points of the argument in [22], pp. 76-90. For 1), see [5] or [22].

(i) If $L = \mathcal{O}_{\mathbf{P}^2}(1)|_\Delta$ is the line bundle determined by the (unique) embedding of $\Delta$ as a plane curve of degree $d = 2k+1$, [22], p. 67, and if $\eta$ is the square-trivial line bundle determined by the double cover $\eta : \tilde{\Delta} \to \Delta$, then $L^k \otimes \eta$ defines a "quadratically normal" embedding $\phi : \Delta \to \mathbf{P}^{2k}$; i.e.: $\phi$ induces an isomorphism $\phi^* : H^0(\mathbf{P}^{2k}; \mathcal{O}(2)) \xrightarrow{\simeq} H^0(\Delta; L^{2k})$.

(ii) If $j : \Delta \to \mathbf{P}^2$ is the plane embedding of $\Delta$, and if $\{f = 0\}$ is a plane equation (of degree $d$) for $\Delta$, then $X$ is the base locus of the net of quadrics in $\mathbf{P}^{2k}$ which corresponds, via $\phi^*$ and $j^* : H^0(\mathbf{P}^2;\mathcal{O}(d-1)) \xrightarrow{\simeq} H^0(\Delta;L^{2k})$, to the "polar net" of $\Delta$; i.e.:

$$X = Q_0 \cap Q_1 \cap Q_2, \text{ where } Q_i = (\phi^*)^{-1} \circ j^* \left(\frac{\partial f}{\partial x_i}\right). \qquad \text{q.e.d.}$$

By analogy with the previous example, we are led to pose the "Prym-Torelli" problem: Does the "Prym variety" $P(\eta)$, defined in part 1) of the theorem, determine the double cover $\eta : \tilde{\Delta} \to \Delta$? We will describe the known results on this question in the next three sections.

## §2. *Counterexamples to Prym-Torelli*

First let us rephrase the problem in terms of the fibers of the "Prym map" $P_g : \mathcal{R}_g \to \mathcal{A}_{g-1}$, where

$\mathcal{R}_g$ = {set of double covers $\eta : \tilde{C} \to C$ of stable curves with $\tilde{C}$ and $C$ of genera $2g-1$ and $g$, respectively, and $P(\eta)$ an abelian variety, taken up to isomorphism}

$\mathcal{A}_{g-1}$ = {set of $(g-1)$ dimensional principally polarized abelian varieties, up to isomorphism}

and $P_g(\eta : \tilde{C} \to C) = P(\eta)$, the Prym variety defined in Section 1. The Prym-Torelli problem then becomes the question of the injectivity of the map $P_g$. This question, strictly put, has a negative answer, as follows:

THEOREM.

(i) *If* $g \le 5$, *then the general fiber of* $P_g$ *has positive dimension.*

(ii) *If* $g = 6$, *then the general fiber of* $P_6$ *consists of exactly 27 points.*

(iii) $P_g$ *is never strictly injective for any* $g \ge 1$.

*Proof of (i).* If $g \le 5$, then $\dim(\mathcal{A}_{g-1}) = \frac{1}{2}(g)(g-1) < 3g-3 = \dim(\mathcal{R}_g)$. This was pointed out by Wirtinger [24]. q.e.d.

REMARKS ON (iii). If $C$ is hyperelliptic, of any genus, then the fiber $P_g^{-1}(P_g(\tilde{C} \to C))$ is always positive dimensional (Mumford [15, p. 344]). Counterexamples to Prym-Torelli with $C$ trigonal were given by Beauville for $g \leq 10$ [5, p. 388], based on Recillas [19], but the most comprehensive source of counterexamples, existing in all genera, is Donagi's "tetragonal construction," for $C$ having a $g_4^1$, [8, 3]. In fact, it is conjectured in [8] that all counterexamples to Prym-Torelli arise from the tetragonal construction.

*Proof of (ii).* (This is joint work with Ron Donagi [9], with important assistance from Clemens. In particular, both the problem and the way of attacking it were suggested to us by Clemens.)

We compute the degree of $P_6$ by analyzing the fibers over a neighborhood of general jacobian variety $J(X) \in \mathcal{A}_5$, where $X = Q_0 \cap Q_1 \cap Q_2 \subseteq \mathbf{P}^4$ is a general genus 5 curve, as follows:

> Let $\Delta$ = set of singular quadrics in the net $\{\sum_0^2 t_i Q_i\}$, $\tilde{\Delta}$ = the set of rulings by 2-planes on these singular quadrics, and $X^{(2)}$ = the second symmetric product of $X$. Then we have the following proposition.

PROPOSITION. *Over a neighborhood of* $J(X)$ *in* $\mathcal{A}_5$, *there is a commutative diagram:*

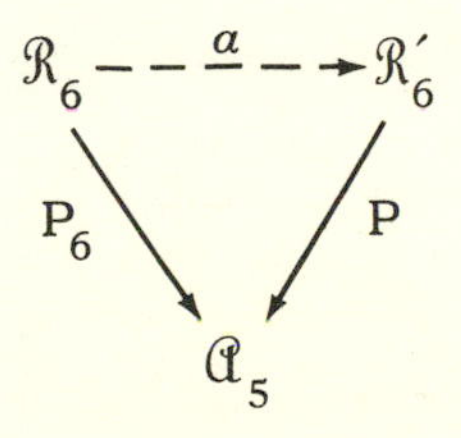


*in which* $a$ *is a bimeromorphic map, and* $P_6$ *and* $P$ *are proper morphisms, such that:*

1) $P^{-1}(J(X)) \cong X^{(2)} \cup \tilde{\Delta} \cup \{*\}$, *where* '$\{*\}$' *denotes an isolated point.*

2) *If* $P^*$ = *the codifferential of* P, *and* $T^*_{J(X)}\mathcal{A}_5 \cong$ {*quadratic polynomials on* $\mathbf{P}^4$}, *is the standard identification of the cotangent space to* $(\mathcal{A}_5)$ [5, p. 380], *then*

(i) *at* $(p,q) \in X^{(2)}$, $P(\ker P^*)$ = *the pencil of quadrics containing* $X \cup \overline{pq}$, *in* $\mathbf{P}^4$, *where* $\overline{pq}$ *is the line in* $\mathbf{P}^4$ *joining* p *and* q.

(ii) *at* $L \in \tilde{\Delta}$, $P(\ker P^*)$ = *the unique singular quadric ruled by* L,

(iii) *at the point* $\{*\}$, $\ker P^* = (0)$.

REMARKS. The properness of $P_6$, a crucial ingredient, and part (iii) of 2), are due to Beauville [4, §6], [5, p. 385]. The structure of $P_6^{-1}(J(X))$, an important preliminary to part 1), is a consequence of the results of Masiewicki and Tjurin (independently) [14], Recillas [19], Wirtinger [24], Mumford [15], and Beauville [4]. This fact was pointed out to us by Clemens.

COROLLARY 1. *If* $\sigma$ *is a general pencil of quadrics through* X *in* $\mathbf{P}^4$, *then*

$$\text{degree}(P) = \#\{w \in P^{-1}(J(X)) : P(\ker P_w^*) \subseteq \sigma\}.$$

COROLLARY 2. $\text{degree}(P) = \text{degree}\, P_6 = 27$.

*Proof of Corollary 1.* If $\tilde{\mathcal{A}}_5$, $\tilde{\mathcal{R}}_6'$ are obtained by blowing up $\mathcal{A}_5$, $\mathcal{R}_6'$ along $\mathcal{J}_5$ (the Jacobi locus), and $P^{-1}(\mathcal{J}_5)$ respectively, then the proposition implies there is a commutative diagram (over a neighborhood of $J(X)$):

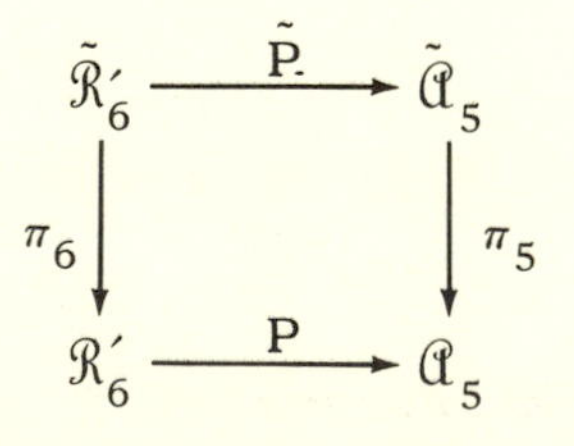

$$\begin{array}{ccc} \tilde{\mathcal{R}}_6' & \xrightarrow{\tilde{P}} & \tilde{\mathcal{A}}_5 \\ \downarrow{\scriptstyle \pi_6} & & \downarrow{\scriptstyle \pi_5} \\ \mathcal{R}_6' & \xrightarrow{P} & \mathcal{A}_5 \end{array}$$

where $\tilde{P}(z) = P(\ker(P \circ \pi_6)_z^*)$ if $\pi_6(z) \in P^{-1}(\mathcal{J}_5)$. The proposition implies that $\dim \ker P_* = \dim P^{-1}(J(X))$, where $P_*$ = (the differential of P). Therefore the normal sheaf to $P^{-1}(\mathcal{J}_5)$, is locally a bundle; and if $\sigma \in \tilde{\mathcal{A}}_5$, with $\pi_5(\sigma) = J(X)$, is represented by a general pencil in the projective conormal space to $\mathcal{J}_5$ at $J(X)$, (i.e., by a general pencil of quadrics through $X$ in $P^4$), then $\sigma$ is a regular value of $\tilde{P}$. Since

$$\pi_6 : \tilde{P}^{-1}(\sigma) \xrightarrow{\approx} \{w \in P^{-1}(J(X)) : P(\ker P_w^*) \subseteq \sigma\}$$

is a bijection, this proves Corollary 1. q.e.d.

*Proof of Corollary 2.* Using the notations and the result of Corollary 1, and referring again to part 2) of the proposition, the local degrees of $P$ at the three components of $P^{-1}(J(X))$ are the following:

(i) (Clemens) $\#\{(p,q) \in X^{(2)} : P(\ker P^*_{(p,q)}) \subseteq \sigma\}$
$= \#\{(p,q) : X \cup \overline{pq} \subseteq \text{base locus of } \sigma\}$
$= 16$. (The base locus of $\sigma$ is a quartic del Pezzo surface, "Segre's surface," which contains exactly 16 lines, [20, p. 141].)

(ii) $\#\{L \in \tilde{\Delta} : P(\ker P_L^*) \subseteq \sigma\} = 2 \cdot \#\{\text{singular quadrics in the pencil } \sigma\}$
$= 2 \cdot \#\{\sigma \cap \Delta\}$
$= 10$.

(iii) $\ker P^*_{\{*\}} = (0)$ implies the local degree of $P$ at $\{*\}$ is one.

q.e.d.

REMARKS. The technique used here to compute the local degree of a generically finite map along a positive dimensional component, using derivatives, is apparently equivalent to the generic case of the method developed earlier by Fulton and MacPherson for computing the multiplicity of an excess intersection, using Segre classes [12]. In the present example, an analysis of Clemens' prediction (on geometric grounds) of the local degree along $X^{(2)}$ led to the method used here. The local degree along $\tilde{\Delta}$, and hence the actual degree of $P$, was correctly predicted as well, by Ken Chakiris.

§3. *The generic Prym-Torelli theorem*

Although we have observed that the *strict* Torelli theorem for the Prym map is false, in this section and the next one we sketch arguments for suitably weakened injectivity statements about the Prym map, which will suffice to prove a *generic* Torelli theorem for intersections of three quadrics.

THEOREM. *If* $g \geq 7$, *then* $P_g : \mathcal{R}_g \to \mathcal{A}_{g-1}$ *is generically injective.*

*Proof.* (This was proven for $g \geq 9$ by Kanev [13], and extended to $g \geq 7$ by Donagi (unpublished).) The independent argument here is joint work with Robert Friedman [10]. The first argument for generic Prym-Torelli was announced in [21], but in view of [23] and [5], p. 388, the problem has been reconsidered.

We prove that $P_g$ is generically injective if $P_{g-1}$ is generically finite, hence if $g \geq 7$, by analyzing the fiber of $P_g$ over a decomposable Prym variety of type $(1, g-2)$, i.e.:

If $\mathcal{A}_1 \times \mathcal{A}_{g-2} = \{E \times A : E \in \mathcal{A}_1, A \in \mathcal{A}_{g-2}\} \subseteq \mathcal{A}_{g-1}$, then we have (after a degree two base change in both $\mathcal{R}_g$ and $\mathcal{A}_{g-1}$, which does not affect the degree of $P_g$) the following theorem.

THEOREM.

1) *If* $E \times A \in P_g(\mathcal{R}_g) \cap (\mathcal{A}_1 \times \mathcal{A}_{g-2})$ *is general, and if* $P_{g-1}^{-1}(A) = \{\eta_i : \tilde{C}_i \to C_i\}$, $i = 1, \cdots, \ell$, (= *a finite set of* $\ell$ *points in* $\mathcal{R}_{g-1}$) *then*

$$P_g^{-1}(E \times A) \cong \bigcup_{i=1}^{\ell} C_i$$ (= *a union of* $\ell$ *curves in* $\mathcal{R}_g$).

2) *If* $\tilde{\mathcal{A}}_{g-1}$ = *the blow-up of* $\mathcal{A}_{g-1}$ *along* $\mathcal{A}_1 \times \mathcal{A}_{g-2}$, *then, over a neighborhood of* $E \times A \in \mathcal{A}_{g-1}$, *there is a commutative diagram*:

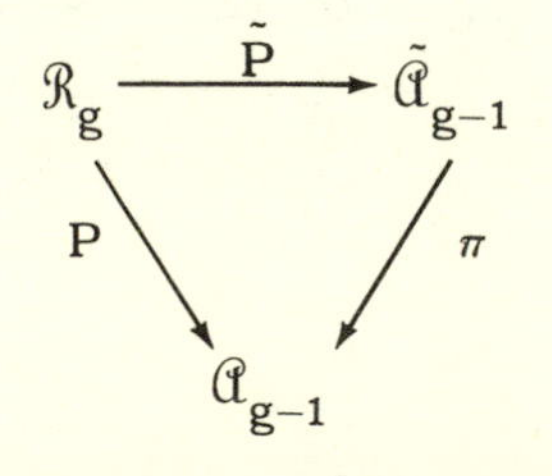

*such that the restriction of* $\tilde{P}$ *to the fibers of* $P$ *and* $\pi$ *over* $E \times A$ *looks as follows*:

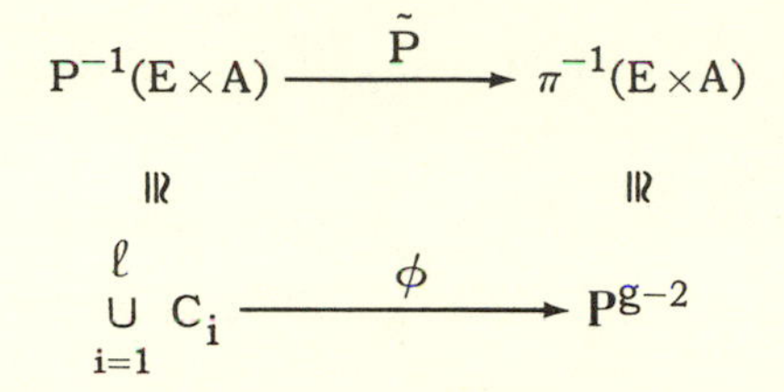


where $\phi|_{C_i} = \phi_{\omega_{C_i} \otimes \eta_i}$, *the "Prym-canonical" map of* $(C_i, \eta_i)$.

COROLLARY. *If* $g \geq 7$, $P_g$ *is generically injective.*

*Proof of 1).* If $(\eta : \tilde{C} \to C) \in P_{g-1}^{-1}(A)$, then for every $p \in C$, we have

$$(1_E \cup \eta : E \underset{p^+}{\cup} \tilde{C} \underset{p^-}{\cup} E \to C \cup E) \in P_g^{-1}(E \times A),$$

where $\eta^{-1}(p) = \{p^+, p^-\}$.

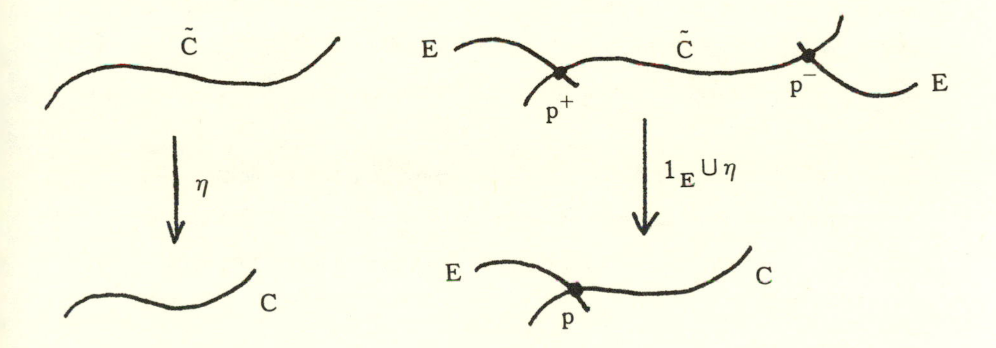


By checking Beauville's list of curves whose Prym variety has a theta-divisor which is singular in codimension one [4], one sees that this construction gives all of $P_g^{-1}(E \times A)$. q.e.d.

*Proof of 2).*

(i) Since

$$T_{E\times A}\mathcal{A}_{g-1} = \mathrm{Hom}^{sym}(H^{1,0}(E)\oplus H^{1,0}(A), H^{0,1}(E)\oplus H^{0,1}(A)),$$

and $T_{E\times A}(\mathcal{A}_1 \times \mathcal{A}_{g-2}) = \mathrm{Hom}^{sym}(H^{1,0}(E), H^{0,1}(E)) \oplus \mathrm{Hom}^{sym}(H^{1,0}(A), H^{0,1}(A))$, and $H^{1,0}(A) = H^0(C;\omega_C \otimes \eta)$ when $A = P(C,\eta)$, we have

$$\begin{aligned} \pi^{-1}(E\times A) &= P(T_{E\times A}\mathcal{A}_{g-1}/T_{E\times A}(\mathcal{A}_1\times\mathcal{A}_{g-2})) \\ &\cong P(\mathrm{Hom}(H^{1,0}(A), H^{0,1}(E))) \\ &\cong P(H^0(C;\omega_C\otimes\eta)^*) = P^{g-2}. \end{aligned}$$

(ii) If $p \in C$, and $\vec{n}$ is the unique vector normal to the boundary of $\mathcal{R}_g$ at $(C \cup E, \eta \cup 1_E)$, then

$$\tilde{P}(p) = \{P_*(\vec{n})\} \in P(T_{E\times A}\mathcal{A}_{g-1}/T_{E\times A}(\mathcal{A}_1\times\mathcal{A}_{g-2})).$$

Moreover it can be shown that $\{P_*(\vec{n})\}$ corresponds, by the isomorphisms above, to the homomorphism with kernel $\{\lambda : \lambda(p) = 0\} \subseteq H^0(C;\omega_C\otimes\eta)$ [10], $= \phi_{\omega_C\otimes\eta}(p)$, by definition of the Prym-canonical map on $C$. q.e.d.

*Proof of Corollary.* Since the $C_i$ are general curves of genus $\geq 6$, their Prym-canonical maps are embeddings [5, Prop. (7.7)], [9, Lemma III. (2.4)], and since these curves are distinct and finite in number, the map $\phi$ on their union $\bigcup_{i=1}^{\ell} C_i$, is generically injective. Thus, since $\tilde{P}$ also has injective derivative along $P^{-1}(E\times A)$, and both $\tilde{P}$ and $P_g$ are proper, they are generically injective for $g \geq 7$. q.e.d.

§4. *The Torelli problem for the intersection of three general quadrics in* $P^{2k}$, $k \geq 3$.

By the result of Beauville, Reid, and Tjurin in example 2, Section 1, we only need to prove generic injectivity of the restriction of $P_g$ to the set of double covers $\eta : \tilde{\Delta} \to \Delta$ arising from nets of quadrics, hence (by the remark following Lemma 2) to the set $\pi^+_{2k+1}$ of "even" double covers of smooth plane curves $\Delta$ of degree $2k+1$, $k \geq 3$, [5, Lemma (6.12)]. (One says $\eta : \tilde{\Delta} \to \Delta$ is "even" if the corresponding line bundle $\eta$, with $\eta^2 = \mathcal{O}_\Delta$ satisfies $h^0(\Delta; \mathcal{O}(k-1) \otimes \eta) \equiv 0 \pmod 2$.) Note that $\pi^+_{2k+1} \subseteq \mathcal{R}_g$, where $g = \frac{1}{2}(2k-1)(2k)$. The following discussion is a preliminary account of joint work with Robert Friedman which we intend to publish soon in full. I want to emphasize that although most difficulties in the proof appear to have been overcome, the full computations needed for the central properness result, Lemma 4, have not at this point been completely carried out.

Proposed Theorem. *If* $k \geq 3$, *then the restricted Prym map* $P_g : \pi^+_{2k+1} \to \mathcal{A}_{g-1}$, *where* $g = \frac{1}{2}(2k-1)(2k)$, *is generically injective.*

*Outline of proof.* We will partially compactify the map $P_g : \mathcal{R}_g \to \mathcal{A}_{g-1}$, and then prove that the compactified map is injective on a (Zariski) neighborhood of some point which is in the closure of the set $\pi^+_{2k+1}$. Since $\pi^+_{2k+1}$ is irreducible, this will prove the theorem.

To be precise, we make the following definitions:

$\mathcal{R}_g$ = {set of étale double covers $\eta : \tilde{C} \to C$, with $\tilde{C}$ and $C$ smooth, connected curves of genera $2g-1$ and $g$ respectively}.

$\tilde{\mathcal{R}}_g$ = {set of étale double covers $\eta : \tilde{C} \to C$, with $\tilde{C}$ and $C$ irreducible, stable curves, of genera $2g-1$ and $g$ respectively}.

Call an involution $\iota : \tilde{C} \to \tilde{C}$ of a stable curve "admissible" if no smooth points of $\tilde{C}$ are fixed by $\iota$, and if at a fixed node the two branches are not exchanged. Then let

$\overline{\mathcal{R}_g}$ = {set of double covers $\eta : \tilde{C} \to C$, with $\tilde{C}$, $C$ stable curves of genera $2g-1$ and $g$, and $(\eta : \tilde{C} \to \tilde{C}/\iota = C)$ the natural double cover associated to an admissible involution of $\tilde{C}$}.

If $(\eta : \tilde{C} \to C) \in \overline{\mathcal{R}_g}$, define the "étale normalization" $(N\eta : N\tilde{C} \to NC)$, to be the induced double cover after normalizing $\tilde{C}$ and $C$ at precisely those nodes where $\eta$ is étale. Finally, denote by $\mathcal{A}_g^*$ and $\overline{\mathcal{A}_g}$ respectively, the Satake and the "Voronoi" compactifications of $\mathcal{A}_g$, [16, 17]. Recall, as a set $\mathcal{A}_g^* = \mathcal{A}_g \cup \mathcal{A}_{g-1} \cup \cdots \cup \mathcal{A}_1 \cup \mathcal{A}_0$, and $\overline{\mathcal{A}_g}$, an example of one of Mumford's compactifications, may be thought of as a modification of $\mathcal{A}_g^*$, "blown up" along the boundary, i.e.: there is a natural bimeromorphic morphism $\sigma : \overline{\mathcal{A}_g} \to \mathcal{A}_g^*$ which is isomorphic over $\mathcal{A}_g$, and $\sigma^{-1}(\mathcal{A}_g^* - \mathcal{A}_g)$ has codimension one in $\overline{\mathcal{A}_g}$. The steps in the proof of the theorem can now be outlined in the following sequence of lemmas:

LEMMA 1. *There exists a commutative diagram of regular morphisms*:

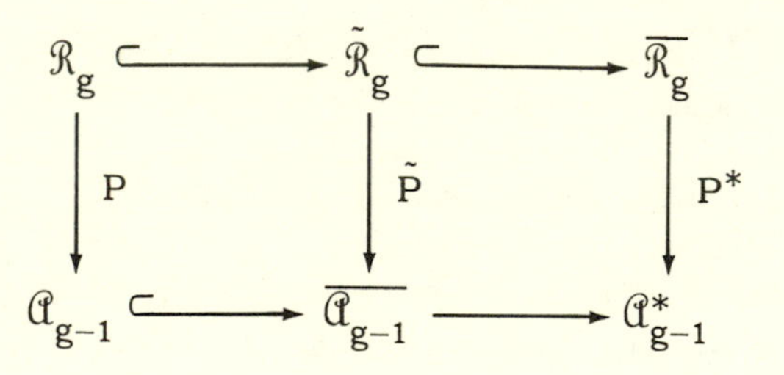


*in which*:

(i) $\eta \in \mathcal{R}_g \Rightarrow P(\eta)$ = *the classical Prym variety* $(\ker \eta_*)^0$;

(ii) $\eta \in \overline{\mathcal{R}_g} \Rightarrow P^*(\eta) = (\ker (N\eta)_*)^0$ = *the abelian variety obtained by modding out all* $C^*$'s *in the generalized Prym variety* $(\ker \eta_*)^0$;

(iii) $\eta \in \tilde{\mathcal{R}}_g \Rightarrow \tilde{P}(\eta)$ *determines a group extension of form* $1 \to (C^*)^r \to (\ker \eta_*)^0 \to P^*(\eta) \to 0$;

(iv) $P^*$ *is a proper morphism* (*but* $P$ *and* $\tilde{P}$ *are not*).

*Proof*. The characterization of $P^*(\eta)$ is stated in [5] and may be proved by checking that when a single étale node on $C$ is normalized (along with its preimages on $\tilde{C}$ ), the effect on the generalized Prym is either to leave it unchanged, or to mod out a $\mathbb{C}^*$. The extension $\tilde{P}$ of $P$ is constructed using the monodromy criterion in [17, Thm. (7.29)]. The properness of $P^*$ is equivalent to the compactness of $\overline{\mathcal{R}_g}$ which is proved by checking that a double cover of stable curves deforms to an étale double cover of smooth curves precisely if it arises from an admissible involution, and then using the compactness of the space of stable curves. q.e.d.

LEMMA 2. *If* $\tilde{\pi}^+_{2k+1}$ *denotes the closure in* $\tilde{\mathcal{R}}_g$ *of the set* $\pi^+_{2k+1}$, *and if* $k \geq 3$, *then for some* $g' \geq 7$, *the general point of* $\mathcal{R}_{g'}$ *occurs as the normalization of some double cover* $(\eta : \tilde{\Delta} \to \Delta)$ *in* $\tilde{\pi}^+_{2k+1}$. *Moreover,* $\tilde{\pi}^+_{2k+1}$ *is irreducible.*

*Proof.* For the first part, one uses the Brill-Noether theorem to prove the existence of (infinitely many) plane models of degree $2k+1$, $k \geq 3$, for generic curves of genus $g' \geq 7$, and then a theorem of Arbarello and Cornalba [2] on the existence of plane nodal deformations of these models, to prove the existence of an irreducible nodal plane curve $\Delta$ of arithmetic genus $g$, with normalization a general curve $N\Delta$ of genus $g' \geq 7$. Any étale double cover $N\tilde{\Delta} \to N\Delta$, can then be lifted to a finite number of étale covers $\tilde{\Delta} \to \Delta$, half of which are even, of the nodal curve $\Delta$. The irreducibility of $\pi^+_{2k+1}$ (and hence of its closure) has been proved by Joe Harris and by Fabrizio Catanese (to appear). q.e.d.

REMARK. Since the double covers coming from nets of quadrics form a Zariski-open subset of the irreducible set $\pi^+_{2k+1}$ ([5, Prop. (6.23)]), the theorem above will indeed imply the generic Torelli theorem for nets of quadrics in $\mathbf{P}^{2n}$, $n \geq 3$.

LEMMA 3. *If* $(\eta : \tilde{\Delta} \to \Delta) \in \tilde{\pi}^+_{2k+1}$ *is a double cover whose normalization is generic in* $\mathcal{R}_{g'}$, $g' \geq 7$ *as in Lemma 2, then* $\tilde{P}^{-1}(\tilde{P}(\eta)) = \{\eta\}$, *where* $\tilde{P}$ *is the map* $\tilde{P} : \tilde{\mathcal{R}}_g \to \overline{\mathcal{A}_{g-1}}$ *in Lemma 1.*

*Proof.* If $N\eta : N\tilde{\Delta} \to N\Delta$ is the normalization of the double cover $\eta$, $\{p_i, q_i\} \subseteq N\Delta$ is the pair of marked points lying over the $i^{th}$ node of $\Delta$, and $N\eta^{-1}(p_i) = \{p_i^+, p_i^-\}$, and $N\eta^{-1}(q_i) = \{q_i^+, q_i^-\}$ are the preimages of these points in $N\tilde{\Delta}$, then the group extension associated to the Voronoi point $\tilde{P}(\eta)$ determines the following data:

(i) the Satake point $P^*(\eta) = P(N\eta)$,

(ii) the "extension classes"

$$\{\pm([p_i^+ - q_i^+] - [p_i^- - q_i^-]\}_{i=1,\cdots,r} \in \bigoplus_{i=1}^{r} P^*(\eta) .$$

Since the normalized double cover $N\eta$ is assumed generic in $\mathcal{R}_{g'}$, $g' \geq 7$, the generic Prym Torelli theorem says $N\eta$ can be recovered from the Satake point $P^*(\eta)$. Moreover since generic curves of genus 7 have no $g_4^1$'s, the extension classes actually determine the pairs of marked points of $N\tilde{\Delta}$ which, when re-identified, allow one to recover $\eta : \tilde{\Delta} \to \Delta$.

q.e.d.

LEMMA 4. *Let* $(\eta : \tilde{\Delta} \to \Delta)$ *again be a double cover with generic normalization as described in Lemmas 2 and 3. Then there is an open neighborhood* U *of* $\tilde{P}(\eta)$ *in* $\overline{\mathcal{A}_{g-1}}$, *such that the restriction* $\tilde{P} : \tilde{P}^{-1}(U) \to U$, *is proper when regarded as a map onto its image.*

*Proposed Proof.* Since $\overline{\mathcal{R}_g}$ is compact, it suffices to show that if an étale double cover $\eta_t$ of irreducible curves specializes to an admissible double cover $(\eta_0 : \tilde{C}_0 \to C_0)$ of stable curves, while the Voronoi point $\tilde{P}(\eta_t)$ specializes to that of the double cover $\eta : \tilde{\Delta} \to \Delta$ described in Lemmas 2 and 3, then $N'\tilde{C}_0 = N\tilde{\Delta}$, and $N'\eta_0 = N\eta$. This differs from Lemma 3 because we don't know that $\tilde{C}_0$ is irreducible, so $N'\tilde{C}_0$ may not be connected. Hence the equality $P^*(\eta_0) = P^*(\eta)$ only implies that $N'\tilde{C}_0$ has one component isomorphic to $N\tilde{\Delta}$, (and on which $N'\eta_0 = N\eta$ ), but there could be other components which contribute nothing to $P^*(\eta_0)$. To rule that out, observe that our hypothesis on $\eta_0$ implies the equivalence of the two group extensions:

(1) $1 \to (\mathbb{C}^*)^r \to (\ker \eta_*)^0 \to P^*(\eta) \to 0$

(2) $1 \to (\mathbb{C}^*)^r \to (\ker (\eta_0)_*)^0 \to P^*(\eta_0) \to 0$

and hence the equality of their extension classes. As we noted in the proof of Lemma 3 the extension classes in case (1) determine, up to sign, a set of r divisors each of degree 4, supported at the set of 4r distinct marked points of $N\tilde{\Delta}$ which lie over the nodes of $\tilde{\Delta}$. In case (2) one should again find r divisors of degree 4 supported at marked points lying on that component of $N\tilde{C}_0'$ which is isomorphic to $N\tilde{\Delta}$. If $\tilde{C}_0$ is reducible, then some extension classes arise from nodes which have only one of their branches on $N\tilde{\Delta}$, and hence only one marked point on this component. It follows that there are fewer than 4r distinct points occurring in the supports of these divisors. This contradiction finishes the argument.

LEMMA 5. *$\tilde{P} : \tilde{\mathfrak{R}}_g \to \overline{\mathfrak{A}_{g-1}}$ is injective on a Zariski-open neighborhood of the point $(\eta : \tilde{\Delta} \to \Delta)$ described in Lemmas 2, 3, and 4.*

*Proof.* Since P is an algebraic map, [4], it suffices by Lemmas 3 and 4 to show that $\tilde{P}$ has injective derivative at the point $\eta$.

Let $(\tilde{P})_*$ be the differential of $\tilde{P}$ and let $\vec{v}$ be a tangent vector to $\tilde{\mathfrak{R}}_g$ at $\eta$. One checks first, by an explicit computation in the normal coordinates to $\partial(\overline{\mathfrak{A}}_{g-1})$, (= the boundary of $\overline{\mathfrak{A}}_{g-1}$), that $\tilde{P}_*(\vec{v})$ is tangent to $\partial(\overline{\mathfrak{A}}_{g-1})$ only if $\vec{v}$ is tangent to $\partial(\tilde{\mathfrak{R}}_g)$. Using this and "Wirtinger's theorem" [5, Cor. (7.11)], (generic injectivity of the differential of $P_{g'}$ for $g' \geq 6$), the assumption $\tilde{P}_*(\vec{v}) = 0$ implies that $\vec{v}$ must be tangent to the locus in $\partial(\tilde{\mathfrak{R}}_g)$ where $\eta : \tilde{\Delta} \to \Delta$ has constant normalization. Since then only the marked points over the nodes are varying on the curve $\tilde{\Delta}$, and only the extension classes are varying on the generalized Prym variety, and since the directions corresponding to different nodes are independent in both spaces, we are reduced to showing that the map

$$(N\tilde{\Delta})^2 \xrightarrow{\delta} P(N\eta)$$

$$(p,q) \to [p - \iota(p)] - [q - \iota(q)]$$

(to which $\tilde{P}$ restricts when only one node varies) is an immersion at $(p,q)$ if $p \neq q$ and $p \neq \iota(q)$. If $\frac{\partial}{\partial p}$ and $\frac{\partial}{\partial q}$ denote the "natural" basis of $T_{(p,q)}(N\tilde{\Delta})^2$, we must show the two vectors $\delta_*\left(\frac{\partial}{\partial p}\right)$ and $\delta_*\left(\frac{\partial}{\partial q}\right)$ are independent in $T_0P(N\eta)$. But the lines through these two vectors represent, in $PT_0(PN\eta)$, the Prym-canonical images of the points $\eta(p)$ and $\eta(q)$, [9, p. 39] so it suffices to know that the Prym-canonical map on $N\Delta$ is injective. Since this is true for any smooth curve without any $g^1_4$'s [5, Prop. (7.7)], [9, Lemma III, (2.4)] and hence for a generic curve of genus $g' \geq 7$, it holds for $N\Delta$. q.e.d.

REMARKS. The inspiration for this argument comes from many directions. The general plan of the proof resembles an unpublished one by Alberto Collino for the case of cubic threefolds, and the one in the 1981 Harvard Ph.D. thesis of Robert Friedman for K–3 surfaces. The wisdom of considering nodal plane curves was emphasized by Ken Chakiris, and the suggestion to prove properness only for irreducible curves in Lemma 4 is due to David Mumford. The proof of Lemma 3 imitates arguments by Namikawa [18] and Carlson [6] for Torelli for singular curves, and the first part especially of the proof of Lemma 5 follows Collino's calculation of the derivative of the compactified period map of cubic threefolds. The proof of Lemma 1 leans heavily on Beauville's analysis of generalized Prym varieties [4], while Lemma 2, as noted there, is taken bodily from work of Arbarello-Cornalba, Catanese, and Harris. q.e.d.

OPEN PROBLEM. Is in fact every smooth complete intersection of three quadrics in $P^{2k}$, $k \geq 3$, determined by its intermediate Jacobian?

## REFERENCES

[1] A. Andreotti, "On a theorem of Torelli," Am. J. Math., 80(1958), no. 4, pp. 801-828.

[2] E. Arbarello and M. Cornalba, Su una proprieta notevole dei morfismi di una curva, Rend. Sem. Mat. Torino 38(1980), no. 2, pp. 87-99.

[3] A. Beauville, "Sous-varietes speciales des varietes de Prym," preprint from l'Ecole Polytechnique, Plateau de Palaiseau, France.

[4] A. Beauville, "Prym varieties and the Schottky problem," Inv. math. 41 (1977), pp. 149-196.

[5] ________, "Variétés de Prym et Jacobiennes intermediares," Ann. Sci. Ecole Norm. Sup. 10(1977), pp. 309-391.

[6] J. Carlson, "Extensions of mixed Hodge structures," in Géométrie algébrique d'Angers 1979, Sijthoff and Noordhoff, Alphen aan den Rijn, The Netherlands, 1980, pp. 107-127.

[7] R. Donagi, "Group Law on the intersection of two quadrics," Ann. Scuola Norm. Sup. Pisa Cl. Sci. (4) 7 (1980), no. 2, pp. 217-239.

[8] ________, "The tetragonal construction," Bull. A.M.S. 4(2) (1981), pp. 181-185.

[9] R. Donagi and R. Smith, "The Structure of the Prym Map," Acta Math. 146 (1981), pp. 25-102.

[10] R. Friedman and R. Smith, "The generic Torelli theorem for the Prym map," Invent. Math., 67 (1982), pp. 473-490.

[11] ________, "The generic Torelli theorem for the intersection of three quadrics," in preparation.

[12] W. Fulton and R. MacPherson, "Intersecting cycles on an algebraic variety," in Real and complex singularities, Oslo 1976, Sijthoff and Noordhoff, Alphen aan den Rijn, The Netherlands, pp. 179-197.

[13] V. Kanev, "The degree of the Prym map is equal to one," preprint from Steklov Institute (1981).

[14] L. Masiewicki, Universal properties of Prym varieties with an application to algebraic curves of genus five, Trans. Am. Math. Soc., 222 (1976), pp. 221-240.

[15] D. Mumford, "Prym Varieties I," in Contributions to Analysis, N.Y., Academic Press, 1974.

[16] Y. Namikawa, "A new compactification of the Siegel space ...," I, Math. Ann., 221 (1976), pp. 97-141; II, *ibid.*, pp. 201-241.

[17] ________, Toroidal Compactification of Siegel Spaces, Lecture Notes in Math. no. 812, Springer-Verlag, Berlin, 1980.

[18] ________, "On the canonical holomorphic map from the moduli space of stable curves to the Igusa monoidal transform," Nagoya Math. J., vol 52 (1973), pp. 197-259.

[19] S. Recillas, "Jacobians of curves with a $g_4^1$ are Prym varieties of trigonal curves," Bol. Soc. Mat. Mexicana 19 (1974), pp. 9-13.

[20] J.G. Semple and L. Roth, "Introduction to Algebraic Geometry," Oxford Univ. Press, 1949.

[21] A. Tjurin, "The geometry of the Poincaré theta-divisor of a Prym variety," Math. USSR Izvestija 9(1975), no. 5, pp. 951-986.

[22] A. Tjurin, "On the intersections of quadrics," Russian Math. surveys 30(1975), pp. 51-105.

[23] ________, "Correction to the paper 'The geometry of the Poincaré theta-divisor of a Prym variety,'" Math. USSR Izvestija 12(1978), no. 2, p. 438.

[24] W. Wirtinger, "Untersuchungen über Theta functionen," Teubner, Berlin, 1895.

ROY SMITH
MATHEMATICS DEPARTMENT
UNIVERSITY OF GEORGIA
ATHENS, GA 30602

Chapter XII

# INFINITESIMAL VARIATION OF HODGE STRUCTURE AND THE GENERIC GLOBAL TORELLI THEOREM

Phillip Griffiths
Written by Loring Tu

In Chapter III we defined an infinitesimal variation of Hodge structure $V = \{H_{\mathbf{Z}}, H^{p,q}, Q, T, \delta\}$. It is given by a polarized Hodge structure $\{H_{\mathbf{Z}}, H^{p,q}, Q\}$ and a linear map

$$\delta : T \to \oplus \operatorname{Hom}(H^{p,q}, H^{p-1,q+1})$$

satisfying a symmetry condition and preserving the bilinear form $Q$. Associated to an infinitesimal variation are a number of invariants with geometric meaning. We recall two:

a) (Kernel of the co-iterate of the differential.) Let

$$\delta^{n*} : \operatorname{Sym}^2 H^{n,0} \to \operatorname{Sym}^n T^*$$

be the n-th co-iterate of the differential. The first invariant is $\mathcal{Q} = \ker \delta^{n*}$, which we think of as a linear system of quadrics on $PH^{n,0*}$.

b) (Annihilator of a Hodge class.) Consider the case of weight $n = 2$. Let

$$\lambda \in H^{1,1}_{\mathbf{Z}} = H_{\mathbf{Z}} \cap H^{1,1}$$

be a Hodge class. If $\xi \in T$, then $\delta(\xi) : H^{2,0} \to H^{1,1}$. We set

$$H^{2,0}(-\lambda) = \{\omega \in H^{2,0} : Q(\delta(\xi)\omega, \lambda) = 0 \text{ for all } \xi \in T\} .$$

Given a collection of Hodge classes $\{\lambda_i\}_{i \in I}$, we may also consider $\cap_{i \in I} H^{2,0}(-\lambda_i)$.

By the generic global Torelli theorem we mean the following statement: *Let* $\mathfrak{M}$ *be a moduli space and* $\phi : \mathfrak{M} \to \Gamma \backslash D$ *a period map. Then* $\phi : \mathfrak{M} \to \phi(\mathfrak{M})$ *has degree one onto its image.*

The infinitesimal variation of Hodge structure may be used to prove the generic global Torelli theorem. We state this as a general principle below. As a matter of notation, for a point $s \in \mathfrak{M}$ let $X_s$ be the variety parametrized by $s$.

GENERAL PRINCIPLE 1. Suppose that for a general point $s \in \mathfrak{M}$ the infinitesimal variation of Hodge structure associated to $X_s$ uniquely determines $X_s$. Then the generic global Torelli theorem holds for $\mathfrak{M}$.

*Proof.* Let $p$ be a regular value. Locally around $p$ factor $\phi$ into a submersion followed by an immersion. If $s_1$ and $s_2$ are two distinct points in $\phi^{-1}(p)$, then $X_{s_1}$ and $X_{s_2}$ are distinct varieties with the same infinitesimal variation of Hodge structure, contradicting the hypothesis. Therefore, $\phi : \mathfrak{M} \to \phi(\mathfrak{M})$ has degree one. q.e.d.

EXAMPLE 2. Let $\mathfrak{M} = \mathfrak{M}_g$ be the moduli space of smooth curves of genus $g \geq 5$. As we saw in Chapter III, the canonical image $\phi_K(C)$ of a general curve of genus $g \geq 5$ may be reconstructed from its infinitesimal variation of Hodge structure:

$$\phi_K(C) = \bigcap_{Q \in \mathcal{Q}} Q .$$

By the general principle above, the generic global Torelli theorem follows.

Using these ideas we shall outline the proof of the following two theorems.

THEOREM A. *The generic global Torelli theorem holds for smooth cubic hypersurfaces* $X$ *in* $\mathbf{P}^{3m+1}$.

REMARK. For $m \geq 2$ the period matrix space $D$ is not a bounded symmetric domain.

REMARK ADDED IN PROOF. This result is now subsumed in Donagi's theorem reported on in the next chapter. It may be interesting to see exactly the nature of the difficulty in extending the proof of Theorem A to the general case (cf. the discussion following example 7 below), and then to see how Donagi resolved this difficulty as explained in Chapter XIII.

THEOREM B. *Let* $F \subset \mathbf{P}^3$ *be a smooth surface of degree* $d \geq 5$ *containing at least* $d-5$ *sets of* $d$ *coplanar lines and let* $T \simeq H^1(F,\Theta)$ *be the tangent space to* $\phi(F)$ *in* $D$. *Then* $F$ *is uniquely determined by its infinitesimal variation of Hodge structure* $V(F) = \{H^2_{prim}(F,\mathbf{Z}), H^{p,q}_{prim}(F), Q, T, \delta\}$.

REMARK. We shall outline a proof of Theorem B in case

$$F = F_d = \{x_0^d + x_1^d + x_2^d + x_3^d = 0\}$$

is the Fermat hypersurface of degree $d$. The proof in [5] shows that

$$\mathrm{Aut}\, V(F_d) \simeq \mathrm{Aut}(F_d) .$$

If $\mathcal{H}(F_d) = \{H^2_{prim}(F_d,\mathbf{Z}), H^{p,q}_{prim}(F_d), Q\}$, then we suspect that

$$\mathrm{Aut}\, \mathcal{H}(F_d) \underset{\neq}{\supset} \mathrm{Aut}(F_d) ,$$

but we have not been able to prove this.

*Sketch of a proof of Theorem A* (see Carlson and Griffiths [2, pp. 51-76] for details). Let $X \subset \mathbf{P}^{n+1} = \mathbf{P}V^*$, where $V^* \simeq \mathbf{C}^{n+2}$, be a smooth hypersurface of degree $d$ with defining equation

$$F(x_0, \cdots, x_{n+1}) = 0, \qquad x_i \in V .$$

Set

$$S = SV \simeq \mathbf{C}[x_0, \cdots, x_{n+1}], \qquad S = \oplus S^q = \oplus S^q V ,$$

and let $J_F$ be the Jacobian ideal of $F$ in $S$:

$$J_F = (\partial F/\partial x_0, \cdots, \partial F/\partial x_{n+1}), \qquad J_F = \oplus J_F^q .$$

The first point is the following proposition.

PROPOSITION 3. *If* X *is smooth, then the homogeneous piece* $J_F^{d-1}$ *of the Jacobian ideal of* F *in degree* d–1 *uniquely determines* X *up to* PGL(V).

*Proof when* X *is general* (i.e., Aut X = {e} ). Let $U \subset PS^d$ be a neighborhood of F. By Mumford's geometric invariant theory if X is smooth and has no automorphism, then $\mathfrak{M} = U/PGL(V)$ exists and is smooth. Let $\pi : U \to U/PGL(V)$ be the projection. We claim that the vertical tangent space at $P \in U$ is $J_P^d$, for if $A \in \mathfrak{gl}(V)$, then

$$t \mapsto P(e^{tA}x)$$

is a curve in a fiber of $\pi$ and

$$\frac{\partial}{\partial t}\Big|_{t=0} P(e^{tA}x) = \Sigma \frac{\partial P}{\partial x_i} a_{ij} x_j .$$

Now suppose $F_0$ and $F_1$ are two polynomials of degree d such that $J_{F_0}^{d-1} = J_{F_1}^{d-1}$. Then $J_{F_0}^d = J_{F_1}^d$. Let $F_t = (1-t)F_0 + tF_1$ be the segment joining $F_0$ and $F_1$. We have

$$\frac{\partial F_t}{\partial t} = -F_0 + F_1 .$$

By Euler's theorem, $F_0 \in J_{F_0}^d = J_{F_t}^d$ and $F_1 \in J_{F_1}^d = J_{F_t}^d$. So

$$\frac{\partial F_t}{\partial t} \in J_{F_t}^d .$$

Since the tangent vector field of $F_t$ is always vertical, $F_t$ lies in a fiber of $\pi$. It follows that all the $F_t$, in particular $F_0$ and $F_1$, are projectively equivalent. q.e.d.

We recall below three facts from residue theory.

i) (Griffiths [4]). Let $\Omega = \Sigma(-1)^i x_i dx_0 \cdots \widehat{dx_i} \cdots dx_{n+1}$. This is a non-zero section of $\Omega^{n+1}_{P^{n+1}}(n+2)$. If $A$ is a polynomial of degree

$$\sigma = d(q+1) - (n+2),$$

then the Poincaré residue map induces an isomorphism

$$(*) \qquad S^\sigma/J_F^\sigma \xrightarrow{\sim} H^{n-q,q}_{prim}(X)$$

$$A \longmapsto \operatorname{res} \frac{A\Omega}{F^{q+1}},$$

where $S^\sigma$ denotes the homogeneous polynomials of degree $\sigma$ in the variables $x_0, \cdots, x_{n+1}$. Furthermore, the diagram

$$\begin{array}{ccccc} S^\sigma/J_F^\sigma & \otimes & S^{\sigma'}/J_F^{\sigma'} & \xrightarrow{\text{Res}} & C \\ \wr\| & & \wr\| & & \| \\ H^{n-q,q} & \otimes & H^{q,n-q} & \xrightarrow{Q} & C \end{array}$$

is commutative, where Res denotes the Grothendieck residue map.

ii) (Carlson and Griffiths [2]). Suppose we fix a homomorphism

$$\tau : H^{n-q,q}_{prim}(X) \to S^\sigma$$

such that

$$\tau(\operatorname{res} A\Omega/F^{q+1}) = A .$$

Then $\tau$ induces a map $S^2\tau : S^2H^{n-q,q} \to S^2(S^\sigma)$ and the basic diagram (Chapter II(12)) induces the diagram

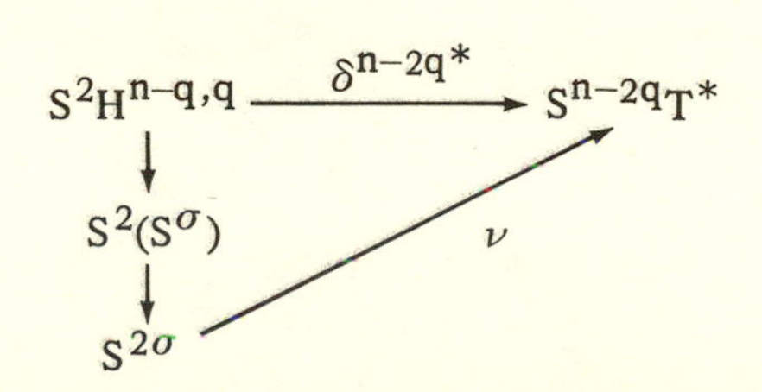

$$\begin{array}{ccc} S^2H^{n-q,q} & \xrightarrow{\delta^{n-2q*}} & S^{n-2q}T^* \\ \downarrow & & \\ S^2(S^\sigma) & & \nearrow \nu \\ \downarrow & & \\ S^{2\sigma} & & \end{array}$$

In this diagram

$$\ker \nu = J_F^{2\sigma} .$$

iii) (Carlson and Griffiths [2]). It follows from Macaulay's theorem that $J_F^{2\sigma}$ determines $J_F$.

To continue the proof of Theorem A we introduce the notion of the *level* of a Hodge structure.

DEFINITION 4. A Hodge structure $\{H_{\mathbf{Z}}, H^{p,q}, Q\}$ of weight $n$ is said to have *level* $a$ if $H^{a,n-a} \neq 0$ but $H^{p,q} = 0$ for $p > a$.

REMARK 5. A Hodge structure of weight $n$ and level $a$ really looks like a Hodge structure of weight $2a-n$:

$$H = H^{a,n-a} \oplus H^{a-1,n-a+1} \oplus \cdots \oplus H^{n-a,a} .$$

If $X \subset \mathbf{P}^{n+1}$ is a smooth hypersurface of degree $d$, then

$$\begin{aligned} h^{n,0} &= h^0(\mathbf{P}^{n+1}, \Omega^{n+1}(X)) \cong h^0(\mathbf{P}^{n+1}, \mathcal{O}(d-n-2)) \\ &= \binom{d-1}{n+1} . \end{aligned}$$

Therefore,

$$h^{n,0} \neq 0 \text{ if and only if } d-1 \geq n+1 .$$

So for $d \geq n+2$, level $H^n(X) = n$ and for $d \leq n+1$, level $H^n(X) < n$.

EXAMPLE 6 (Cubic hypersurface in $\mathbf{P}^{3m+1}$). In this case

$$\sigma = 3(q+1) - (3m+2) \geq 0 \text{ iff } n-q \leq 2m + \frac{1}{3} .$$

So the first nonzero Hodge piece in $H^{3m}(X)$ is $H^{2m,m}(X)$. Furthermore, since

$$\sigma = 3(m+1) - (3m+2) = 1 ,$$

the cohomology classes in $H^{2m,m}(X)$ correspond to linear polynomials:

$$H^{2m,m}(X) \simeq V .$$

EXAMPLE 7. $d \geq n+2$.

As noted above, the level of $H^n(X)$ here is $n$ and

$$H^{n,0} \simeq S^{d-n-2} .$$

We return now to the proof of Theorem A. Let $X$ be a cubic hypersurface in $\mathbf{P}^{3m+1}$. Fix an isomorphism

$$\tau : H^{2m,m}(X) \xrightarrow{\sim} V .$$

From the infinitesimal variation of Hodge structure we get $J_F^2 \subset S^2V$, since $\ker \delta^{2m*} = J_F^2$. By iii) and Proposition 3, $J_F^2$ determines $X$ up to the action of $PGL(V)$. If we had chosen another isomorphism $\tau' : H^{2m,m}(X) \xrightarrow{\sim} V$, the polynomial ideal $J_F$ and the equation for $X$ would be altered by an element of $PGL(V)$. So this construction is independent of the isomorphism $\tau$. This proves Theorem A.

Next consider the case $d \geq n+2$. In order for $J^{2\sigma}$ to be nonzero, it is necessary that $2\sigma \geq d-1$, which is equivalent to $d \geq 2n+3$. Under this hypothesis, the same argument goes through except for the fact that the isomorphism

$$\tau : H^{n,0}(X) \xrightarrow{\sim} S^{d-n-2}$$

may not be induced from an automorphism of $V$. This problem, except in a small number of cases, has now been resolved by Ron Donagi (see Chapter XIII).

*Sketch of proof of Theorem B* (for details see Griffiths and Harris [5]). Let $F$ be a smooth surface of degree $d$ in $\mathbf{P}^3$ and $\omega$ the hyperplane class on $F$. If $\lambda$ is the fundamental class of a line on $F$, then

$$\lambda \cdot \omega = 1 \qquad \text{and} \qquad \lambda^2 = 2-d .$$

[The self-intersection number of $\lambda$ may be computed as follows. Let $H$ be a hyperplane section of $F$ containing $\Lambda$. Then $H = \Lambda + R$, where the residual intersection $R$ is a plane curve of degree $d-1$. So $\Lambda \cdot R = d-1$. Since $1 = \Lambda \cdot H = \Lambda \cdot (\Lambda + R)$, we have $\Lambda \cdot \Lambda = 2-d$.]

DEFINITION 8. A cohomology class $\lambda \in H^{1,1}(F, Z)$ is called a *Hodge line* if

$$\lambda \cdot \omega = 1 \qquad \text{and} \qquad \lambda^2 = 2-d \ .$$

REMARK 9. It can be shown that every Hodge line is the fundamental class of a unique line on $F$. By Lefschetz' theorem on the (1,1)-class there is a divisor $\Lambda$ whose fundamental class is $\lambda$. So the problem is to show the effectiveness of $\Lambda$. For $d = 4, 5$ this follows from the Riemann-Roch theorem for surfaces. For $d \geq 6$ a special argument is necessary; we refer the reader to the paper cited above.

We call a set of $d$ coplanar concurrent lines on $F$ a *star*. (Here "concurrent" means that the lines all meet at the same point.) If $x$ and $y$ are the affine coordinates of the plane of a star, then the lines in the star have equations

$$y = A_i x \ , \qquad i = 1, \cdots, d \ ,$$

for some constants $A_i$. In case the $A_i$'s can be taken to be the $d$-th roots of unity, the star is said to be a *special star*. On the Fermat surface there are $6d$ sets of special stars; moreover, each line belongs to exactly two special stars (Shioda [7], Ziv Ran [6]). Thus a Fermat surface has $3d^2$ lines.

We now outline an argument that the Fermat surface $F_d$ can be reconstructed from its infinitesimal variation of Hodge structure. First note that the lines and their pairwise intersection numbers on $F_d$ may be recovered from the polarized Hodge structure. For suppose a surface $F$ has the same polarized Hodge structure as the Fermat $F_d$. Then $F$ has $3d^2$ Hodge lines. By Remark 9 these Hodge lines are the fundamental classes of $3d^2$ lines on $F$. Moreover, from the intersection form $Q$ we get the intersection number of any pair of lines on $F$. In fact we can even conclude that these $3d^2$ lines fall into $6d$ sets of $d$ coplanar lines. Consider three mutually intersecting lines $\Lambda_1$, $\Lambda_2$, and $\Lambda_3$. They form either Configuration (i) or (ii) in the figure below.

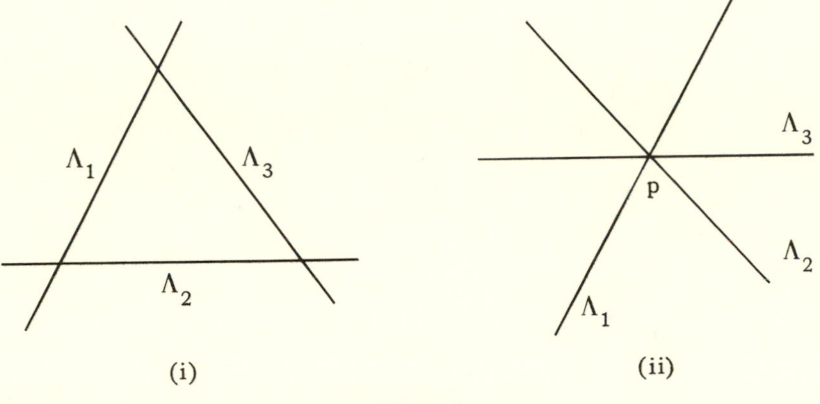


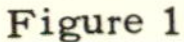
Figure 1

Case (i) is clearly coplanar, but so is Case (ii), since in (ii) each line lies in the tangent plane $T_p(F)$ at the common intersection point $p$. So the lines on $F$ must satisfy the same coplanarity relations as those on $F_d$.

However, the polarized Hodge structure alone cannot distinguish (i) from (ii). For this we need the infinitesimal variation of Hodge structure.

Let $V(F) = \{H_{\mathbf{Z}}, H^{p,q}, Q, \delta, T\}$ be the infinitesimal variation of Hodge structure of $F$. Recall that for $\xi \in T$ the differential

$$\delta(\xi) : H^{2,0} \to H^{1,1}$$

can be given as the cup product with the Kodaira-Spencer class $\rho(\xi) \in H^1(\Theta)$. Let $\Lambda$ be a cycle with fundamental class $\lambda \in H^{1,1}_{\mathbf{Z}}$ and let

$$H^0(F, K(-\Lambda)) = \{\psi \in H^0(K) : \psi \text{ vanishes on } \Lambda\}.$$

Then

$$H^0(F, K(-\Lambda)) \subseteq H^{2,0}(-\lambda),$$

since

$$Q(\delta(\xi)\psi,\lambda) = \int_{\Lambda} \rho(\xi)\psi = 0 .$$

MAIN LEMMA 10. *If* F *is a smooth surface in* $\mathbf{P}^3$ *and* $\Lambda$ *is a line on* F *with fundamental class* $\lambda$, *then*

$$H^0(F, K(-\Lambda)) = H^{2,0}(-\lambda) .$$

Using this lemma one can distinguish between (i) and (ii): in Case (i)

$$H^{2,0}(-\lambda_1) + H^{2,0}(-\lambda_2) + H^{2,0}(-\lambda_3) = H^{2,0} ,$$

while in Case (ii)

$$H^{2,0}(-\lambda_1) + H^{2,0}(-\lambda_2) + H^{2,0}(-\lambda_3) \neq H^{2,0} ;$$

in this last inequality if $H^{2,0}$ is identified with the polynomials of degree $d-4$ on $\mathbf{P}^3$, then the left-hand side consists of those polynomials vanishing at p. Indeed, one can even recover the line $\Lambda \subset F$ from the infinitesimal variation of Hodge structure, for $\Lambda$ is the base locus of the linear system $H^{2,0}(-\lambda) \subset H^0(F, K)$.

Thus if F has the same infinitesimal variation of Hodge structure as the Fermat $F_d$, then F also has 6d special stars. Further arguments concerning the automorphisms of F show that the configuration of lines on F is projectively equivalent to that on $F_d$. So by a projective transformation we can line up the lines of F with those of $F_d$. But then

$$\deg F \cap F_d \geq 3d^2 \geq d^2 + 1 .$$

By Bezout's theorem F and $F_d$ are identical. q.e.d.

## REFERENCES

[1] J. Carlson, M. Green, P. Griffiths, and J. Harris, Infinitesimal variation of Hodge structures, to appear in Comp. Math.

[2] J. Carlson and P. Griffiths, Infinitesimal variation of Hodge structure and the global Torelli problem, Journées de géométrie algébrique d'Angers, Sijthoff and Noordhoff, 1980, pp. 51-76.

[3] R. Donagi, Generic Torelli and variational Schottky, Chapter XIII, this volume.

[4] P. Griffiths, Periods of certain rational integrals, Ann. of Math. 90 (1969), 460-541.

[5] P. Griffiths and J. Harris, Infinitesimal variation of Hodge structures (II): an infinitesimal invariant of Hodge classes, to appear in Comp. Math.

[6] Z. Ran, Cycles on Fermat hypersurfaces, Comp. Math. 42 (1980/81), 121-142.

[7] T. Shioda, The Hodge conjecture for Fermat varieties, Math. Ann. 245 (1979), 175-184.

PHILLIP GRIFFITHS
MATHEMATICS DEPARTMENT
HARVARD UNIVERSITY
CAMBRIDGE, MA 02138

LORING TU
MATHEMATICS DEPARTMENT
JOHNS HOPKINS UNIVERSITY
BALTIMORE, MD 21218

Chapter XIII

# GENERIC TORELLI AND VARIATIONAL SCHOTTKY

Ron Donagi*

*Introduction*

Our aim in this work is to give a simple treatment of the generic Torelli theorem for projective hypersurfaces, proved in [5], and of related results. The main result of [5] asserts the generic injectivity of the period map for $n$-dimensional projective hypersurfaces of degree $d$ with the following possible exceptions:

(0) $n = 2$, $d = 3$ (cubic surfaces).

(1) $d$ divides $n + 2$.

(2) $d = 4$, $n = 4m$ or $d = 6$, $n = 6m + 1$ $(m \geq 1)$.

The proof of that result uses a melange of methods. Here we focus attention on one of these methods, the use of symmetrizers. At the cost of a few more exceptions we achieve greater clarity and unity of exposition. We feel that this is the technique which is most likely to work in more general situations, such as for sufficiently ample divisors on arbitrary varieties ([3], §IIIb). In this work we add only one harmless analogue, the generic Torelli theorem for branched double covers of $\mathbf{P}^n$ (§6).

In the last section we discuss Schottky's problem of determining which Hodge structures arise from geometry. We formulate the variational version of this problem, solve it for hypersurfaces using the symmetrizer construction, and muse on the variational Schottky problem for curves and its relation to classical Schottky.

*Research partially supported by the NSF.

Contents



## §1. *Infinitesimal variation of Hodge structure*

We start by recapitulating the idea of an infinitesimal variation of Hodge structure and its use in proving generic Torelli results.

Let

$$p : M \to D$$

be a rational map of varieties or an almost-everywhere defined map of manifolds. Let TD denote the tangent bundle of D, let d be the rank of p at a generic point of M, and let G(d, TD) be the Grassmannian-bundle over D of d-dimensional subspaces of TD. The *prolongation* of p is the map

$$v : M \to G(d, TD)$$

defined by

$$v(x) = (p(x), p_* T_x M) .$$

The *principle of prolongation* says that p and v have the same degree, so in particular if v is generically injective then so is p. The principle becomes evident when we consider a dense open subset of M on which the degree of p is constant, possibly infinite.

In our application, p is the period map for a family of hypersurfaces and v is the infinitesimal variation of Hodge structure. Each of these maps consists of the specification of certain algebraic data (a filtered vector space and various bilinear maps) and some transcendental data (the lattice). Naturally, the algebraic part is more accessible to manipulations. However, the algebraic part of $p(X)$ has no invariants; thus any method for recovering information directly from $p(X)$ would have to be essentially transcendental. On the other hand, as we shall see, the algebraic portion of $v(X)$ is rich enough that the hypersurface $X$ can be recovered from it using simple algebraic constructions.

In fact, let $v_a(X)$ be the algebraic part of $v(X)$. It consists of a filtered, polarized vector space together with certain bilinear maps giving the infinitesimal variation (§2). We wish to suggest that these bilinear maps might be considered as (algebraic) analogues of the (transcendental) lattice in $p(X)$. In particular this leads to formulation of variational analogues to the Torelli and Schottky problems:

1.1. VARIATIONAL TORELLI. Is the map $v_a$ (from the moduli space of some varieties $X$, to an appropriately constructed "variational period space" parametrizing the data of type $v_a(X)$ ) injective?

1.2. VARIATIONAL SCHOTTKY. Describe (by equations and presumably also some inequalities) the image of $v_a$.

The principle of prolongation gives the implication:

$$\text{Variational Torelli} \Rightarrow \text{Generic Torelli}.$$

We only see a relationship between variational and standard Schottky in the presence of a theta function, cf. §7. In §7, we solve problem 1.2 for hypersurfaces (satisfying certain numerical conditions). The arguments in §§3-5, together with a recent strengthening of the Symmetrizer Lemma (cf. Remark 3.2) give an affirmative answer to problem 1.1 for hypersurfaces (cf. Remark 5.3).

## §2. *Hodge theory for non-singular hypersurfaces*

In this section we review Griffiths' description of the Hodge filtration of a hypersurface and of its infinitesimal variation in terms of vector spaces of polynomials on $\mathbf{P}^{n+1}$. The idea is to generalize the classical adjunction formula identifying $H^{n,0}(X) = H^0(\Omega_X^n)$ with the vector space of polynomials of degree $d-n-2$ on $\mathbf{P}^{n+1}$. For complete proofs we refer the reader to [6].

We let $S = S^*V$ denote the symmetric algebra on the vector space $V = H^0(\mathbf{P}^{n+1}, \mathcal{O}(1))$, let $f \in S^dV$ be a homogeneous polynomial of degree $d$ defining a nonsingular hypersurface $X \subset \mathbf{P}^{n+1}$, $J = J(f)$ the ideal generated by the partials of $f$, and $R = S/J$ the quotient ring, called the Jacobian ring of $f$. We let $S^a, J^a, R^a$ denote the a-th homogeneous piece of $S$, $J$, $R$ respectively.

THEOREM 2.1. *There are natural isomorphisms*

$$\lambda_a : R^{t_a} \xrightarrow{\sim} F^a/F^{a+1}$$

*from certain graded pieces of the Jacobian ring to the successive quotients of the Hodge filtration*

$$H^n(X) = F^0 \supset F^1 \supset \cdots \supset F^n \supset F^{n+1} = (0)\,.$$

*Here* $t_a = (n-a+1)d - (n+2)$.

The main ingredient of the proof is the residue map

$$\mathrm{Res} : H^{n+1}(\mathbf{P}^{n+1} \setminus X) \to H^n(X)\,.$$

This can be defined purely algebraically, but is most transparent over $\mathbf{C}$ where it is the adjoint of the tube map

$$H_n(X) \to H_{n+1}(\mathbf{P}^{n+1} \setminus X)$$

sending an n-cycle in $X$ to the boundary of a normal tube over it in $\mathbf{P}^{n+1}$. The image of Res is the primitive part $H_0^n(X)$ of $H^n(X)$.

The cohomology of the affine variety $\mathbf{P}^{n+1} \setminus X$ can be computed using the algebraic deRham complex. Thus any cohomology class in $H^{n+1}(\mathbf{P}^{n+1} \setminus X)$ (hence also in $H_0^n(X)$ ) can be represented by a meromorphic form

$$\alpha = \frac{A \cdot \Omega}{f^{k+1}}$$

where

$$\Omega = \sum_{i=1}^{n+2} (-1)^i x_i dx_1 \wedge \cdots \wedge \widehat{dx_i} \wedge \cdots \wedge dx_{n+2}$$

is the standard section of $\omega_{P^{n+1}}(n+2)$, $k$ is a non-negative integer, and $A$ is a polynomial on $P^{n+1}$ chosen to make $\alpha$ homogeneous of degree $0$, i.e.,

$$\deg(A) = (k+1)d - (n+2) .$$

The marvelous thing is that any class in $H^n(X)$ can be represented by an $\alpha$ with $k \leq n$, and in fact a class in $H^{n-p,p}$ can be represented by a form with $k \leq p$, determined uniquely modulo classes with $A \in J$. The details are in [6].

Next we consider the infinitesimal variation of the Hodge filtration as $f$ varies. It is given by a map

$$v : H^1(\theta_X) \to T_{p(X)}\mathcal{F}$$

where $H^1(\theta_X)$ is the tangent space to the deformation space of $X$, and $p(X)$ is the period of $X$ considered as a point of the appropriate flag space $\mathcal{F}$, i.e., $p(X)$ is given by a filtration

$$H^n(X) = F^0 \supset F^1 \supset \cdots \supset F^n \supset (0) .$$

Now $\mathcal{F}$ can be embedded in a product

$$\mathcal{F} \subset \prod_{i=1}^{n} G_i$$

where $G_i$ is the Grassmannian of subspaces of $F^0$ whose dimension equals $\dim(F^i)$, hence there is a natural inclusion

$$T_{p(X)}\mathcal{F} \subset \bigoplus_{i=1}^{n} T_{F^i} G_i = \bigoplus_{i=1}^{n} \operatorname{Hom}(F^i, F^0/F^i) .$$

We let

$$v_i : H^1(\theta_X) \to \mathrm{Hom}(F^i, F^0/F^i)$$

be the i-th component, under this decomposition, of the infinitesimal variation $v(X)$.

THEOREM 2.2. (1) $\mathrm{Im}(v_i) \subset \mathrm{Hom}(F^i/F^{i+1}, F^{i-1}/F^i)$.

(2) *There is a natural identification*

$$\lambda_d : R^d \xrightarrow{\sim} H^1(\theta_X)\,.$$

(3) *Via the identifications of part (2) and of Theorem 2.1,* $v_i$ *becomes (up to some universal constants) the multiplication map*

$$R^d \times R^{t_i} \to R^{t_i+d}\,, \qquad t_i = (n-i+1)d - (n+2)\,.$$

Part (1) expresses the infinitesimal period relations, and will follow from part (3). The isomorphism $\lambda_d$ is defined as follows. For $\bar{g} \in R^d$, choose $g \in S^d$ representing $\bar{g}$, and let $\lambda_d(\bar{g})$ be the deformation of $f$ given by $f + tg$, $t$ a parameter. One sees easily that $g \in J^d$ if and only if the deformation is trivial (to first order), i.e., $f + tg$ is the linear part of the Taylor expansion for $T_t f$ where $\{T_t\}$ is a one-parameter group of projective transformations.

To prove part (3), choose $\alpha \in F^i$ and represent it via the residue isomorphism $\lambda_i$ as

$$\alpha = \frac{A \cdot \Omega}{f^{n-i+1}}\,.$$

When $f$ is deformed to $f + tg$, $\alpha$ becomes

$$\alpha_t = \frac{A \cdot \Omega}{(f+tg)^{n-i+1}}$$

so by differentiation with respect to $t$ we find:

$$v_i(X)(\overline{g})(\alpha) = \frac{\partial}{\partial t}\Big|_{t=0} \alpha_t = \text{constant} \cdot \frac{g \cdot A \cdot \Omega}{f^{n-i+2}}$$

This proves that, up to a constant,

$$v_i(X)(\overline{g}) : R^{t_i} \to R^{t_i+d}$$

sends

$$A \mapsto g \cdot A$$

as claimed.

We need two more results on the structure of the ring $R$.

LEMMA 2.3. $\mathrm{Dim}(R^i)$ *depends only on* $i$, $d$, $n$, *for non-singular hypersurfaces of degree* $d$ *in* $P^{n+1}$.

*Proof.* This is a straightforward counting argument, based on the exactness of the Koszul complex

$$0 \to S^{\binom{n+2}{n+2}} \to \cdots \to S^{\binom{n+2}{2}} \to S^{n+2} \to S \to R \to 0 .$$

(Compare [7], page 689.)

THEOREM 2.4: LOCAL DUALITY. $R$ *is an Artinian ring of top degree* $\sigma = (d-2)(n+2)$. $R^\sigma$ *is* 1-*dimensional, and for* $0 \leq a \leq \sigma$ *the pairing*

$$R^a \times R^{\sigma-a} \to R^\sigma$$

*given by multiplication is perfect.*

The proof is in [7], page 659 ff. (The partials of $f$ form a regular sequence precisely when $X = \{f = 0\}$ is non-singular.)

COROLLARY 2.5: MACAULAY'S THEOREM. *The pairing*

$$R^a \times R^b \to R^{a+b}$$

*is non-degenerate in each factor if* $a + b \leq \sigma = (d-2)(n+2)$.

§3. *Symmetrizers*

In this section we describe the main constructions needed to prove the generic Torelli theorem. Given a bilinear map

$$B : U \times V \to W$$

of vector spaces, we construct the vector space

$$T = \{P : U \to V \mid \forall \ell, m \in U, B(\ell, P(m)) = B(m, P(\ell))\} \subset \mathrm{Hom}(U,V)$$

and the natural bilinear map

$$B_{-} : T \times U \to V$$

$$B_{-}(P, \ell) = P(\ell) .$$

We refer to either $T$ or $B_{-}$ as the symmetrizer of $B$. Let

$$B_{a,b} : R^a \times R^b \to R^{a+b}$$

be the multiplication map of graded pieces of $R$.

LEMMA 3.1. *Let* $f$ *be a generic polynomial of degree* $d$ *in* $n+2$ *variables. Assume that* $(d-2)(n-1) \geq 3$, $a \leq d-1$, $b \leq d$. *Then the symmetrizer of* $B_{a,b}$ *is* $B_{b-a,a}$.

*Sketch of proof.* Let $T_{a,b}$ be the symmetrizer vector space of

$$B_{a,b} : R^a \times R^b \to R^{a+b} .$$

We observe that there is a natural map

$$i : R^{b-a} \to T_{b-a}$$

sending $p \in R^{b-a}$ to the transformation $P = i(p)$,

$$P : R^a \to R^b$$

$$P(\ell) = p \cdot \ell .$$

By local duality (Theorem 2.4), $i$ is injective. Hence

$$\dim(T_{a,b}) \geq \dim(R^{b-a}) .$$

By Lemma 2.3, the dimension on the right-hand side is independent of the non-singular polynomial $f$. On the other hand, $\dim(T_{a,b})$ is a semi-continuous function on the parameter space $S^dV$ of polynomials $f$, since $T_{a,b}$ is a linear subspace of $\mathrm{Hom}(R^a, R^b)$ (which has constant dimension by Lemma 2.3) defined by a continuously varying family of linear equations. We conclude that equality

$$\dim(R^{b-a}) = \dim(T_{b-a})$$

for one non-singular polynomial $f$ implies equality for generic $f$. Verification of the lemma when $f$ is the Fermat polynomial

$$f = \sum_{i=1}^{n+2} x_i^d$$

is a rather routine exercise in counting. (Details are in [5]. The numerical hypotheses on $a$, $b$, $d$, $n$ are needed to make this counting work.)

q.e.d.

REMARK 3.2. In collaboration with M. Green, we have now developed a more sophisticated argument for proving the Symmetrizer Lemma, based on the systematic use of various Koszul-like complexes and their hypercohomology spectral sequences, in analogy with the work of Peters [10] and Lieberman, Wilsker and Peters [8] on local Torelli. One crucial advantage of the new method is that we can now remove the generic restriction on $f$ (cf. §7). We hope that these formal arguments might also extend to yield generic Torelli results in other situations, such as complete intersections or sufficiently ample divisors on arbitrary varieties.

## §4. *Recovery of a function from its Jacobian ideal*

PROPOSITION 4.1. *Let* $f \in S^dV$ *be a generic polynomial of degree* $d \geq 4$. *Then any* $g \in S^dV$ *with*

$$J(f) = J(g) \subset S$$

*is proportional to* $f$.

*Proof.* Consider the bilinear map

$$B_g : V^* \times J^{d-1}(g) \to S^{d-2}$$

given by

$$(x, h) \mapsto \frac{\partial h}{\partial x} .$$

It fits into a commutative diagram

$$\begin{array}{ccccc} B_{1,1} : V^* \times & V^* & \longrightarrow & S^2V^* \\ \| & \downarrow \alpha_g & & \downarrow \beta_g \\ B_g \quad : V^* \times & J^{d-1}(g) & \longrightarrow & S^{d-2} \end{array}$$

where $B_{1,1}$ is symmetric multiplication, $\alpha_g$ is the isomorphism

$$x \mapsto \frac{\partial g}{\partial x} ,$$

and $\beta_g$ sends a bivector to the corresponding second partial of $g$.

We claim that for generic $f$, $\beta_g$ must be injective. Indeed $B_g$ depends only on $J(g) = J(f)$ (and not on $g$ itself), hence

$$\mathrm{Im}(\beta_g) = \mathrm{Im}(B_g) = \mathrm{Im}(B_f) .$$

The injectivity of $\beta_g$ is thus equivalent to the non-degeneracy of $B_f$, which is clear, for instance, for

$$f = \sum_{i,j} x_i^2 x_j^{d-2} \qquad (d \geq 4) .$$

By Lemma 3.1, the symmetrizer of $B_{1,1}$ is $S^0V^*$, i.e., consists only of the scalar multiplications of $V^*$. By the injectivity of $\beta_g$, the same is true for $B_g = B_f$. Since both $\alpha_g$ and $\alpha_f$ are in this symmetrizer, they are proportional. By Euler's formula:

$$\sum x_i^* \cdot \alpha_g(x_i) = d \cdot g$$

we see that $f$ and $g$ are proportional. q.e.d.

REMARK 4.2. Various improvements are possible, e.g., the assumption $d \geq 4$ is not necessary, and the genericity of $f$ can be dropped if we settle for $f$, $g$ being related via a (not necessarily scalar) projective automorphism. For details cf. [5] or [3].

## §5. *Generic Torelli for hypersurfaces*

THEOREM 5.1. *The period map for* n-*dimensional hypersurfaces of degree* d *is generically injective unless* $d = 3$, $n = 2$ *or* d *divides* $2(n+2)$.

*Proof.* As explained in §1, our plan is to recover a generic hypersurface $X \subset P^{n+1}$, given by $f = 0$, from the algebraic part of the infinitesimal variation of its Hodge structure $v(X)$. By 4.1, it suffices to recover the subspace $J^{d-1}(f) \subset S^{d-1}$ (or equivalently, the ideal $J(f) \subset S$). By Macaulay's Theorem 2.5, it suffices to recover $J^d(f)$. (Take $a = d-1$, $b = 1$ in 2.5.) We shall construct the epimorphism

$$S^d \twoheadrightarrow R^d$$

whose kernel is $J^d$.

From $v(X)$ we use only the data of the variation* of the first non-zero Hodge piece. In the notation of §2, this piece is $F^a/F^{a+1}$ where $a$ is as large as possible keeping $t_a \geq 0$. (More precisely, $a = \left[n+1-\frac{n+2}{d}\right]$,

*One checks easily that $d = 3$, $n = 2$ is the only case where no variation occurs.

but we do not need this.) We have the isomorphism 2.1:

$$\lambda_a : R^{t_a} \xrightarrow{\approx} F^a/F^{a+1}$$

where $t_a$ is as small as possible, hence

$$0 \leq t_a \leq d-1 ,$$

i.e., $t_a$ is the remainder of $-(n+2)$ modulo $d$. By Theorem 2.2(3), our data amounts to the bilinear map

$$B'_{t_a,d} : R^d \times R^{t_a} \to R^{t_a+d} .$$

More precisely, we are given vector spaces $W^i (i=d,\ t_a,\ t_a+d)$ and a bilinear map

$$B_{t_a,d} : W^d \times W^{t_a} \to W^{d+t_a} .$$

We are also told that there exist isomorphisms

$$\mu_i : R^i \xrightarrow{\approx} W^i \qquad (i=d,\ t_a,\ d+t_a)$$

which make the diagram

(1) 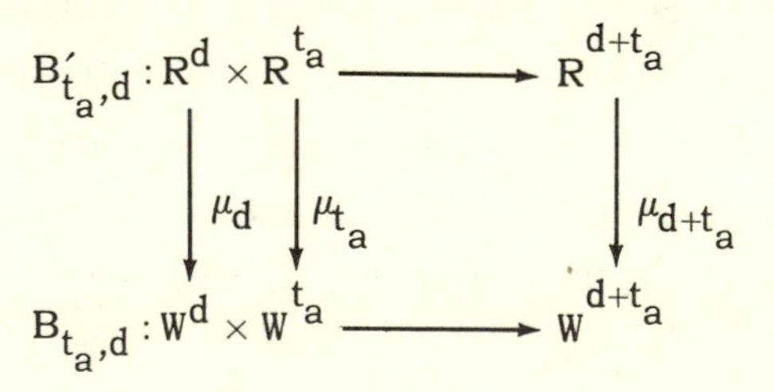

$$\begin{array}{ccc} B'_{t_a,d} : R^d \times R^{t_a} & \longrightarrow & R^{d+t_a} \\ \downarrow \mu_d \quad \downarrow \mu_{t_a} & & \downarrow \mu_{d+t_a} \\ B_{t_a,d} : W^d \times W^{t_a} & \longrightarrow & W^{d+t_a} \end{array}$$

commute, but of course we are given neither the $R^i$ nor the $\mu_i$.

We define a sequence of bilinear maps

$$B_{a_i,b_i} : W^{a_i} \times W^{b_i} \to W^{a_i+b_i}$$

inductively, starting with $a_1 = t_a$, $b_1 = d$. The pairs $(a_i, b_i)$ are obtained by following the Euclidean algorithm, beginning with the pair $(a_1, b_1)$. For each $i \geq 1$ we define $W^{b_{i-1}-a_{i-1}}$ as the symmetrizer vector space, and $B_{a_i,b_i}$ as the symmetrizer map, of the bilinear map $B_{a_{i-1},b_{i-1}}$. The sequence stops when $a_i = b_i = k$, where $k$ is the greatest common divisor of $(d, n+2)$. For each $i$ we let

$$B'_{a_i,b_i} : R^{a_i} \times R^{b_i} \to R^{a_i+b_i}$$

be the multiplication map between the corresponding graded pieces of $R$. The Symmetrizer Lemma (Lemma 3.1) asserts that for each $i$ we have a commutative diagram

$$\begin{array}{ccc} B'_{a_i,b_i} : R^{a_i} \times R^{b_i} & \longrightarrow & R^{a_i+b_i} \\ \downarrow \mu_{a_i} \quad \downarrow \mu_{b_i} & & \downarrow \mu_{a_i+b_i} \\ B_{a_i,b_i} : W^{a_i} \times W^{b_i} & \longrightarrow & W^{a_i+b_i} \end{array} \tag{2}$$

where the $\mu$'s are isomorphisms. Indeed, for $i = 1$ this is the diagram (1). For each successive $i$ we are given the maps $\mu_{a_i+b_i}$ and $\mu_{b_i}$. These define $\mu_{a_i}$ uniquely. $\mu_{a_i}$ is injective by Macaulay (2.5) and surjective by the Symmetrizer Lemma 3.1.

At the last stage our diagram becomes

$$\begin{array}{ccc} B'_{k,k} : R^k \times R^k & \longrightarrow & R^{2k} \\ \downarrow \mu_k \quad \downarrow \mu_k & & \downarrow \mu_{2k} \\ B_{k,k} : W^k \times (W^k)' & \longrightarrow & W^{2k} . \end{array} \tag{3}$$

A priori we do not have an identification of the two spaces denoted $W^k$, but this situation would be remedied in the next step, with the diagram:

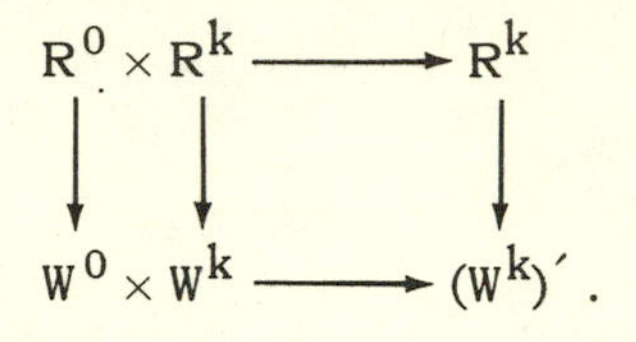

Since $R^0$ is one-dimensional, so is $W^0$, so choosing a generator of $W^0$ we get the required identification of the two $W^k$'s.

Note that by our assumptions on $d$ and $n$, we have $2k \leq d-1$. ($k$ divides $d$, and equals neither $d$ nor $d/2$.) If $2k < d-1$ we may replace the diagram (3) by the equivalent diagram

$$
(4) \qquad \begin{array}{ccc}
B''_{k,k} : S^k \times S^k & \longrightarrow & S^{2k} \\
\downarrow \nu_k \quad \downarrow \nu_k & & \downarrow \nu_{2k} \\
B_{k,k} : W^k \times W^k & \longrightarrow & W^{2k}
\end{array}
$$

where the $\nu$'s are still isomorphisms. The key observation is that $\nu_k$ can be reconstructed from $B_{k,k}$, as follows.

LEMMA 5.2. *Let* $Y \subset P((S^k)^*)$ *denote the Veronese, or image of* $P((V)^*) = P((S^1)^*)$ *under the* $k$*-uple embedding. Then the ideal of* $Y$ *in* $S^{\cdot}(S^k)$ *is generated by* $\ker(B) \subset S^2(S^k)$, *where*

$$B : S^2(S^k) \to S^{2k}$$

*is the multiplication map.*

The lemma is an easy exercise, left to the reader. (Actually what we need is a bit weaker, namely that $Y$ is defined, set theoretically, by $\ker(B)$.) In our case, the lemma says that the map $B_{k,k}$ (plus the identification of the two $W^k$'s) determines the Veronese $Y \subset P((W^k)^*)$. This determines $\nu_k$, as the inverse of the restriction map

$$W^k = H^0(P((W^k)^*),\mathcal{O}(1)) \xrightarrow{\sim} H^0(Y,\mathcal{O}(1)) \xrightarrow{\sim} H^0(P(V^*),\mathcal{O}(k)) = S^k$$

depending only on the choice of isomorphism $P(V^*) \xrightarrow{\sim} Y$, i.e., determined up to automorphism of $V$. If $2k = d-1$ then $k = 1$, so $\nu_k$ can be chosen arbitrarily. (Any two isomorphisms $S^1 \xrightarrow{\sim} W^1$ differ by an automorphism of $V$.)

By decreasing induction on $i$ starting from diagram (4) we now get a sequence of diagrams

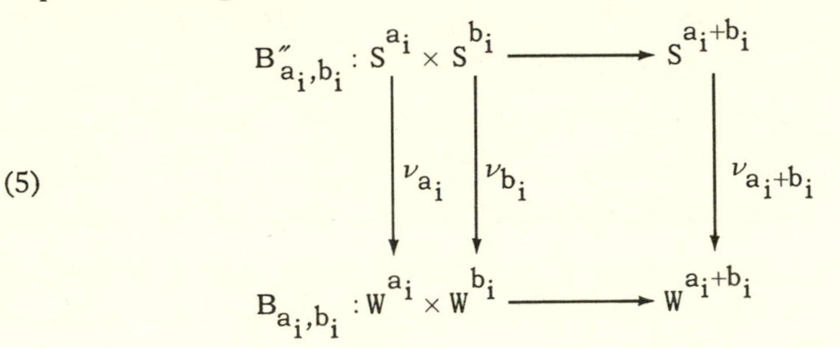

(5)

$$\begin{array}{ccc} B''_{a_i,b_i} : S^{a_i} \times S^{b_i} & \longrightarrow & S^{a_i+b_i} \\ \downarrow \nu_{a_i} \quad \downarrow \nu_{b_i} & & \downarrow \nu_{a_i+b_i} \\ B_{a_i,b_i} : W^{a_i} \times W^{b_i} & \longrightarrow & W^{a_i+b_i} \end{array}$$

where $B''_{a_i,b_i}$ is multiplication between graded pieces of the ring $S^{\cdot}$, and where each $\nu_{a_i+b_i}$ is induced from $\nu_{a_i}, \nu_{b_i}$ and is surjective. (The existence of the $\mu_{a_i}$ etc. in (2) and of the natural quotient maps $S^{a_i} \to R^{a_i}$ etc. guarantees that $\nu_{a_i+b_i}$ is well-defined.)

Going all the way to $i = 1$, we have the map

$$\nu_d : S^d \twoheadrightarrow R^d$$

which was constructed purely in terms of the infinitesimal variation of the Hodge structure of the original hypersurface $X$. Its kernel is $J^d \subset S^d$, so we are done. q.e.d.

REMARK 5.3. Using Remark 3.2 instead of the Symmetrizer Lemma 3.1 and Remark 4.2 instead of Proposition 4.1 in the above argument one can prove the Variational Torelli Theorem (1.1) for hypersurfaces.

§6. *Generic Torelli for double spaces*

Let $X$ be a double cover of $\mathbf{P}^n$, given by the equation $y^2 = f(\underline{x})$ where $f$ is a polynomial of degree $d = 2m$ in the variables $\underline{x} = (x_1, \cdots, x_{n+1})$. $X$ is non-singular if and only if the branch locus $\{f=0\}$ in $\mathbf{P}^n$ is non-singular. In this section we show that the technique of §5 extends almost verbatim to give a generic Torelli result for such varieties.

Double solids are a special case of k-sheeted cyclic branched covers of $\mathbf{P}^n$, given by $y^k = f(\underline{x})$ with $f$ of degree $d = k \cdot m$. These in turn are examples of hypersurfaces in weighted projective spaces, a notion studied in [9], [11] and [4]. We expect our method to generalize and apply to such weighted hypersurfaces as well.

To convert the infinitesimal variation of Hodge structure $v(X)$ into ring data, we need the analogue of Griffiths' results discussed in §2. These were worked out by Steenbrink [11] for "quasismooth" weighted hypersurfaces. Here we specialize his results to the case of double spaces. For $X$ defined by $y^2 = f(\underline{x})$ we let $R$ be the Jacobian ring of the homogeneous polynomial $f$. Let

$$(0) \subset F^n \subset \cdots \subset F^0 = H^n(X)$$

be the Hodge filtration of $H^n(X)$.

THEOREM 6.1 ([11]). *For a non-singular double space $X$ there are natural isomorphisms*

$$\lambda_{s_a} : R^{s_a} \xrightarrow{\sim} F^a/F^{a+1} ,$$

*where* $s_a = (n-a+1)d - (n+m+1) = (2n-2a+1)m - (n+1)$, *and*

$$\lambda_d : R^d \xrightarrow{\sim} T$$

*where* $T \approx H^1(X, \theta_X)$ *is the tangent space to the deformation space of* $X$. *In terms of these identifications the variation maps*

$$v_i : T \times F^a/F^{a+1} \to F^{a-1}/F^a$$

*become the multiplication maps of graded pieces of* R,

$$R^d \times R^{s_a} \to R^{d+s_a} .$$

(For a general weighted hypersurface we must replace R by the appropriate weighted Jacobian ring, and extend the Symmetrizer Lemma 3.1 to that case. For our equation $y^2 = f(\underset{\sim}{x})$, this ring is

$$C[y,x_1,\cdots,x_{n+1}]/\left(y, \frac{\partial f}{\partial x_1}, \cdots, \frac{\partial f}{\partial x_{n+1}}\right) \approx C[x_1,\cdots,x_{n+1}]/J(f)$$

so we get precisely the Jacobian ring of f.)

THEOREM 6.2: GENERIC TORELLI FOR DOUBLE SPACES. *The period map for double covers of* $P^n$ *branched along a non-singular hypersurface of degree* 2m *is generically injective unless* m *divides* n + 1.

The proof is identical to that of Theorem 5.1.

## §7. *Variational Schottky*

In the case of hypersurfaces, the variational Schottky problem 1.2 is: determine which sequences of bilinear maps

$$B_{t_a,d} : R^d \times R^{t_a} \to R^{t_{a-1}}$$

come from non-singular projective hypersurfaces. For simplicity we restrict our discussion here to the first map, where $a = \left[n+1 - \frac{n+2}{d}\right]$. (This map determines all the others.)

The Symmetrizer Lemma 3.1 imposes a set of conditions on $B_{t_a,d}$, namely that the dimension of each of its iterated symmetrizers should equal a specified integer, dimension of some graded piece of R. When these conditions are written down explicitly, they give various equations and inequalities (expressing the vanishing of all minors of a certain size and non-vanishing of some larger minor).

Assuming that all these conditions are satisfied, we can construct inductively the bilinear maps

$$B_i : R^{a_i} \times R^{b_i} \to R^{a_i+b_i}$$

as in the proof of Theorem 5.1. The dependence of these maps on $B_{t_a,d}$ is rational, and could be made explicit if we wished to do so. Following the proof of Theorem 5.1, we see that the polynomial structure

$$S^* \twoheadrightarrow R^* \qquad (* \geq d-1)$$

and hence the Jacobian ideal $J^* \subset S^*$ is determined explicitly by $B_{t_a,d}$, at least under the extra assumption that the greatest common divisor

$$k = g\cdot c\cdot d\cdot (d, n+2)$$

equals $1$. In this case we can impose one extra condition on $B_{t_a,d}$, namely that the bilinear map

$$B_g : V^* \times J^{d-1} \to S^{d-2}$$

defined in §4 (and depending, in our case, rationally on $B_{t_a,d}$) should have a non-zero symmetrizer. The last condition assures the existence of $f \in S^d$ such that

$$J^{d-1}(f) = J^{d-1} ,$$

so by the variational Torelli theorem for hypersurfaces (5.3) the original variational Hodge structure is $v_a(f)$. Thus, the following impressive-sounding theorem is really a trivial corollary of our proof of variational Torelli:

THEOREM 7.1: VARIATIONAL SCHOTTKY FOR HYPERSURFACES. *If* $(n+2, d) = 1$ *then an abstract (algebraic piece of an) infinitesimal variation of Hodge structure* $v_a$ *comes from a non-singular hypersurface of degree* $d$

*if and only if the symmetrizers of all the* $B_i$ *have the expected dimensions (each* $B_i$ *can be constructed once we know that* $\mathrm{Symm}(B_{i-1})$ *has the right dimension) and* $\mathrm{Symm}(B_g) \neq (0)$.

We conclude by pointing out that the variational Schottky problem for curves which are neither hyperelliptic, nor trigonal, nor plane quintics is equivalent to the following question:

7.2. Characterize the ideals $I(C)$ of canonical curves. Equivalently, characterize those bilinear maps

$$B : S^2V \to W, \quad \dim(V) = g, \quad \dim(W) = 3g-3$$

which represent the restriction

$$S^2H^0(C, \omega_C) \twoheadrightarrow H^0(C, \omega_C^2)$$

on a curve of genus $g$.

All we know in this case are the dimensions of the graded pieces $I^k(C)$ ( $I$ is the ideal in $S^{\cdot}V$ generated by $\ker(B)$ ) and the fact that the symmetrizer of each map

$$I^k \times I^\ell \to I^{k+\ell}$$

is $I^{\ell-k}$. This does not seem to be sufficient to force $I$ to come from a curve.

One may think of the theta function of a curve as a way of converting its Hodge structure to an algebraic variation of Hodge structure. Indeed, by the work of Andreotti and Mayer [1], the theta function determines the quadrics of rank 4 in $\ker(B)$, and for a generic curve these actually span, thus determine, $\ker(B)$. Therefore any identities which must be satisfied by $\ker(B)$, or equivalently by the ideal $I$, must imply similar identities satisfied by the theta functions of Jacobians. To make this correspondence more precise, one would like to know the following.

CONJECTURE 7.3. The linear system of quadrics through *any* canonical curve is spanned by quadrics of rank 4.

This conjecture is discussed in [2] where it is proved for $g \leq 6$. It is also known for various special families of curves in all genera, but the general question appears to be difficult.

*Added in proof*: Conjecture 7.3 has recently been proved by Mark Green.

## REFERENCES

[1] A. Andreotti and A. Mayer, On period relations for Abelian integrals on algebraic curves, Ann. Scu. Norm. Sup. di Pisa 21 (1967), pp. 189-238.

[2] E. Arbarello and J. Harris, Canonical curves and quadrics of rank four, Comp. Math. 43 (1981), pp. 145-179.

[3] J. Carlson, M. Green, P. Griffiths and J. Harris, Infinitesimal variation of Hodge structures, to appear in Comp. Math., 1983.

[4] I. Dolgachev, Weighted projective varieties, preprint.

[5] R. Donagi, Generic Torelli for projective hypersurfaces, to appear in Comp. Math.

[6] P. Griffiths, On the periods of certain rational integrals, Ann. of Math. 90 (1969), pp. 460-541.

[7] P. Griffiths and J. Harris, Principles of algebraic geometry, John Wiley and Sons, New York (1978).

[8] D. Lieberman, R. Wilsker and C. Peters, A theorem of local-Torelli type, Math. Ann. 231 (1977), pp. 39-45.

[9] S. Mori, On a generalization of complete intersections, J. Math. Kyoto Univ. 15 (1975), pp. 619-646.

[10] C. Peters, The local Torelli theorem: I. Complete intersections, Math. Ann. 217 (1975), pp. 1-16.

[11] J. Steenbrink, Intersection forms for quasihomogeneous singularities, Comp. Math. 34 (1977), pp. 211-223.

RON DONAGI
MATHEMATICS DEPARTMENT
NORTHEASTERN UNIVERSITY
BOSTON, MA 02115

Chapter XIV

# INTERMEDIATE JACOBIANS AND NORMAL FUNCTIONS

Steven Zucker

The meaning of the term "normal function" has changed slightly since it first (?) appeared in [10]. We begin by discussing the classical situation in fairly classical terms.

Let $X$ be a surface in $\mathbf{P}^3$. One takes a sufficiently general pencil of planes, which cuts out the pencil of curves $\{X_t\}$ on $X$. One chooses affine coordinates $(x,y,z)$ on $\mathbf{P}^3$ so that $y = t$ defines the planes. Let $f(x,y,z) = 0$ be the equation of $X$. Let $\Sigma = \{t : X_t \text{ is singular}\}$.

As $P$ ranges over a suitable space of polynomials,

$$\omega(t) = P(x,y,z)\,dx/(\partial f/\partial z) \tag{1}$$

gives the space of holomorphic 1-forms (differentials of the first kind) on the non-singular $X_t$. Let $C$ be a curve on $X$ in general position, and let $d$ be its degree. Then we write

$$C \cap X_t = \{p_1(t), \cdots, p_d(t)\}\,, \tag{2}$$

where each $p_j(t)$ has coordinates that are multi-valued analytic functions of $t$. Let $p_0$ be one of the base points of the pencil $\{X_t\}$, and consider the Abelian sums

$$\sigma(t) = \sum_{j=1}^{d} \int_{p_0}^{p_j(t)} \omega(t)\,. \tag{3}$$

One can select paths of integration so that $\sigma(t)$ is a multi-valued analytic function of $t$, with singularities on $\Sigma$. We observe that (3) depends linearly on $\omega$.

From a more modern point of view, we can reformulate the above as follows. One blows up the base points of the pencil to obtain the fibered variety

$$h : \tilde{X} \to \mathbf{P}^1 , \tag{4}$$

with $h^{-1}(t) \simeq X_t$. Then the Poincaré residues of the rational differentials

$$P(x,y,z)\, dx \wedge dz / f(x,y,z) \tag{5}$$

determine the 1-forms (1) (cf. [5, §8]), and a basis of the polynomials defines a frame for some extension of the Hodge bundle $\mathcal{F}^1$ associated to $R^1 h_* \mathbf{C}$. An invariant formulation of (3) is that $\sigma$ defines a section $\nu$ of the family of Jacobians over $\mathbf{P}^1 - \Sigma$, associated to (the smooth fibers of) $h$. We say that $\nu$ is the *normal function* over $\mathbf{P}^1 - \Sigma$ associated to the algebraic cycle $C - dE$ on $\tilde{X}$ —where $E \simeq \mathbf{P}^1$ is the exceptional curve over $p_0 \in X$ —whose intersections with the fibers of $h$ are homologous to zero.

I had always wondered what the "normal" in normal function signified. If one goes back to [10, p. 75], one sees that Poincaré intended it to describe the mild nature of the singularities of (3) on $\Sigma$. The most serious property is that for any $t \in \Sigma$, one can subtract a period from $\sigma(t)$ so that the resulting function is regular at $t$. If $C$ does not pass through the singular points of the singular $X_t$ (as one assumes), then since $h$ is a fiber bundle "away from the singular points," one can choose the paths of integration in (3) to be continuously varying near $t$. In other words, that the $\sigma$'s be "normal" is tantamount to saying that $\nu$ have limiting values in the generalized Jacobians of the singular fibers! This idea plays an important role in the general theory to come.

Thus, a curve $C$ on $X$ gives a normal function $\nu$, defined on all of $\mathbf{P}^1$. Conversely, given any normal function $\nu$, one can construct a curve $C$ on $X$ such that $\nu$ is associated to it, by means of Jacobi inversion (for which $d = g(X_t)$ ) [10, IV]. It was later observed by Lefschetz that one can also read off the homology class of $C$ from $\nu$. This led to the first proof of the Hodge conjecture for (1,1) classes on a surface [8, p. 72]. It was Griffiths' idea to try to generalize these techniques so that they might apply in higher dimensions.

Let $(H_{\mathbf{Z}}, \{F^p\})$ be a polarized Hodge structure of odd weight $2k-1$. Then $H_{\mathbf{Z}}$ defines a lattice in $(F^k)^\vee$, so

$$H_{\mathbf{Z}} \backslash (F^k)^\vee \simeq H_{\mathbf{Z}} \backslash H/F^k \tag{6}$$

is naturally a complex torus. In case

$$H_{\mathbf{Z}} = H^{2k-1}(X,\mathbf{Z}) \qquad \text{(modulo torsion)}$$

for some non-singular projective variety $X$, the torus (6) is called the k-*th intermediate Jacobian* of $X$, denoted $J^k(X)$. In this case, we can use Poincaré duality to write

$$J^k(X) \simeq H_{2n-2k+1}(X,\mathbf{Z}) \backslash (F^{n-k+1}H^{2n-2k+1}(X,\mathbf{C}))^\vee, \tag{7}$$

where $n = \dim X$. When $k = n = 1$, the preceding is just the usual Jacobian of $X$.

Let $\mathcal{Z}_h^k(X)$ denote the group of codimension $k$ algebraic cycles on $X$ that are homologically equivalent to zero. One defines the *Abel-Jacobi homomorphism*

$$u : \mathcal{Z}_h^k(X) \to J^k(X) \tag{8}$$

as follows (see [6, p. 166]). Given $Z \in \mathcal{Z}_h^k(X)$, choose a geometric $(2n-2k+1)$-chain $\Gamma$ on $X$ such that $\partial\Gamma = Z$. Then the formula

$$\omega \mapsto \int_\Gamma \omega, \qquad \omega \in F^{n-k+1}H^{2n-2k+1}(X)$$

determines, according to (7), a unique point of $J^k(X)$, which defines $u(Z)$. When $k = n$, (8) is the familiar Albanese mapping.

Let now $f: X \to S$ be a smooth projective morphism, and put $X_s = f^{-1}(s)$ for $s \in S$. Since the Hodge filtration varies holomorphically with $s$, so does $J^k(X_s)$. More precisely, one gets a holomorphic fiber space of tori

$$\pi : J^k(X/S) \to S , \tag{9}$$

such that $\pi^{-1}(s) \simeq J^k(X_s)$. Letting $\mathcal{F}^k$ denote the Hodge filtration bundle (locally free sheaf), and $\mathcal{J}^k$ the sheaf of holomorphic sections of (9), we may write the basic exact sequence

$$0 \to R^{2k-1}f_*Z \text{ (mod torsion)} \to (\mathcal{F}^k)^\vee \to \mathcal{J}^k \to 0 . \tag{10}$$

We will continue to abuse Poincaré and call $H^0(S, \mathcal{J}^k)$ the group of *normal functions* over $S$.

REMARK. The construction of an intermediate Jacobian was first done by Weil [12, p. 82]. In contrast to (6), he took the real torus $H_Z \backslash H_R$, and used the negative of the (real) now-called Weil operator, which is the direct sum of the multiplications by $i^{p-q}$ on $H^{p,q}$, to define a complex structure on it. Equivalently, one is replacing $F^0/F^k$ by $\bigoplus_{q-p\equiv 1 \pmod 4} H^{p,q}$ in the definition (6). The two constructions coincide for $k = 1$. One can define an Abel-Jacobi homomorphism for these tori as well [9, p. 1179]. Weil's intermediate Jacobians are naturally polarized, hence are abelian varieties (whereas Griffiths' are not in general), though they do not depend holomorphically on parameters.

The connecting homomorphism $\delta$ in the long exact sequence of cohomology associated to (10) associates to each $\nu \in H^0(S, \mathcal{J}^k)$ the element $\delta(\nu) \in H^1(S, R^{2k-1}f_*Z)$, which we call the *cohomology class* of the normal function $\nu$; it is, by definition, the obstruction to lifting $\nu$ to a (single-valued) section of the bundle $(\mathcal{F}^k)^\vee$ of Lie algebras. We remark in passing that (10) makes sense for an abstract variation of Hodge structure (defined over $Z$).

The Abel-Jacobi homomorphism (8) varies holomorphically in the following sense. Let $\mathcal{Z}_h^k(X/S)$ denote the group of algebraic cycles $Z$ on $X$ such that $Z \cdot X_s \in \mathcal{Z}_h^k(X_s)$ for every $s \in S$. Then the values $u(Z \cdot X_s) \in J^k(X_s)$ determine a normal function

$$\nu_Z \in H^0(S, \mathcal{J}^k) .$$

The cohomology class $\delta(\nu_Z)$ and the topological cycle class $(Z) \in H^{2k}(X, Z)$ are compatible [13, (3.9)]. To explain the meaning of this statement, let $\{L^p\}$ denote the Leray filtration on $H^*(X)$ associated to $f$. One has a natural homomorphism

$$\text{(11)} \quad L^1H^{2k}(X,Q) \simeq \ker\{H^{2k}(X,Q) \longrightarrow H^0(S,R^{2k}f_*Q)\} \xrightarrow{\rho} H^1(S,R^{2k-1}f_*Q) .$$

(The right-hand side is isomorphic to $L^1/L^2$, by degeneration of the Leray spectral sequence.) By assumption, $(Z) \in L^1$. Then

$$\rho((Z)) = \delta(\nu_Z) .$$

There is a differential equation satisfied by the normal functions associated to algebraic cycles. The connection on the cohomology bundle $\mathcal{H}$ induces

$$\nabla : (\mathcal{F}^k)^\vee \simeq (\mathcal{H}/\mathcal{F}^k) \to \Omega_S^1 \otimes (\mathcal{H}/\mathcal{F}^{k-1}) .$$

Since $R^{2k-1}f_*C$ gives the flat sections of $\mathcal{H}$, one can define

$$\text{(12)} \qquad \nabla : \mathcal{J}^k \to \Omega_S^1 \otimes (\mathcal{H}/\mathcal{F}^{k-1}) .$$

We denote by $\mathcal{J}_h^k$ the kernel of $\nabla$ in (12), the sheaf of germs of *horizontal* normal functions. If $Z \in \mathcal{Z}_h^k(X/S)$, then $\nu_Z$ is horizontal [6, (A.8)].

For the remainder of this chapter, we take $S$ to be a non-singular algebraic curve. We let $\overline{S}$ denote its smooth completion, $j : S \to \overline{S}$ the inclusion, $\Sigma = \overline{S} - S$; and let $\overline{f} : \overline{X} \to \overline{S}$ be a completion of $f$. There is a natural extension of (10) to all of $\overline{S}$ :

(13) $$0 \to j_* R^{2k-1} f_* \mathbf{Z} \to \mathcal{H}_e / \mathcal{F}_e^k \to \bar{\mathcal{J}}^k \to 0 ,$$

where $\mathcal{H}_e$ and $\mathcal{F}_e$ are as in Chapter VII, §2. One can associate a space to the sheaf $\bar{\mathcal{J}}^k$ [13, p. 191]:

(14) $$\bar{\pi} : \bar{J}^k(\bar{X}/\bar{S}) \to \bar{S} ,$$

which is evidently independent of the choice of $\bar{X}$ and which agrees with (9) over $S$. The fiber of (13) over $s \in \Sigma$ is a generalized torus, isomorphic to $\Lambda \backslash \mathbf{C}^g$, where $g = \mathrm{rk}(\mathcal{F}^k)$ and $\Lambda$ is some (partial) lattice of rank at most $2g$.

REMARK. It has *not* been proved in general that $\bar{J}^k$ is a separated space. To prove that, it is equivalent to show the following. Let $\Delta^*$ be a small punctured disc about a point of $\Sigma$. Let $M(s)$ be the (multi-valued) period matrix for $X_s$ $(s \in \Delta^*)$ with respect to a local frame for $\mathcal{H}_e/\mathcal{F}_e^k$. Does there exist $\varepsilon > 0$ such that

(15) $$|M(s)v| > \varepsilon \quad \text{for all} \quad 0 \neq v \in \mathbf{Z}^{2g}, \quad s \in \Delta^* ?$$

This has been checked if the monodromy $T$ satisfies $(T-1)^2 = 0$ [13, (2.9)], and for an $SL_2$-orbit by Walter Parry. Unfortunately, the inequality (15) is unstable under small perturbation, so the general case does not seem to follow from a coarse reading of [11].

The differential operator $\nabla$ of (12) extends to give

(16) $$\bar{\nabla} : \bar{\mathcal{J}}^k \to \Omega^1_{\bar{S}}(\log \Sigma) \otimes (\mathcal{H}_e/\mathcal{F}_e^{k-1}) .$$

As before, we let $\bar{\mathcal{J}}^k_h$ denote the kernel of $\bar{\nabla}$ in (16), the sheaf of horizontal normal functions over $\bar{S}$.

Note that for $Z \in \mathcal{Z}^k_h(X/S)$, one has a priori only that $\nu_Z$ is given as element of $H^0(S, \mathcal{J}^k_h)$. It has been recently proved [4] that $\nu_Z$ extends to all of $\bar{S}$, i.e., is the restriction of an element (unique, of course) of $H^0(\bar{S}, \bar{\mathcal{J}}^k_h)$, if $Z$ satisfies the following condition: over a punctured disc

$\Delta^*$ about each point of $\Sigma$, the cohomology class of $Z$ restricts to zero. Also, Clemens has shown that under certain hypotheses the Abel-Jacobi mapping for families of cycles extends to a mapping into a "Néron model" [1]. It is a local question at each $s \in \Sigma$. His assumptions are that the local monodromy $T$ satisfy $(T-1)^2 = 0$, and that the image of $N = (T-1)$ be of pure type $(k-1, k-1)$ in the limit mixed Hodge structure; these are satisfied, for example, by a generic pencil of hyperplane sections of a smooth variety. The fiber over $s$ in the Néron model is an Abelian Lie group whose identity component is $\overline{\pi}^{-1}(s)$ (14).

We remark that elements of $H^0(\overline{S}, \overline{\mathcal{J}}^k)$ are the true normal functions, in the sense of Poincaré. As before, though restricting to the subgroup of horizontal normal functions, we have a homomorphism

$$\overline{\delta} : H^0(\overline{S}, \overline{\mathcal{J}}^k_h) \to H^1(\overline{S}, j_* R^{2k-1} f_* Z) \tag{17}$$

(if we pass to $Q$ coefficients, the right-hand side is isomorphic to $H^1(\overline{S}, R^{2k-1}\overline{f}_* Q)$ ), compatible with cycle classes. The main result on normal functions is:

THEOREM ([15, §9]). *The image of* $\overline{\delta}$ *is exactly the set of classes of type* $(k,k)$.

The above statement depends on the assertion that $H^1(\overline{S}, j_* R^{2k-1} f_* Z)$ underlies a Hodge structure of weight $2k$, induced by that of $H^{2k}(\overline{X}, Z)$ through the Leray filtration. The main work in [15] was the construction of the Hodge structure by means of a cohomological Hodge complex on $\overline{S}$, after which the proof of the above theorem became a triviality. For earlier versions of the theorem, see [7] and [13].

We recall the relation between the above theorem and the Hodge conjecture. There is a commutative diagram

$$\begin{array}{ccc} \mathcal{Z}^k(\overline{X}) \supset \mathcal{Z}^k(X/S) & \overset{u}{\dashrightarrow} & H^0(\overline{S}, \overline{\mathcal{J}}^k_h) \\ \downarrow c & & \downarrow \overline{\delta} \\ H^{2k}(\overline{X}, Z) & \overset{\rho}{\dashrightarrow} & H^1(\overline{S}, j_* R^{2k-1} f_* Z) \end{array}$$

where $\mathcal{Z}^k(\overline{X})$ denotes the group of all algebraic cycles of codimension k on $\overline{X}$, and the dotted arrows are defined only on appropriate subgroups. The Hodge conjecture asserts that, after passing to $\mathbf{Q}$ coefficients, c maps onto the classes of types (k,k). Now, the theorem on normal functions asserts the corresponding thing for $\overline{\delta}$. Arguing inductively (to dispose of the "rest" of the Leray filtration), one can see that it would suffice to know that u is surjective. Although this is true for $k = 1$, it is unfortunately the case that the image of the Abel-Jacobi homomorphism (8) is in general, very small [5, §13], because of conditions imposed by horizontality. Thus, Jacobi inversion tends to be false for the higher intermediate Jacobians. This is the chief drawback to the use of normal functions in studying the Hodge Conjecture.

There are some special cases in which the Abel-Jacobi homomorphism is surjective, for example cubic hypersurfaces in $\mathbf{P}^4$ [2]. One can use this to obtain the Hodge conjecture for the cubic fourfold, by applying the theorem on normal functions to a pencil of hyperplane sections [14, §3]. One also sees here a second drawback to the method of normal functions: that more direct arguments happen to be available (e.g., [14, App. A]). Likewise, the Hodge conjecture for (1,1) classes and divisors (the only general case known) is now proved by a well-known sheaf-theoretic argument. One gets the impression that whenever there is enough geometric information to prove Jacobi inversion, then one should be able to find a more direct route to the Hodge conjecture. The last statement is not, however, a theorem.

In closing, I wish to point out that when $\overline{f}: \overline{X} \to \overline{S}$ is an elliptic surface (and $k=1$), a normal function is essentially the same thing as a section of the elliptic surface, so it has direct geometric meaning. For applications of the kind of techniques discussed in this article to this very concrete case, see [3].

## REFERENCES

[1] C.H. Clemens, "On extending Abel-Jacobi mappings," 1981.

[2] C.H. Clemens and P. Griffiths, "The intermediate Jacobian of the cubic threefold," Annals of Math. 95 (1972), pp. 281-356.

[3] D. Cox and S. Zucker, "Intersection numbers of sections of elliptic surfaces," Inv. math. 53 (1979), pp. 1-44.

[4] F. El Zein and S. Zucker, "Extendability of normal functions associated to algebraic cycles," this volume, Chapter XV, pp. 269-288.

[5] P. Griffiths, "On the periods of certain rational integrals," Annals of Math. 90 (1969), pp. 460-541.

[6] ———, "Periods of integrals on algebraic manifolds, III," Publ. Math. IHES 38 (1970), pp. 125-180.

[7] ———, "A theorem concerning the differential equations satisfied by normal functions associated to algebraic cycles," Amer. J. Math 101 (1979), pp. 94-131.

[8] S. Lefschetz, "L'Analysis situs et la géometrie algébrique," Gauthier-Villars, Paris, 1924.

[9] D. Lieberman, "Higher Picard varieties," Amer. J. Math. 90 (1968), pp. 1165-1199.

[10] H. Poincaré, "Sur les courbes tracées sur les surfaces algébriques," Ann. Sci. ENS (série 3) 27 (1910), pp. 55-108.

[11] W. Schmid, "Variation of Hodge structure: the singularities of the period mapping," Inv. Math. 22 (1973), pp. 211-319.

[12] A. Weil, "Variétés Kählériennes," Hermann, Paris, 1971.

[13] S. Zucker, "Generalized intermediate Jacobians and the theorem on normal functions," Inv. math. 33 (1976), pp. 185-222.

[14] ———, "The Hodge conjecture for cubic fourfolds," Compositio Math. 34 (1977), pp. 199-209.

[15] ———, "Hodge theory with degenerating coefficients: $L_2$ cohomology in the Poincaré metric," Ann. of Math. 109 (1979), pp. 415-476.

STEVEN ZUCKER
MATHEMATICS DEPARTMENT
JOHNS HOPKINS UNIVERSITY
BALTIMORE, MD 21218

Chapter XV

# EXTENDABILITY OF NORMAL FUNCTIONS ASSOCIATED TO ALGEBRAIC CYCLES

Fouad El Zein* and Steven Zucker**

Let $f: X \to \Delta$ be a projective (or Kähler) holomorphic mapping from the complex manifold $X$ to the disc, smooth over the punctured disc $\Delta^*$. For any analytic cycle $Z$ on $X$ that is homologous to zero on the smooth fibers of $f$, there is a normal function $\nu_Z$ defined over $\Delta^*$ via the Abel-Jacobi homomorphism (see [7, §16]). There remains the question of deciding whether $\nu_Z$ extends to the generalized intermediate Jacobian [8, §2], [16, §2], [17, §9] over the origin. (See also Chapter XIV.)

We give here a natural condition on the cycle $Z$ that is sufficient to imply that $\nu_Z$ extends. Specifically, we prove:

THEOREM. *Let* $Z$ *be of codimension* $k$ *in* $X$. *If the cycle class of* $Z$ *restricts to zero in* $H^{2k}(f^{-1}(\Delta^*), \mathbb{Z})$, *then* $\nu_Z$ *extends across the origin.*

Our condition on $Z$ seems very reasonable, since the generalized intermediate Jacobian is also constructed from data in the deleted neighborhood. In particular, the result is independent of the singular fiber $f^{-1}(0)$, so it is useful to assume, after Hironaka, that the latter is a divisor with normal crossings.

The main technique in the proof is the use of Deligne's algebraic construction of the Abel-Jacobi homomorphism, which suppresses the integration. In fact, he gave a construction of a cycle class for all algebraic

*Supported by CNRS (ERA589) and NSF.

**Supported in part by NSF grant MCS 81-08814.

cycles on a projective manifold, coinciding with the Abel-Jacobi homomorphism for cycles homologically equivalent to zero. We learned of this from [2]. Since there is no published account of it in the literature, we present one here in §2, with the permission of all involved.

We point out that J. King has independently obtained results on the extendability of normal functions, and that H. Clemens has done related work [3].

We wish to thank S. Bloch, P. Deligne, and P. Griffiths for helpful conversations.

§1. *Preliminaries*

1.1. Let $K^{\cdot}$ be a complex of sheaves, with differential $d$, on the space $X$. One lets $\{F^pK^{\cdot}\}$ denote the "*filtration bête*"

$$F^pK^q = \begin{cases} K^q & \text{if} \quad q \geq p, \\ 0 & \text{otherwise.} \end{cases} \tag{1}$$

For the corresponding quotient complexes, we set

$$\sigma_p K^{\cdot} = K^{\cdot}/F^pK^{\cdot} . \tag{2}$$

Given a morphism of complexes

$$\Phi : K^{\cdot} \to L^{\cdot} , \tag{3}$$

one defines (see [6] or [15]) the *mapping cone* of $\Phi$, denoted $C^{\cdot}(\Phi)$, to be the complex with terms

$$C^q(\Phi) = K^{q+1} \oplus L^q , \tag{4}$$

and differential

$$d(k,\ell) = (-dk, d\ell + \Phi(k)) . \tag{5}$$

There is the basic exact sequence

(6) $$0 \to L^{\cdot} \to C^{\cdot}(\Phi) \to K^{\cdot}[1] \to 0 ,$$

such that the connecting homomorphism in hypercohomology:

(7) $$H^i(X,K^{\cdot}) \simeq H^{i-1}(X,K^{\cdot}[1]) \to H^i(X,L^{\cdot})$$

is the usual mapping induced by $\Phi$.

Suppose that one is given, more generally, a system of morphisms of complexes

(8) $$\begin{array}{ccc} K^{\cdot} & \xrightarrow{\kappa} & L^{\cdot} & & \\ & & \uparrow \lambda & & \\ & & M^{\cdot} & \xrightarrow{\mu} & N^{\cdot} \end{array} ,$$

in which $\lambda$ is a quasi-isomorphism. (This defines a morphism $\Phi : K^{\cdot} \to N^{\cdot}$ in the derived category.) We may use (8) to construct a mapping cone $C^{\cdot}(\Phi)$ as follows. One defines

$$\Psi : (K^{\cdot} \oplus M^{\cdot}) \to (L^{\cdot} \oplus N^{\cdot})$$

by

(9) $$\Psi(k,m) = (\kappa(k) + \lambda(m), \mu(m)) .$$

In analogy with (6), $C^{\cdot}(\Psi)$ contains $N^{\cdot}$ as a subcomplex, and the quotient is quasi-isomorphic to $K^{\cdot}[1]$. One therefore puts

$$C^{\cdot}(\Phi) = C^{\cdot}(\Psi) .$$

1.2. Let $X$ be a projective manifold over $\mathbb{C}$. The k-th intermediate Jacobian

(10) $$J^k(X) = H^{2k-1}(X,\mathbb{Z})\backslash H^{2k-1}(X,\mathbb{C})/F^kH^{2k-1}(X,\mathbb{C})$$

can be expressed as

(11) $$J^k(X) = H^{2k-1}(X,\mathbb{Z})\backslash H^{2k-1}(X,\sigma_k\Omega^{\cdot}_X)\,.$$

This isomorphism is deduced from the short exact sequence

$$0 \to F^kH^{2k-1}(X,\mathbb{C}) \to H^{2k-1}(X,\mathbb{C}) \to H^{2k-1}(X,\sigma_k\Omega^{\cdot}_X) \to 0\,,$$

which follows from the degeneration of the spectral sequence of $F$ on $\Omega^{\cdot}_X$ at the $E_1$ term.

1.3. Let $Z$ be an element of $\mathcal{Z}^k(X)$, the group of algebraic cycles of codimension $k$ in $X$. The topological homology class of $Z$

(12) $$c(Z) \in H_{2(n-k)}(|Z|,\mathbb{Z})\,,$$

where $n$ is the dimension of $X$, is defined classically by means of a triangulation of $|Z|$. By duality, it induces a class, also denoted $c(Z)$, in $H^{2k}_{|Z|}(X,\mathbb{Z})$. There are many ways of constructing the latter; we refer the reader to [1], [10], and [14].

One has also a cycle class for $Z$ in de Rham cohomology:

(13) $$c_{DR}(Z) \in H^{2k}_{|Z|}(X,F^k\Omega^{\cdot}_X)\,,$$

constructed in [5].

1.4. We recall the Abel-Jacobi homomorphism

(14) $$u : \mathcal{Z}^k_h(X) \to J^k(X)\,,$$

where $\mathcal{Z}^k_h(X)$ is the subgroup of $\mathcal{Z}^k(X)$ consisting of those cycles $Z$ for which $c(Z) = 0$.

REMARK. If $Z \in \mathcal{Z}^k_h(X)$, then $u(Z)$ can be realized in the following manner. Write $Z = \partial\Gamma$. We have an exact sequence of mixed Hodge structures:

(15) $$0 = H^{2k-1}_{|Z|}(X,\mathbb{Z}) \to H^{2k-1}(X,\mathbb{Z}) \to H^{2k-1}(X-|Z|,\mathbb{Z}) \to H^{2k}_{|Z|}(X,\mathbb{Z}) \to H^{2k}(X,\mathbb{Z})\,.$$

By assumption, the cycle class $c(Z)$ in $H^{2k}_{|Z|}(X,Z)$ goes to zero in $H^{2k}(X,Z)$; and $\Gamma$ has a cycle class in $H^{2k-1}(X-|Z|,Z)$. Since $H^{2k}_{|Z|}(X,Z)$ is generated by the cycle classes of the components of $Z$, it is pure of type $(k,k)$. We then obtain from (15)

$$H^{2k-1}(X,C)/F^kH^{2k-1}(X,C) \xrightarrow{\sim} H^{2k-1}(X-|Z|,C)/F^kH^{2k-1}(X-|Z|,C)\ ,$$

by strictness. Thus, we can identify in (10)

$$(16) \qquad J^k(X) \simeq H^{2k-1}(X,Z)\backslash H^{2k-1}(X-|Z|,C)/F^kH^{2k-1}(X-|Z|,C)\ .$$

Then $u(Z)$ is just the image of the cycle class of $\Gamma$.

§2. *Deligne groups*

We give an exposition of unpublished ideas of Deligne, after the appendix of [2].

2.1. Let $X$ be a smooth projective variety of dimension $n$.

DEFINITION 1. For any positive integer $k$,

i) the $k$-*th Deligne complex* $\underline{D}_X(k)^{\cdot}$ is the complex starting in degree $0$:

$$Z_X \to \mathcal{O}_X \to \Omega^1_X \to \cdots \to \Omega^{k-1}_X \to 0\ ;$$

ii) the $k$-*th Deligne group* is

$$D^k(X) = H^{2k}(X,\underline{D}_X(k)^{\cdot})\ .$$

We also write

$$D_p(X) = D^{n-p}(X)\ .$$

The short exact sequence of complexes

$$(17) \qquad 0 \to \sigma_k\Omega^{\cdot}_X[-1] \to \underline{D}_X(k)^{\cdot} \to Z_X \to 0$$

induces an exact sequence of hypercohomology

$$H^{2k-1}(X,Z) \to H^{2k-1}(X,\sigma_k\Omega^{\cdot}_X) \to D^k(X) \to H^{2k}(X,Z) \to H^{2k}(X,\sigma_k\Omega^{\cdot}_X) .$$

Let $H^{k,k}(X,Z)$ denote the group of integral cohomology classes of type $(k,k)$ on $X$. Using (11), we can read off of the above the following short exact sequence:

$$0 \longrightarrow J^k(X) \xrightarrow{\iota} D^k(X) \xrightarrow{c} H^{k,k}(X,Z) \longrightarrow 0 . \tag{18}$$

2.2. Let $Z \in \mathcal{Z}^k(X)$. We want to define a cycle class $v(Z) \in D^k(X)$. Consider the complex $\underline{D}^{C}_X(k)^{\cdot}$ :

$$0 \to C_X \to \mathcal{O}_X \to \Omega^1_X \to \cdots \to \Omega^{k-1}_X \to 0 . \tag{19}$$

Using the quasi-isomorphism

$$C_X \xrightarrow{\sim} \Omega^{\cdot}_X$$

and the exact sequence

$$0 \to F^k\Omega^{\cdot}_X \to \Omega^{\cdot}_X \to \sigma_k\Omega^{\cdot}_X \to 0 ,$$

we get an isomorphism in the derived category:

$$\underline{D}^{C}_X(k)^{\cdot} \simeq F^k\Omega^{\cdot}_X . \tag{20}$$

This induces an isomorphism

$$H^{2k}_{|Z|}(X,\underline{D}^{C}_X(k)^{\cdot}) \simeq H^{2k}_{|Z|}(X,F^k\Omega^{\cdot}_X) . \tag{21}$$

From the exact sequences

$$0 \to Z_X \to C_X \to C_X/Z_X \to 0 ,$$

$$0 \to \underline{D}_X(k)^{\cdot} \to \underline{D}^{C}_X(k)^{\cdot} \to C_X/Z_X \to 0 ,$$

and (21), one deduces the following commutative diagram with exact rows:

$$
(22)\qquad
\begin{array}{ccccccc}
0 = H^{2k-1}_{|Z|}(X, C/Z) & \longrightarrow & H^{2k}_{|Z|}(X, \underline{D}_X(k)^{\cdot}) & \longrightarrow & H^{2k}_{|Z|}(X, F^k\Omega^{\cdot}_X) & \longrightarrow & H^{2k}_{|Z|}(X, C/Z) \\
\downarrow & & \downarrow & & \downarrow & & \downarrow = \\
0 = H^{2k-1}_{|Z|}(X, C/Z) & \longrightarrow & H^{2k}_{|Z|}(X, Z) & \longrightarrow & H^{2k}_{|Z|}(X, C) & \longrightarrow & H^{2k}_{|Z|}(X, C/Z)
\end{array}
$$

Now, $Z$ has cycle classes in both $H^{2k}_{|Z|}(X, F^k\Omega^{\cdot}_X)$ and $H^{2k}_{|Z|}(X, Z)$ (see (1.3)), such that their images in $H^{2k}_{|Z|}(X, C)$ coincide. From (22), one may then assign to $Z$ the unique element in $H^{2k}_{|Z|}(X, \underline{D}_X(k)^{\cdot})$ which maps to the two cycle classes. Its image in $D^k(X)$ will be called the *Deligne class* of $Z$, and is denoted $v(Z)$. It is not hard to see that $v(Z)$ depends only on the rational equivalence class of $Z$.

EXAMPLE. Let $k = 1$. The exponential exact sequence

$$0 \to Z_X \to \mathcal{O}_X \to \mathcal{O}^*_X \to 0$$

gives that $\underline{D}_X(1)^{\cdot}$ is quasi-isomorphic to $\mathcal{O}^*_X[-1]$. It follows that

$$D^1(X) \simeq H^1(X, \mathcal{O}^*_X).$$

One readily checks that $v : \mathcal{Z}^1(X) \to D^1(X)$ is identified with the usual divisor class mapping under the above isomorphism.

We consider now the diagram

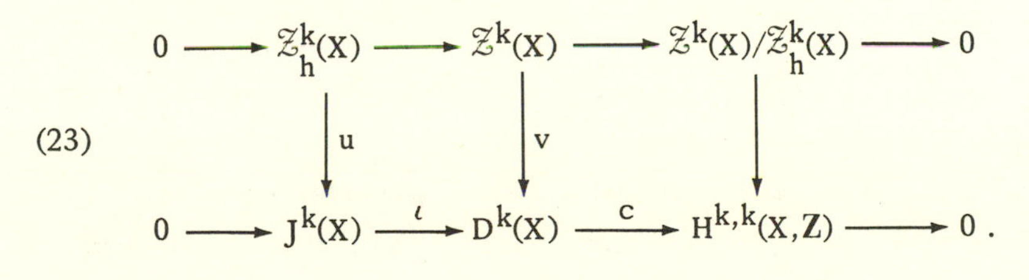

$$
(23)\qquad
\begin{array}{ccccccccc}
0 & \longrightarrow & \mathcal{Z}^k_h(X) & \longrightarrow & \mathcal{Z}^k(X) & \longrightarrow & \mathcal{Z}^k(X)/\mathcal{Z}^k_h(X) & \longrightarrow & 0 \\
 & & \downarrow u & & \downarrow v & & \downarrow & & \\
0 & \longrightarrow & J^k(X) & \xrightarrow{\iota} & D^k(X) & \xrightarrow{c} & H^{k,k}(X, Z) & \longrightarrow & 0 .
\end{array}
$$

PROPOSITION 1. *In (23), we have:*

i) $c(v(Z))$ *is the cohomology class of* $Z$ *in* $X$,

ii) *if* $Z \in \mathcal{Z}^k_h(X)$, *i.e., if* $c(v(Z)) = 0$, *then* $v(Z) = \iota u(Z)$.

*Proof.* Of course, (i) follows immediately from the construction of $v(Z)$, so it remains to prove (ii). We will represent $v(Z)$ by a cocycle in a certain cochain complex, so as to make (ii) apparent. To do this, we choose compatible resolutions of $\mathbf{Z}_X$ and $\Omega^{\cdot}_X$; namely, let $\mathcal{I}^i_X$ denote the sheaf of locally integral currents (see [10]) of real codimension $i$ on $X$, and $\mathcal{D}'^{\,p,q}_X$ the sheaf of distributions (i.e., currents) of type $(p,q)$ on $X$. Then $\mathcal{I}^{\cdot}_X$ is a soft resolution of $\mathbf{Z}_X$, and $\mathcal{D}'^{\,p,\cdot}_X$ is a fine resolution of $\Omega^p_X$ for each $p$. On the single complex of distributions $\mathcal{D}'^{\,\cdot}_X$, there is an evident extension of $F$ from $\Omega^{\cdot}_X$, obtained by truncation with respect to $p$ only. Then $\sigma_p \mathcal{D}'^{\,\cdot}_X$ resolves $\sigma_p \Omega^{\cdot}_X$. The natural inclusion of $\mathcal{I}^{\cdot}_X$ in $\mathcal{D}'^{\,\cdot}_X$ induces a mapping

$$\tilde{\Phi}^k : \mathcal{I}^{\cdot}_X \to \sigma_k \mathcal{D}'^{\,\cdot}_X$$

for any $k$, and we put

$$\tilde{\underline{D}}_X(k)^{\cdot} = C^{\cdot}(\tilde{\Phi}^k)[-1]. \tag{24}$$

Then clearly:

LEMMA 1. *The inclusion of* $\underline{D}_X(k)^{\cdot}$ *in* $\tilde{\underline{D}}_X(k)^{\cdot}$ *induces an isomorphism on cohomology. Thus,*

$$D^k(X) \simeq H^{2k}(X, \tilde{\underline{D}}_X(k)^{\cdot}).$$

Let $Z \in \mathcal{Z}^k(X)$. Then $Z$ defines a current

$$\langle Z \rangle \in H^0_{|Z|}(X, \mathcal{I}^{2k}_X) \subset H^0_{|Z|}(X, \tilde{\underline{D}}_X(k)^{2k}) \tag{25}$$

(see [10]). Since $\langle Z \rangle$ is of type $(k,k)$, it projects to zero in $H^0(X, \sigma_k \mathcal{D}'^{\,\cdot}_X)$, and hence is a cocycle for $\tilde{\underline{D}}_X(k)^{\cdot}$. It is also clear that $\langle Z \rangle$ represents the usual cycle classes of $Z$. Therefore, from (22), $\langle Z \rangle$ represents $v(Z)$ in $D^k(X)$. Suppose now that $Z \in \mathcal{Z}^k_h(X)$. As in the definition of the Abel-Jacobi homomorphism, let

$$\Gamma \in H^0(X, \mathcal{I}^{2k-1}_X)$$

satisfy $\partial\Gamma = \langle Z \rangle$. By (5), we have in $\tilde{\underline{D}}_X(k)^{\cdot}$ the relation

$$d(-\Gamma,0) = (\langle Z \rangle,0) - (0,\tilde{\Gamma}) , \tag{26}$$

where $\tilde{\Gamma} = \tilde{\Phi}^k(\Gamma)$. Therefore, $(0,\tilde{\Gamma})$ also represents $v(Z)$ in $D^k(X)$. That $\tilde{\Gamma}$ represents $u(Z)$ in $J^k(X)$ follows immediately from the definition of $u$, so we are done.

2.3. The Deligne groups and Deligne classes satisfy the usual functoriality:

PROPOSITION 2. *Let* X *and* Y *be non-singular projective varieties, and let* $f: X \to Y$ *be a morphism. Then there are induced mappings*

$$f^*: D^k(Y) \to D^k(X) ,$$

$$f_*: D_p(X) \to D_p(Y) ,$$

*making* $D^k$ *(resp.* $D_p$ *) a contravariant (resp. covariant) functor. Furthermore, the following squares commute:*

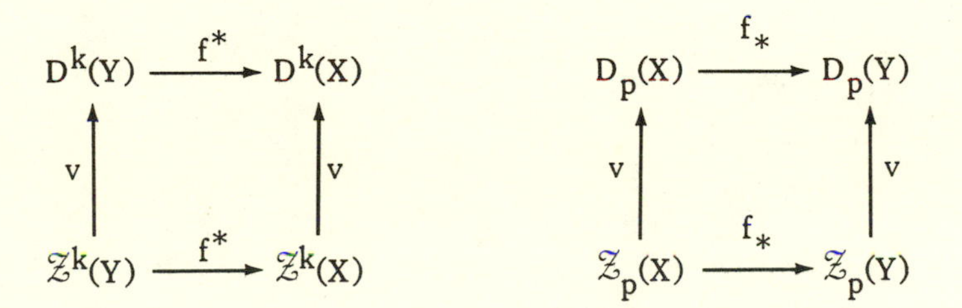

$$\begin{array}{ccc} D^k(Y) & \xrightarrow{f^*} & D^k(X) \\ \uparrow v & & \uparrow v \\ \mathcal{Z}^k(Y) & \xrightarrow{f^*} & \mathcal{Z}^k(X) \end{array} \qquad \begin{array}{ccc} D_p(X) & \xrightarrow{f_*} & D_p(Y) \\ \uparrow v & & \uparrow v \\ \mathcal{Z}_p(X) & \xrightarrow{f_*} & \mathcal{Z}_p(Y) \end{array}$$

*(Here,* $\mathcal{Z}_p$ *denotes cycles of dimension* p.*)*

The proof of Proposition 2 is straightforward, so we omit the details. One uses the Cech resolution of $\underline{D}_X(k)^{\cdot}$ and (22) in discussing $f^*$, whereas one uses Lemma 1 and (25) for the treatment of $f_*$.

We follow [2] in defining products on the Deligne groups. Let

$$\mu: \underline{D}_X(p)^{\cdot} \otimes_{\mathbf{Z}} \underline{D}_X(q)^{\cdot} \to \underline{D}_X(p+q)^{\cdot} \tag{27}$$

denote the morphism of complexes defined, for any $p$ and $q$, by the rule: if $\phi \in \underline{D}_X(p)^i$ and $\psi \in \underline{D}_X(q)^j$, then

$$\mu(\phi \otimes \psi) = \begin{cases} \phi \cdot \psi & \text{if} \quad i = 0 \\ \phi \wedge d_0 \psi & \text{if} \quad i > 0,\ j = q \\ 0 & \text{otherwise,} \end{cases} \tag{28}$$

where for $\psi \in \underline{D}_X(q)^q = \Omega_X^{q-1}$, $d_0\psi$ is its usual exterior derivative (with values in $\Omega_X^q$ ).

PROPOSITION 3. i) *The multiplication* $\mu$ *is associative and homotopy graded-commutative. Thus, it induces on*

$$D^{\cdot}(X) = \bigoplus_{k \geq 0} D^k(X)$$

*the structure of a commutative ring with unit.*

ii) *The image under* $\iota$ *of*

$$J^{\cdot}(X) = \bigoplus_{k \geq 0} J^k(X)$$

*is a square-zero ideal in* $D^{\cdot}(X)$.

iii) *The morphism*

$$c : D^{\cdot}(X) \to H^{even}(X, Z)$$

*is a ring homomorphism.*

*Proof.* For i), one defines the homotopy operator

$$H : \underline{D}_X(p)^{\cdot} \otimes_Z \underline{D}_X(q)^{\cdot} \to \underline{D}_X(p+q)^{\cdot}$$

by the formula

$$H(\phi \otimes \psi) = \begin{cases} (-1)^i \phi \wedge \psi & \text{if} \quad i,j \neq 0 \\ 0 & \text{otherwise.} \end{cases} \tag{29}$$

One checks that

(30) $$(-1)^{ij}\mu(\psi\otimes\phi) = \mu(\phi\otimes\psi) + (dH+Hd)(\phi\otimes\psi)\,.$$

To see (ii), we recall that $\iota(J^k)$ is the image of the mapping

(31) $$H^{2k}(X,\sigma_k\Omega^{\cdot}_X[-1]) \to D^k(X)\,,$$

so it consists of those classes represented via Lemma 1 by cocycles whose component in $H^0(X,\mathcal{I}^{2k}_X)$ is zero. By (27), this property is preserved under taking the product with anything, so $J^{\cdot}(X)$ is an ideal. One can see that the product of two elements of $J^{\cdot}(X)$ is zero by representing the image of (31) by d-closed currents, namely harmonic $(2k-1)$-forms.

We write $\phi\cdot\psi$ for $\mu(\phi\otimes\psi)$ when $\phi$ and $\psi$ are elements of $D^{\cdot}(X)$. The following properties of the product are easily verified:

PROPOSITION 4. i) $f^*(\phi\cdot\psi) = f^*(\phi)\cdot f^*(\psi)$,

ii) $f_*(\eta\cdot f^*\psi) = (f_*\eta)\cdot\psi$,

iii) *For a pair of cycles* $Z$ *and* $Z'$,

$$v(Z\cdot Z') = v(Z)\cdot v(Z')$$

*whenever the intersection cycle* $Z\cdot Z'$ *is defined.*

It follows that $v$ commutes with correspondences defined by algebraic cycles, in the following sense. Let $T\in\mathcal{Z}^{\cdot}(X\times Y)$ define the correspondence

$$\tau : H^{\cdot}(X) \to H^{\cdot}(Y)\,;$$

i.e., $\tau(\xi) = (\pi_Y)_*(\pi_X^*(\xi)\cdot c(T))$. Then $v(T)$ induces a correspondence

$$\tau : D^{\cdot}(X) \to D^{\cdot}(Y)\,,$$

and if $Z\in\mathcal{Z}^{\cdot}(X)$, then

(32) $$\tau(v(Z)) = v(\tau(Z))\,.$$

This completes the verification that the functorial properties of the Abel-Jacobi homomorphism (see [11]) extend to the Deligne class mapping.

## §3. *Relative Deligne groups*

3.1. Let $f : X \to S$ be a smooth projective morphism, with n-dimensional fibers, onto a reduced analytic space $S$. The sheaf of germs of cross-sections of the associated family of k-th intermediate Jacobians can be written

$$(33) \qquad \mathcal{J}^k_{X/S} \simeq R^{2k-1}f_*\sigma_k\Omega^{\cdot}_{X/S}/\mathrm{im}\, R^{2k-1}f_*\mathbf{Z}\,.$$

This follows from (11) and the fact that the sheaf $R^{2k-1}f_*\sigma_k\Omega^{\cdot}_{X/S}$ is locally free, and at each point $s \in S$

$$R^{2k-1}f_*\sigma_k\Omega^{\cdot}_{X/S}\otimes_{\mathcal{O}_S} k(s) \simeq H^{2k-1}(X_s,\sigma_k\Omega^{\cdot}_{X_s})\,,$$

where $X_s = f^{-1}(s)$ and $k(s)$ is the residue field $\mathbf{C}$ at $s$.

3.2. Denote by $\underline{D}_{X/S}(k)^{\cdot}$ the complex starting in degree zero

$$(34) \qquad \mathbf{Z}_X \to \sigma_k\Omega^{\cdot}_{X/S}\,.$$

Via the natural projection of $\underline{D}_X(k)^{\cdot}$ onto $\underline{D}_{X/S}(k)^{\cdot}$, we get a homomorphism

$$(35) \qquad v : \mathcal{Z}^k(X) \to H^0(S, R^{2k}f_*\underline{D}_{X/S}(k)^{\cdot})\,.$$

It follows from slicing theory (see [10, §3]) that (35) restricts, for each $s \in S$, to

$$v : \mathcal{Z}^k(X_s) \to D^k(X_s)$$

on the subset of $\mathcal{Z}^k(X)$ consisting of those cycles $Z$ for which $Z \cdot X_s$ is defined.

Let $\mathcal{Z}^k_h(X/S)$ denote the group of cycles relatively homologous to zero; i.e., $Z \in \mathcal{Z}^k_h(X/S)$ if and only if $c(Z)$ is zero in $H^0(S, R^{2k}f_*\mathbf{Z}_X)$.

This condition merely asserts that the restriction of $c(Z)$ to $H^{2k}(X_s,\mathbf{Z})$ is zero for all $s \in S$, or equivalently that the homological intersection $Z \cdot X_s$ is zero in $H_{2(n-k)}(X_s,\mathbf{Z})$. We see from Proposition 1 that the restriction of (35) to $\mathcal{Z}_h^k(X/S)$,

$$u : \mathcal{Z}_h^k(X/S) \to H^0(S,\mathcal{J}^k_{X/S}), \tag{36}$$

gives the Abel-Jacobi homomorphism (when defined).

3.3. Let now $f : X \to \Delta$ be a projective morphism from a smooth variety to the disc $\Delta$, smooth over $\Delta^*$, with fiber at $0$ a normal crossing divisor $Y = f^{-1}(0)$. The relative logarithmic complex $\Omega^{\cdot}_{X/\Delta}(\log Y)$ is composed of the locally free sheaves

$$\Omega^p_{X/\Delta}(\log Y) = \Lambda^p \Omega^1_{X/\Delta}(\log Y),$$

where $\Omega^1_{X/\Delta}(\log Y) = \Omega^1_X(\log Y)/f^*\Omega^1_\Delta(\log 0)$.

We define the morphism (in the derived category)

$$\Phi^k : Ri_*\mathbf{Z}_{X^*} \to \sigma_k\Omega^{\cdot}_{X/\Delta}(\log Y), \tag{37}$$

where $X^* = X - Y$ and $i : X^* \to X$ denotes the inclusion, by composing the following morphisms:

$$Ri_*\mathbf{Z}_{X^*} \to Ri_*\mathbf{C}_{X^*} \overset{\gamma}{\simeq} \Omega^{\cdot}_X(\log Y) \to \Omega^{\cdot}_{X/\Delta}(\log Y) \to \sigma_k\Omega^{\cdot}_{X/\Delta}(\log Y)$$

(the quasi-isomorphism $\gamma$ has been constructed in [4, (3.1)]). It can be realized in the form (8) as

$$\begin{array}{ccc} Ri_*\mathbf{Z}_{X^*} & \to & i_*\mathcal{D}'^{\cdot}_{X^*} \\ & & \uparrow \\ & & \Omega^{\cdot}_X(\log Y) \to \sigma_k\Omega^{\cdot}_{X/\Delta}(\log Y). \end{array}$$

DEFINITION 2. i) The *k-th relative Deligne complex* of f is

$$\underline{D}_{X/\Delta}(k)^{\cdot} = C^{\cdot}(\Phi^k)[-1];$$

ii) the *sheaf of k-th relative Deligne groups* on $\Delta$ is

$$D^k(X/\Delta) = R^{2k}f_*\underline{D}_{X/\Delta}(k)^{\cdot}.$$

We remark that the above definition globalizes to the case $f: X \to S$, where S is any smooth analytic curve.

3.4. We give now an interpretation [16], in terms of the Hodge filtration of the limit mixed Hodge structure, of the fiber at 0 of the sheaves

$$\mathcal{J}^k_{X/\Delta} = R^{2k-1}f_*\sigma_k\Omega^{\cdot}_{X/S}(\log Y)/\mathrm{im}\, R^{2k-1}(f\circ i)_*\mathbf{Z}_{X^*}.$$

We recall that the above sheaves can be constructed without assuming that $X_0 = Y$ be a normal crossing divisor, and moreover they depend only on $X^*$ [16, §2].

We use the following result:

PROPOSITION 5 ([12, (2.16)]). *Consider the Poincaré half plane* H *as the universal covering of* $\Delta^*$, *and let*

$$X_\infty = X^* \times_{\Delta^*} H.$$

*Then for all* q,

$$R^qf_*\Omega^{\cdot}_{X/\Delta}(\log Y)\otimes_{\mathcal{O}_\Delta} k(0) \simeq H^q(Y, \Omega^{\cdot}_{X/\Delta}(\log Y)\otimes_{\mathcal{O}_X}\mathcal{O}_Y) \simeq H^q(X_\infty, \mathbf{C}).$$

Let $\{F^p\Omega^{\cdot}_{X/\Delta}(\log Y)\}$ denote the Hodge filtration (1) by forms of degrees at least p. It induces a filtration on $\Omega^{\cdot}_{X/\Delta}(\log Y)\otimes_{\mathcal{O}_X}\mathcal{O}_Y$, and therefore induces a filtration F on the cohomology $H^{\cdot}(X_\infty, \mathbf{C})$.

PROPOSITION 6. *We have*

$$R^{2k-1}f_*\sigma_k\Omega^{\cdot}_{X/\Delta}(\log Y)\otimes_{\mathcal{O}_\Delta}k(0) \simeq H^{2k-1}(X_\infty,C)/F^kH^{2k-1}(X_\infty,C)\,.$$

*Proof.* We deduce from the degeneration at $E_1$ ([13, (2.9)]) of the spectral sequence

$${}_FE_1^{p,q} = R^qf_*\Omega^p_{X/\Delta}(\log Y)$$

the short exact sequence

$$0 \to R^if_*F^k\Omega^{\cdot}_{X/\Delta}(\log Y) \to R^if_*\Omega^{\cdot}_{X/\Delta}(\log Y) \to R^if_*\sigma_k\Omega^{\cdot}_{X/\Delta}(\log Y) \to 0\,.$$

Tensoring with $k(0)$ and using Proposition 5, we get

$$0 \to F^kH^i(X_\infty,C) \to H^i(X_\infty,C) \to R^if_*\sigma_k\Omega^{\cdot}_{X/\Delta}(\log Y)\otimes_{\mathcal{O}_\Delta}k(0) \to 0\,,$$

which gives the desired isomorphism.

PROPOSITION 7. *The generalized intermediate Jacobian at* $0$ *is isomorphic to*

$$\mathrm{im}\, H^{2k-1}(X^*,Z)\backslash H^{2k-1}(X_\infty,C)/F^kH^{2k-1}(X_\infty,C)\,.$$

*Proof.* Let $g$ denote the restriction of $f$ to $X^*$, and $j$ the inclusion of $\Delta^*$ in $\Delta$. We show that

$$\mathrm{im}\,\{H^{2k-1}(X^*,Z) \to H^{2k-1}(X_\infty,C)/F^kH^{2k-1}(X_\infty,C)\}$$

is the lattice $\mathrm{im}\, H^0(\Delta,j_*R^{2k-1}g_*Z)$ used in [16, (2.5)]. We claim that the canonical morphism

$$R^q(f\circ i)_*Z_{X^*} = R^q(j\circ g)_*Z_X \to j_*R^qg_*Z_{X^*}$$

is surjective, or equivalently

$$H^q(X^*,Z) \xrightarrow{\psi} H^0(\Delta^*,R^qg_*Z) \to 0$$

is exact. This follows from the degeneration at $E_2$ of the Leray spectral sequence for $g$, which one has because

$$H^i(\Delta^*, R^q g_* Z) = 0 \quad \text{for} \quad i \neq 0,1 \quad (\text{any } q).$$

The kernel of $\psi$ consists of those cohomology classes vanishing in

$$H^q(X_\infty, Z) \simeq H^q(X_s, Z).$$

Thus, $H^{2k-1}(X^*, Z)$ and

$$H^0(\Delta, j_* R^{2k-1} g_* Z) \simeq H^0(\Delta^*, R^{2k-1} g_* Z)$$

have the same image even in $H^{2k-1}(X_\infty, Z)$, so we are done.

## §4. *Extension of the normal function*

4.1. With the notations of (3.3), consider the long exact sequence associated to the cone $\underline{D}_{X/\Delta}(k)^{\cdot}$ :

$$\to R^q(f \circ i)_* Z_{X^*} \to R^q f_* \sigma_k \Omega^{\cdot}_{X/\Delta}(\log Y) \to R^{q+1} f_* \underline{D}_{X/\Delta}(k)^{\cdot} \to R^{q+1}(f \circ i)_* Z_{X^*} \to .$$

Put

$$(38) \qquad H^{k,k}(X/\Delta, Z) = \ker \{R^{2k}(f \circ i)_* Z_{X^*} \to R^{2k} f_* \sigma_k \Omega^{\cdot}_{X/\Delta}(\log Y)\},$$

which one might call the sheaf of relative Hodge classes, extending to $\Delta$ the subsheaf of $R^{2k} g_* Z$ consisting of cohomology classes that are of type $(k,k)$ on all fibers of $g$. Then we have the short exact sequence

$$(39) \qquad 0 \longrightarrow \mathcal{J}^k_{X/\Delta} \longrightarrow D^k(X/\Delta) \xrightarrow{c} H^{k,k}(X/\Delta) \longrightarrow 0 .$$

4.2. *Relative Deligne class.* Let $Z \in \mathcal{Z}^k(X)$ be a cycle of codimension $k$ in $X$. We deduce a relative cycle class $v(Z)$ in $H^0(\Delta, D^k(X/\Delta))$ by pushing the absolute cycle class $v(Z) \in D^k(X)$, constructed in 2.2, via the following morphism of cones:

(40) $\underline{D}_X(k)^{\cdot} \simeq C^{\cdot}(Z \to \sigma_k \Omega^{\cdot}_X) \to C^{\cdot}(Ri_* Z_{X^*} \to \sigma_k \Omega^{\cdot}_{X/\Delta}(\log Y)) = \underline{D}_{X/\Delta}(k)^{\cdot}$.

4.3. We now give our main result.

THEOREM. *Let* $f : X \to \Delta$ *be a projective morphism over a disc, and* $Z \in \mathcal{Z}^k(X)$ *be an analytic cycle whose restriction* $Z^*$ *to* $X^*$ *lies in* $\mathcal{Z}^k_h(X^*/\Delta^*)$. *Suppose that* $c(Z^*) = 0$ *in* $H^{2k}(X^*, \mathbf{Z})$, *or equivalently that* $Z$ *is homologous to a cycle supported on* $X_0$. *Then* $v(Z)$ *lifts to a class in* $H^0(\Delta, \mathcal{J}^k_{X/\Delta})$; *in other words, the normal function* $\nu_Z$ *associated to* $Z$ *by the Abel-Jacobi homomorphism extends to* $\Delta$.

*Proof.* We may assume without loss of generality that $X_0 = Y$ is a normal crossing divisor. Using the exact sequence (39), we see that the hypothesis of the theorem says that $c(v(Z)) = 0$; then $v(Z)$ lifts to a section of $\mathcal{J}^k_{X/\Delta}$, inducing $\nu_Z$ on $\Delta^*$ (see (36)).

REMARK. i) If one uses the complex

$$Z_X \to \sigma_k \Omega^{\cdot}_{X/\Delta}(\log Y)$$

instead of Definition 2(i), one would have to assume that $Z$ is cohomologous to zero on all fibers, i.e., on $X_0$. This condition appears already in [9].

ii) James King has also obtained conditions on the cycle $Z$ in order to extend the normal function $\nu_Z$, using a notion of currents with logarithmic singularities.

4.4. We look now a little more closely at the hypothesis of our theorem and the condition in the above remark. Let $X$ be now a projective variety of dimension $n+1$ and $Y \subset X$ a normal crossing divisor with irreducible components $\{Y_\lambda\}_{\lambda \in \Lambda}$. Then we have:

PROPOSITION 8. *Let* $Z \in \mathcal{Z}^k(X)$.

i) *The restriction of the cycle class* $c(Z)$ *to* $H^{2k}(X-Y, \mathbf{Q})$ *is zero if and only if it is in the image of*

$$\bigoplus_{\lambda\epsilon\Lambda} H^{2k}_{Y_\lambda}(X,Q) \simeq \bigoplus_{\lambda\epsilon\Lambda} H_{2(n+1-k)}(Y_\lambda,Q) .$$

ii) *The restriction of* $c(Z)$ *to* $H^{2k}(Y,Q)$ *is zero if and only if its restriction to* $H^{2k}(Y_\lambda,Q)$ *is zero for all* $\lambda \epsilon \Lambda$.

*Proof.* Consider the morphism of mixed Hodge structures

$$a : H^{2k}(X,Q) \to H^{2k}(Y,Q) ,$$

and the associated

$$Gr^W_{2k}a : Gr^W_{2k}H^{2k}(X,Q) \simeq H^{2k}(X,Q) \to Gr^W_{2k}H^{2k}(Y,Q) .$$

Then by strictness,

$$\ker a = \ker Gr^W_{2k}a .$$

Since $Gr^W_{2k}H^{2k}(Y,Q)$ can be identified as a subspace of

$$\bigoplus_{\lambda\epsilon\Lambda} H^{2k}(Y_\lambda,Q) ,$$

one sees that

$$\ker a = \ker \{H^{2k}(X,Q) \to \bigoplus_{\lambda\epsilon\Lambda} H^{2k}(Y_\lambda,Q)\} .$$

This gives (ii). One obtains (i) by the dual argument for the morphism

$$H^{2k}_Y(X,Q) \to H^{2k}(X,Q) .$$

REMARK. The above discussion can be carried out in a neighborhood of $Y$, using the mixed Hodge structures mentioned in Chapter VII, Theorem 5.

4.5. PROBLEM (Interpretation as periods). Consider the generalized intermediate Jacobian at $0$ (see Proposition 7). Suppose that the cycle $Z$ satisfies the condition of the theorem in 4.3. Then there exists a chain $\Gamma$ (of infinite supports) in $X^*$, of dimension $(2n+3-2k)$, such that $\partial\Gamma = Z|_{X^*}$.

i) Can the value of $u(Z)$ at $0$ be expressed via integrals over $\Gamma$ of suitable $C^\infty$ differential forms on $X^*$ or $X_\infty$?

ii) Show that for $\omega$ a relatively closed $C^\infty$ $(2n-2k+1)$-form on $X$ with logarithmic poles along $Y$, and of the appropriate Hodge filtration level,

$$\lim_{s\to 0} \int_{\Gamma\cdot X_s} \omega$$

exists, and this determines the value of $u(Z)$ at $0$.

## REFERENCES


[1] M. Atiyah and F. Hirzebruch, "Analytic cycles on compact manifolds," Topology 1 (1962), pp. 25-45.

[2] S. Bloch and P. Griffiths, "A theorem about normal functions associated to Lefschetz pencils on algebraic varieties," 1971.

[3] H. Clemens, "On extending Abel-Jacobi mappings," 1981.

[4] P. Deligne, "Théorie de Hodge II," Pub. Math. IHES 40 (1972), pp. 5-57.

[5] F. El Zein, "Complexe dualisant," Bul. Soc. Math. France, Memoire No. 58 (1978).

[6] ———, "Mixed Hodge structures," Trans. AMS 275 (1983).

[7] P. Griffiths, "On the periods of certain rational integrals," Annals of Math. 90 (1969), pp. 460-541.

[8] ———, "A theorem concerning the differential equations satisfied by normal functions associated to algebraic cycles," Amer. J. Math. 101 (1979), pp. 94-131.

[9] ———, "An observation on normal functions," preprint.

[10] J. King, "The currents defined by analytic varieties," Acta Math. 127 (1971), pp. 185-220.

[11] D. Lieberman, "Higher Picard varieties," Amer. J. Math. 90 (1968), pp. 1165-1199.

[12] J. Steenbrink, "Limits of Hodge structures," Inventiones Math. 31 (1976), pp. 229-257.

[13] J. Steenbrink, "Mixed Hodge structure on the vanishing cohomology," in: Real and Complex Singularities, Oslo, 1976, pp. 525-563.

[14] J.-L. Verdier, "Classe d'homologie d'un cycle," Séminaire de Géometrie Analytique, Astérisque 36-37 (1976), pp. 101-151.

[15] ———, "Catégories dérivées," SGA 4½, Lecture Notes in Math. 569, Springer-Verlag, Berlin-Heidelberg-New York, 1977, pp. 262-311.

[16] S. Zucker, "Generalized intermediate Jacobians and the theorem on normal functions," Inventiones Math. 33 (1976), pp. 185-222.

[17] ———, "Hodge theory with degenerating coefficients: $L_2$ cohomology in the Poincaré metric," Annals of Math. 109 (1979), pp. 415-476.

FOUAD EL ZEIN
UNIVERSITÉ de PARIS VII
UER de MATHÉMATIQUES et INFORMATIQUE
TOUR 45-55
2, PLACE JUSSIEU
75251 PARIS CEDEX 05
FRANCE

STEVEN ZUCKER
MATHEMATICS DEPARTMENT
JOHNS HOPKINS UNIVERSITY
BALTIMORE, MD 21218

## Chapter XVI

## SOME RESULTS ABOUT ABEL-JACOBI MAPPINGS

Herbert Clemens

### §1. *The Abel-Jacobi mapping*

Let V be a smooth irreducible complex projective manifold. One defines

(1) $\mathfrak{G}^r(V)$ = (group of algebraic cycles of codimension r in V which are *homologous* to zero modulo those which are *rationally* equivalent to zero).

The classical result of Jacobi is:

THEOREM 1. *For* $r = \dim V = 1$, *the natural map*

$$\mathfrak{G}^1(V) \to J^1(V) = \frac{(H^{1,0}(V))^*}{H_1(V)_Z}$$

$$Z'-Z'' \mapsto \int_{Z''}^{Z'}$$

*is an isomorphism.*

In [13; App. A], Griffiths defined for arbitrary V his *intermediate Jacobians*

$$J^r(V) = \frac{(H^{2\hat{r}+1,0}(V) + \cdots + H^{\hat{r}+1,\hat{r}}(V))^*}{H_{2\hat{r}+1}(V)_Z}$$

where $r+\hat{r} = \dim V$, and a natural mapping

$$\Phi : \mathfrak{G}^r(V) \to J^r(V)$$

(2)

$$Z'-Z'' \mapsto \int_{Z''}^{Z'} ,$$

which he called the *Abel-Jacobi mapping.* The first seeds of (2) appear on page 333 of Weil's *Foundations* [17], but a detailed analysis of properties of (2) in the case $r \geq 2$ has come only in the past 15 years following the appearance of [12], [14], and [15].

The particular direction of the development which we wish to follow here began with Griffiths surprising use of the mapping (2) to show:

THEOREM 2. *There exist threefolds* V *for which the subgroup*

$$\Phi(\mathfrak{G}^2(V)) \subseteq J^2(V)$$

*is a non-zero countable group* [14; pp. 462-463].

The examples used to establish Theorem 2 are non-singular quintic hypersurfaces in $\mathbf{P}^4$. Griffiths constructs a smooth quintic fourfold

$$W \subseteq \mathbf{P}^5$$

containing two planes, $K'$ and $K''$. Letting

$$H_t \subseteq \mathbf{P}^5$$

be a generic pencil of hyperplanes, he constructs a *Poincaré normal function*

(3)

$$\int_{Z''_t}^{Z'_t} \in J^2(V_t)$$

where

$$V_t = W \cdot H_t\,, \quad Z'_t = K' \cdot H_t\,, \quad \text{etc.}$$

By showing that, if (3) vanishes identically, then $K'$ and $K''$ are homologous in $W$ (which is false), he concludes that, for generic $t$,

$$\int_{Z''_t}^{Z'_t} \neq 0 \tag{4}$$

in $J^2(V_t)$. On the other hand, by the irreducibility of the action of monodromy on the family $\{V_t\}$, there can be no non-zero element

$$\gamma_t \in H_3(V_t)_{\mathbf{Z}}$$

( $t$ generic) such that

$$\int_{\gamma_t} \omega = 0$$

for all $\omega \in H^{3,0}(V)$. This fact, taken together with (4), is already enough to imply Theorem 2. This is because, if $S_t$ is a connected projective variety with

$$S_t \to \mathfrak{G}^2(V_t) ,$$

then the image of the natural mapping

$$\begin{aligned} H_1(S_t)_{\mathbf{Z}} &\to H_3(V_t)_{\mathbf{Z}} \\ \sigma &\mapsto \bigcup_{s \in \sigma} Z_s , \end{aligned} \tag{5}$$

lies in a submodule of $H_3(V_t)_{\mathbf{Z}}$ which is annihilated by $H^{3,0}(V_t)$.

In fact, the mapping (5), which is also sometimes called an Abel-Jacobi mapping, underlies a morphism of type (1,1) of polarized Hodge structures. It is this fact that underlies Grothendieck's correction of the generalized Hodge conjecture in [15].

In the following sections we will outline some recent progress along the lines of Theorem 2. Most of the results apply in the situation

$$\dim V = 2r-1 ,$$

but, to keep the discussion focused on the principal issues involved, we will restrict our attention almost exclusively to the case

$$r = 2 ,$$

that is, the case of algebraic one-cycles on threefolds.

## §2. *Parametrizing abelian subvarieties of* $J^2(V)$

If $V$ is a threefold for which

$$H^{3,0}(V) \neq 0 ,$$

then typically there is no non-zero polarized sub-Hodge-structure

$$H \subseteq H^3(V) \tag{6}$$

which is both defined over $\mathbf{Q}$ and perpendicular to $H^{3,0}(V)$. This fact is fundamental to Theorem 2. Grothendieck's version of the generalized Hodge conjecture states in this context:

CONJECTURE 1. Suppose $H$ as in (6) is a polarized sub-Hodge-structure defined over $\mathbf{Q}$ and perpendicular to $H^{3,0}(V)$. Via cup product view $H_{\mathbf{C}}$ as a subspace of $H^3(V)^*_{\mathbf{C}}$ and let $E$ denote its projection into $J^2(V)$. Then $E$ lies in the image of the Abel-Jacobi mapping

$$\Phi : \mathfrak{G}^2(V) \to J^2(V) .$$

Just as in (5), let us suppose that

$$\begin{array}{ccc} \mathfrak{Z} & \xrightarrow{\nu} & V \\ \downarrow{\scriptstyle \pi} & & \\ S & & \end{array} \tag{7}$$

is an algebraic family of algebraic one cycles with S connected. Then there is induced a morphism

$$\Phi_S : \mathrm{Alb}(S) \to J^2(V)$$

(8)
$$\int_{s''}^{s'} \to \int_{Z''}^{Z'}$$

where $Z' = \nu_*(\pi^*(s'))$, etc. If we set

$$E = \Phi_S(\mathrm{Alb}(S)) ,$$

then E satisfies the hypothesis of Conjecture 1, so the only subtori of $J^2(V)$ that can be parametrized by *algebraic* families of algebraic cycles are those satisfying the hypothesis of Conjecture 1. The question is whether every such E is so parametrized.

There are some special V for which Conjecture 1 holds. Several are mentioned in [15], [1] and [2]. Among these, the case of the cubic threefold, proved by Gherardelli, should be singled out, since the fundamental principle used by Gherardelli has been the basis for the success in dealing with Conjecture 1 in many other cases for which

(9)
$$H^{3,0}(V) = 0 .$$

PRINCIPLE. If V can be covered by rational curves, then Conjecture 1 is true.

In Gherardelli's case, the curves in question were lines. A general treatment is implicit in [11].

Most desirable, under assumption (9), is to find an algebraic family (7) such that the Abel-Jacobi mapping

$$\Phi_S : \mathrm{Alb}(S) \to J(V)$$

given in (8) is actually an isomorphism. Examples are found in [5] for lines on the cubic threefold and [18] for lines on the quartic double solid. This is fundamentally a topological question since it hinges on proving that, modulo torsion, the map

$$\begin{aligned} H_1(S)_{\mathbf{Z}} &\to H_3(V)_{\mathbf{Z}} \\ \sigma &\to \nu_*(\pi^*(\sigma)) \end{aligned} \tag{10}$$

is an isomorphism. A criterion for surjectivity (over $\mathbf{Z}$ ) for the mapping (10) is given by:

THEOREM 3 ([6]). *Suppose* $V$ *is a very ample divisor in a fourfold* $W$ *with* $H_3(W)_{\mathbf{Z}} = 0$. *Suppose that a family* $S_V$ *of algebraic curves on* $V$ *varies algebraically with* $V$ *in such a way that, when* $V$ *specializes to generic* $V_0$ *with one node,* $S_V$ *specializes to a reduced connected family* $S_{V_0}$ *whose generic curve does not pass through the node of* $V_0$. *Suppose further that there exists a curve* $Z_{\bar{s}}$, $\bar{s} \in S_{V_0}$, *such that the node of* $V_0$ *is a simple point of* $Z_{\bar{s}}$. *Finally assume*

$$\mathfrak{S} = \bigcup_{V \to V_0} S_V$$

*is smooth at* $\bar{s}$. *Then the mapping (10) is surjective for all non-singular* $V$.

The most striking example for which $\Phi_S$ in (8) is an isomorphism is the case in which $S$ is the surface of conics on the quartic threefold. This result, proved by M. Letizia of the University of Trento (Italy), should be the model for the remaining Fano threefolds.

In another direction are several cases in which (9) holds and in which

$$\Phi : \mathfrak{G}^2(V) \to J^2(V) \tag{11}$$

is an isomorphism. For example, if $V$ is a "conic bundle", then Beauville showed (11) to be an isomorphism in [1]. Bloch and Murre showed the analogous result for the "principle sequence" of Fano threefolds

$$V_4 \subseteq \mathbf{P}^4, \quad V_3 \cap V_2 \subseteq \mathbf{P}^5, \quad V_2 \cap V_2 \cap V_2 \subseteq \mathbf{P}^6$$

in [2].

The discussion surrounding Theorem 2 and Conjecture 1 implies that, once

$$H^{3,0}(V) \neq 0 ,$$

there is no hope that the mapping (11) is an isomorphism; however, if $V$ is "regular", i.e., $h^{2,0}(V) = h^{1,0}(V) = 0$, it seems possible that (11) is *injective.* In fact, it is reasonable to expect that, if Conjecture 1 is true for all regular threefolds, then (11) will be injective for all regular threefolds. To see why one might expect this, we proceed as follows. Suppose

$$\int_{Z''}^{Z'} = 0 \tag{12}$$

in $J^2(V)$. By smoothing lemmas of Kleiman, we can assume $Z'$ and $Z''$ smooth and irreducible, and, by a result from Bloch's thesis (Columbia University, 1971):

$$Z',\ Z'' \subseteq A \ \text{ very ample, smooth } \subseteq V .$$

If $B_t$ is a one-parameter family of divisors on $V$ specializing to $2A$, we can form a one-parameter family of threefolds

$$W_t \to W_0 \tag{13}$$

where $W_t$ is a double cover of $V$ branched along $B_t$.

If $Z'$ and $Z''$ are homologous in $A$, then they are linearly equivalent in $A$ since $V$ and hence $A$ is regular, and there is nothing to prove. On the other hand, even if $Z' - Z''$ is not homologous to zero in $A$, it is homologous to zero in $V$, and so gives rise to an extension datum in Deligne's mixed Hodge structure on

$$H^3(W_0)\,.$$

Condition (12) says that this extension datum is zero. That is, one has satisfied at least the *necessary* condition for there to exist a family of elliptic curves

(14) $$E_t \subseteq J^2(W_t)$$

such that:

i) $E_t$ is annihilated by $H^{3,0}(W_t)$,

ii) $E_0 \simeq \mathbf{C}^*$, in fact to *that* $\mathbf{C}^*$ in the generalized Jacobian $J^2(W_0)$ corresponding to the class

$$(Z'-Z'') \in H_2(A)^0_{\mathbf{Z}}\,.$$

(See [7; Thm. 5.66].)

The critical missing ingredient is to show, perhaps by a constant count for $A$ sufficiently ample, that there actually exists a one-parameter family (13) giving rise to a family (14) of one-dimensional abelian subvarieties of $J^2(W_t)$. If this can be done, then Conjecture 1 can be applied for generic $t$ in order to parametrize $E_t$ by algebraic one-cycles. Taking the limits of these families of algebraic one-cycles as $t \to 0$, we would obtain a family

$$S \cong \mathbf{C}^*$$

of algebraic one-cycles on $W_0$. But $W_0$ has two components, each isomorphic to $V$. For each $s \in \overline{S}$, we take only the components of the algebraic one-cycle which lie in one component of $W_0$. We thereby obtain a family of algebraic one-cycles

$$Z_s\,,\quad s \in \overline{S} = \mathbf{P}^1\,,$$

such that $(Z_\infty - Z_0)$ lies in $A$ and is homologous, and therefore linearly equivalent to, $Z' - Z''$ in $A$. Thus $(Z'-Z'')$ would have to be rationally equivalent to zero in $V$.

§3. *Cycles not algebraically equivalent to zero*

In this final section, we will outline some results about the part of the image of

$$\Phi : \mathfrak{G}^2(V) \to J^2(V) \tag{15}$$

which does not lie in any abelian subvariety of $J^2(V)$ orthogonal to $H^{3,0}(V)$. A prototype of this situation is provided by the following result, which was arrived at independently by Ceresa-Collino [3] and the author [8]:

THEOREM 4. *Let* $V \subseteq P^4$ *be a generic hypersurface which has only one node singularity. Let* $\tilde{V}$ *be the minimal resolution of* $V$, $E$ *the exceptional surface, and* $Z'$, $Z''$ *lines from the two rulings of* $E$. *Then* $J^2(\tilde{V})$ *contains no abelian subvariety orthogonal to* $H^{3,0}(\tilde{V})$, *and* $Z'$ *and* $Z''$ *are not algebraically equivalent.* (*We assume* $\deg V \geq 5$.)

*Proof.* As we have seen in (8) ff., the fact that $J^2(\tilde{V})$ contains no abelian subvarieties perpendicular to $H^{3,0}(\tilde{V})$ implies that $Z'$ and $Z''$ cannot be *algebraically* equivalent in $\tilde{V}$ unless

$$\int_{Z''}^{Z'} = 0 .$$

However $Z'$ and $Z''$ are homologous. In fact, just as in the case of curves, we can find a 3-chain $\Gamma$ bounding $Z' - Z''$ by forming a family of smooth hypersurfaces

$$V_t \to V , \tag{16}$$

taking a transverse cycle $\Gamma_t$ in $V_t$, and letting $\Gamma$ be the "proper transform" in $\tilde{V}$ of

$$\lim_{t \to 0} \Gamma_t .$$

In fact, this shows that $\int_{Z''}^{Z'}$ is the extension datum associated to Deligne's mixed Hodge structure on

$$H^3(V) .$$

This last fact gives the key to the proof of Theorem 4. For, if $Z'$ and $Z''$ are algebraically equivalent, then there exists an algebraic family (7) and $s', s'' \epsilon S$ such that

$$Z' - Z'' = Z_{s'} - Z_{s''} .$$

Since $J^2(\tilde{V})$ contains no abelian subvarieties perpendicular to $H^{3,0}(\tilde{V})$, the map

$$\nu_* \circ \pi^* : S \to J^2(\tilde{V})$$

must be the zero map. So we conclude that

$$\int_{Z''}^{Z'} \omega = 0$$

for all $\omega \epsilon H^{3,0}(\tilde{V}) + H^{2,1}(\tilde{V})$ and suitably chosen $\Gamma$ with

$$\partial\Gamma = Z' - Z'' .$$

In other words, the mixed Hodge structure on $H^3(V)$ must always be split over $\mathbf{Z}$. It is then not hard to specialize $V$ to a case in which the extension datum has a concrete geometric interpretation, e.g., as a linear system of degree zero on the curve of lines with fourth order contact to $V$ at its node. It is then straightforward to show that such a linear system is not always the trivial one.

Ceresa applied similar techniques to answer a long-standing question:

THEOREM 5 ([4]). *If* $C$ *is a generic curve of genus* $g \geq 3$, *then* $C$ *and* $-C$ *are not algebraically equivalent in* $J^1(C)$.

The trick in Theorem 5 is to specialize to the case

$$C = C' + E \tag{17}$$

with $E$ elliptic and proceed by induction until reaching the genus three case. In this case, one specializes to a curve (17) such that $C'$ and $E$ meet at a point which is *not* a Weierstrass point for the hyperelliptic curve $C'$.

There is an interesting question connected with Theorem 5 which probably bears more looking into. Namely Ceresa shows that for higher symmetric products

$$C^{(r)} \subseteq J^1(C)\,,$$

with $r \leq g-2$, the analogous result holds. An especially interesting case is

$$g \text{ odd}\,, \quad r = \frac{g-1}{2}\,.$$

For, in this case, if $\omega$ is a generator of

$$H^{g,0}(J(C))\,,$$

then the integral

$$\int_{-C^{(r)}}^{C^{(r)}} \omega\,,$$

which would have to vanish if $C^{(r)}$ and $-C^{(r)}$ were algebraically equivalent, does in fact vanish along the divisor in moduli space consisting of those $C$ having a $g_d^1$ with

$$d = \frac{g-1}{2}\,.$$

This is the same divisor which plays such an important role in Harris and Mumford's recent spectacular result on the Kodaira dimension of moduli space.

Finally there is the question as to whether, for all threefolds V, the group $\mathfrak{G}^2(V)$, or at least its image in $J^2(V)$, is finitely generated over the (image) subgroup of cycles algebraically equivalent to zero. Again the quintic threefold serves to dispose of this question just as it did in Theorem 2.

THEOREM 6 ([9]). *If* V *is a generic quintic threefold, the image of the Abel-Jacobi mapping*

$$\Phi : \mathfrak{G}^2(V) \to J(V)$$

*is a totally disconnected group which, when tensored with* $\mathbf{Q}$, *is not a finite dimensional vector space.*

Although the actual proof proceeds differently, let us give a simpler heuristic argument that perhaps will convince the reader that something like Theorem 6 must be true. A constant count shows that a generic quintic threefold V should contain a finite number of rational curves of degree d for each $d \in \mathbf{N}$. (For instance, there are 2875 lines on V.)*
The normal bundle of a rational curve on V must be of the form

$$\mathfrak{O}(p) + \mathfrak{O}(q)$$

where $(p+q) = -2$. So we get, for each d, a generically finite morphism

$$\rho_d : E_d \to \mathbf{P}, \tag{18}$$

where $\mathbf{P}$ is the projective space of all quintics and

$$\rho_d^{-1}(V) = \text{(the algebraic set whose points are reduced rational curves of degree } d \text{ on } V\text{)}.$$

It is not hard to show that $\rho_d$ is etale at a curve $C \subseteq V$ if and only if the normal bundle

$$\mathfrak{N}_{C,V} \cong \mathfrak{O}(-1) + \mathfrak{O}(-1),$$

*Added in proof: S. Katz has recently shown that there are 609,250 conics on V.

and that a Zariski open subset of the ramification divisor $R_d$ is given by those $C \subseteq V$ with

$$\mathfrak{N}_{C,V} \cong \mathfrak{O} + \mathfrak{O}(-2) .$$

By looking at the set of rational curves of degree $d$ on a generic quintic fourfold and counting constants, one sees that

$$D_d = \rho_d(R_d) \not\supseteq \Delta \tag{19}$$

where $\Delta$ = (set of all singular quintic threefolds). For example, a generic quintic fourfold contains only a two-parameter family of lines, so a generic quintic threefold with a node does *not* contain a line passing through the node.*

In fact, one might well expect that the various $D_d$, $d \in N$, have no common components. Let's suppose that this is true. Let

$$L \subseteq P$$

be a generic line and

$$V_t, \quad t \in L ,$$

the corresponding pencil of quintic threefolds. We pick a basepoint

$$\bar{t} \in L$$

such that $\bar{t} \notin \cup_{d \in N} D_d$. Given $d_1, d_2 \in N$ we can picture $L$ as follows:

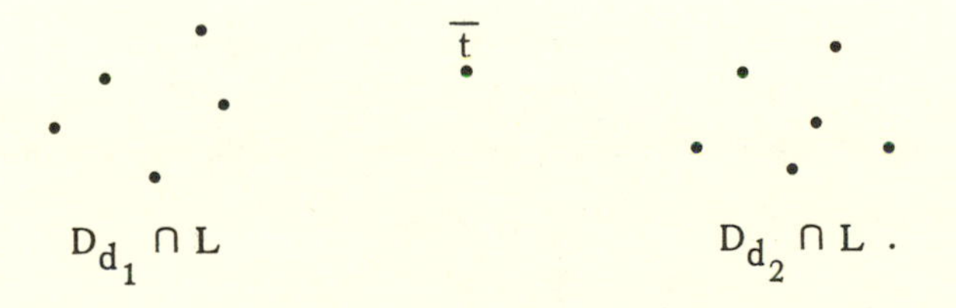


---

*It was this fact about the lines, pointed out to the author by A. Beauville, which led to Theorem 6.

Since $(D_{d_1} \cap L)$ and $(D_{d_2} \cap L)$ are disjoint, the action of $\pi_1(L-(D_{d_1} \cap L), \bar{t})$ on the fibre of $\rho_{d_2}^{-1}(L)$ over $\bar{t}$ must be *trivial* as must the action of $\pi_1(L-(D_{d_2} \cap L), \bar{t})$ on the fibre of $\rho_{d_1}^{-1}(L)$ over $\bar{t}$. Suppose now that we had, for generic $t \in L$, a relation of rational equivalence

$$\Sigma\, a_j L_j(t) + \Sigma\, b_k M_k(t) \equiv 0\,, \tag{20}$$

where the $L_j(t)$ are rational curves of degree $d_1$ on $V_t$ and the $M_k(t)$ are rational curves of degree $d_2$ on $V_t$. If

$$t_0 \in (D_{d_2} \cap L)$$

is, for example, a point where $M_1(t)$ and $M_2(t)$ are interchanged, then monodromy on the path

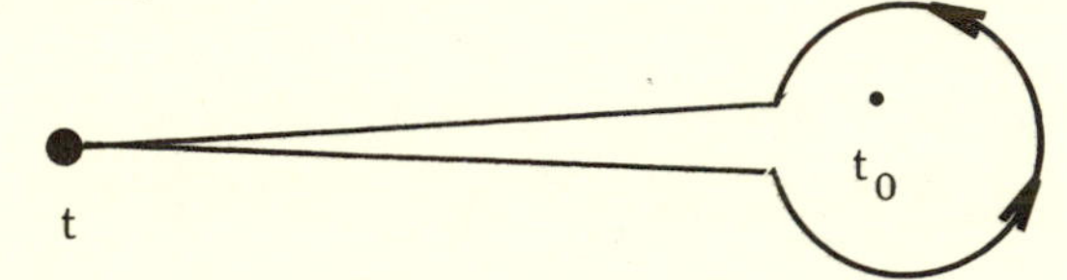


and (20) yield, again for generic $t$, a relation

$$\Sigma\, a_j L_j(t) + b_2 M_1(t) + b_1 M_2(t) + \sum_{k \geq 3} b_k M_k(t) \equiv 0\,. \tag{21}$$

Subtracting (21) from (20), we obtain

$$(b_1 - b_2)(M_1(t) - M_2(t)) \equiv 0\,.$$

But, as in the case $d = 1$ which Griffiths exploited in Theorem 2, one does *not* expect that

$$\int_{M_2(t)}^{M_1(t)} \in J(V_t)$$

is a point of finite order for generic $t$.

This line of reasoning suggests that generically the only relations which exist among rational curves on $V_t$ are the relations of the form

$$5(\Sigma C) + d(\text{plane section of } V_t) \equiv 0$$

where the summation is over *all* curves $C$ in $V_t$ which correspond to a given component of some $E_d$.

Needless to say, the above argument has many gaps. For example, it seems quite difficult to actually compute the branching divisors $D_d$. So the actual proof of Theorem 6 is obtained by exploiting the properties of the Neron model (see [10]) near quintics $V$ such that:

1) $V$ contains a plane $K$ and a quartic surface $Q$;
2) $Q$ contains rational curves of infinitely many degrees which are rigid in $V$.

By varying $K$ we can arrange that exactly *one* predetermined rational curve on $Q$ is in the ramification locus of the corresponding $\rho_d$ above the point $V$. Playing with this fact and properties of the Neron model for degenerations to $V$, one obtains a contradiction to the assumption that the rational equivalence classes (of the deformations to a generic quintic threefold) of the rational curves on $Q \subseteq V$ are finitely generated.

## REFERENCES


[1] A. Beauville, "Variétés de Prym et Jacobiennes intermédiares." Ann. scient. Éc. Norm. Sup. *10*(1977), pp. 309-391.

[2] S. Bloch and J. P. Murre, "On the Chow group of certain types of Fano threefolds." Comp. Math. *39*(1979), pp. 47-105.

[3] G. Ceresa and A. Collino, "Algebraic equivalence of lines on quintic threefolds." Preprint, University of Torino.

[4] G. Ceresa, "$C$ is not algebraically equivalent to $C^-$ in its Jacobian." Annals of Math. (to appear).

[5] H. Clemens and P. Griffiths, "The intermediate Jacobian of the cubic threefold." Annals of Math. *95*(1972), pp. 281-356.

[6] H. Clemens, "On the surjectivity of Abel-Jacobi mappings." Annals of Math. *117*(1983), pp. 71-76.

[7] H. Clemens, "Double Solids." Adv. in Math. *47*(1983), pp. 107-230.

[8] ________, "Applications of mixed Hodge theory to the study of threefolds." Rend. Sem. Mat. Univ. Pol. Torino, *39*(1981).

[9] ________, "Homological equivalence, modulo algebraic equivalence, is not finitely generated." Publ. I.H.E.S. (to appear).

[10] ________, "On extending Abel-Jacobi mappings." Preprint, Univ. of Utah, 1981.

[11] A. Conte and J. P. Murre, "The Hodge conjecture for fourfolds admitting a covering by rational curves." Math. Ann. *238*(1978), pp. 79-88.

[12] F. Gherardelli, "Un' osservazione sulla varietà cubica di $\mathbf{P}^4$." Rend. Sem. Mat. Fis. Milano *37*(1967), pp. 157-160.

[13] P. Griffiths, "Periods of integrals on algebraic manifolds, III." Publ. Math. I.H.E.S. No. 38(1970), pp. 125-180.

[14] ________, "On the periods of certain rational integrals, I, II." Annals of Math. *90*(1969), pp. 460-541.

[15] A. Grothendieck, "Hodge's general conjecture is false for trivial reasons." Topology *8*(1969), pp. 299-303.

[16] V. Iskovskih, "Fano 3-folds, I-II." Math. U.S.S.R. Izvestija *11*(1977), pp. 485-527 and *12*(1978), pp. 469-506.

[17] A. Weil, *Foundations of Algebraic Geometry.* Providence, R.I.: A.M.S. Colloq. Publ. Vol. XXIX (2nd ed., 1962).

[18] G. Welters, "The Fano surface of lines on a double $\mathbf{P}^3$ with 4th order discriminant." Preprint, Univ. of Utrecht, 1979-80.

HERBERT CLEMENS
MATHEMATICS DEPARTMENT
UNIVERSITY OF UTAH
SALT LAKE CITY, UT 84112

## Chapter XVII
# INFINITESIMAL INVARIANT OF NORMAL FUNCTIONS

Phillip Griffiths

§1. As we saw in the previous chapter, to a primitive Hodge class we may associate a global analytic object in the form of its normal function $\nu$. However, due to the failure of Jacobi inversion in higher dimensions this has been of limited use. In this talk I want to discuss a first-order invariant $\delta\nu$ associated to a normal function $\nu$ and to interpret $\delta\nu$ geometrically in case $\nu = \nu_Z$ is the normal function associated to an algebraic cycle. Before doing this I want to motivate things by recalling two examples of the derivative of the Abel-Jacobi map on a fixed variety.

EXAMPLE 1. For a curve $C$ with base point $p_0$ we have the usual Abelian sum mapping

$$u : C^{(d)} \to J(C)$$

$$u(D) = \left( \cdots, \sum_i \int_{p_0}^{p_i} \omega^{\alpha}, \cdots \right) ,$$

where $D = \Sigma p_i$ and the $\omega^{\alpha}$ give a basis for $H^0(\Omega^1_C)$. Then the codifferential

$$\begin{array}{c} u^*(D) : T^*_{(0)}(J(C)) \to T^*_D(C^{(d)}) \\ \cong \\ H^0(\Omega^1_C) \end{array}$$

is simply given by interpreting the differential $du$ and it is represented by the Brill-Noether matrix (cf. Kleiman and Laksov [7], Griffiths and Harris [6])

$$u^*(D) = \begin{bmatrix} \omega^1(p_1) \cdots \omega^g(p_1) \\ \\ \\ \omega^1(p_d) \cdots \omega^g(p_d) \end{bmatrix} .$$

It follows that

$$\ker u^*(D) = H^0(\Omega^1_C(-D)) .$$

By Riemann-Roch,

$$\operatorname{rank} u_*(D) = d - \dim |D| .$$

This simple observation is the key to studying special linear series via $J(C)$ (loc. cit.).

EXAMPLE 2. For a smooth cubic threefold $X \subset PV \simeq \mathbf{P}^4$, we have $h^{3,0}(X) = 0$ and via residues,

$$T^*_{(0)}J(X) \simeq H^{2,1}(X) \simeq V^* .$$

Moreover, there is the Fano surface $S \subset G(2,V)$ of lines $\Lambda$ on $X$ and Abel-Jacobi map

$$u : S \to J(X)$$

with codifferential

$$u^*(\Lambda) : V^* \to T^*_\Lambda(S) .$$

The main fact in the study of $u$ turns out to be

$$\ker u^*(\Lambda) = \Lambda^\perp \subset V^*$$

(Clemens and Griffiths [3]). It follows that the tangent bundle $TS$ is the universal subbundle over $S$.

REMARK. In each case du picks out the differentials vanishing on a cycle. It is this phenomenon that we would like to generalize to families.

§2. We consider a variation of Hodge structure $\{\mathcal{H}_{\mathbf{Z}}, \mathcal{H}^{p,q}, Q, \nabla, S\}$ of odd weight $n = 2m-1$. The sheaf of holomorphic sections of the associated family of intermediate Jacobians is

$$\mathcal{J} = \mathcal{F}^m\backslash\mathcal{H}/\mathcal{H}_{\mathbf{Z}} \simeq \mathcal{F}^{m^*}/\mathcal{H}_{\mathbf{Z}},$$

where the isomorphism $\mathcal{F}^m\backslash\mathcal{H} \simeq \mathcal{F}^{m^*}$ is given by $Q$. The Gauss-Manin connection induces a differential operator

$$\overline{\nabla} : \mathcal{J} \to (\mathcal{H}/\mathcal{F}^{m-1}) \otimes \Omega^1_S \cong \Omega^1_S(\mathcal{F}^{m-1^*}),$$

and a normal function gives a holomorphic section $\nu \in H^0(S, \mathcal{J})$ satisfying

$$\overline{\nabla}\nu = 0 \quad \text{(quasi-horizontality)}.$$

EXAMPLE 3. Suppose we have a family $\pi : \mathfrak{X} \to S$ of smooth varieties $X_s = \pi^{-1}(s)$ with $\dim X_s = 2m-1$ and a codimension $m$ cycle $\mathcal{Z} \subset \mathfrak{X}$ such that the intersections $\mathcal{Z} \cdot X_s = Z_s$ are homologous to zero. Write $Z_s = \partial\Gamma_s$. For $\omega(s) \in \mathcal{F}^m$ a holomorphic section we define $\nu_{\mathcal{Z}} \in \mathcal{F}^{m^*}/\mathcal{H}^*_{\mathbf{Z}}$ by

$$\nu_{\mathcal{Z}}(\omega)(s) = \int_{\Gamma_s} \omega(s).$$

This defines a normal function $\nu_{\mathcal{Z}}$. For curves,

$$Z_s = \sum_i p_i(s) - q_i(s) \in \mathrm{Div}^0(X_s)$$

$$\nu_{\mathcal{Z}}(\omega)(s) = \sum_i \int_{q_i(s)}^{p_i(s)} \omega(s).$$

Intuitively we would like to set

$$\left(\frac{d\nu_{\mathcal{Z}}}{ds}\right)(\omega) = \frac{d}{ds}\left(\sum \int_{q_i(s)}^{p_i(s)} \omega(s)\right)$$

$$= \sum \int_{q_i(s)}^{p_i(s)} \frac{d\omega}{ds}(s) + \sum \omega(s) \Big|_{q_i'(s)}^{p_i'(s)}$$

$$= A(s) + B(s) ,$$

where $A(s)$ is the first sum and $B(s)$ the second sum. Then $B(s) = 0$ if $(\omega) \supset \operatorname{Supp} Z_s$, so in this way $\frac{d\nu_Z}{ds}(\omega)$ "changes character" for those $\omega$ passing through $\operatorname{Supp} Z_s$, generalizing the phenomena noted for du in Examples 1 and 2. Unfortunately, neither $A(s)$ nor $B(s)$ makes sense and so we must proceed differently. Intuitively, the reason for the difficulty is that there is no intrinsic holomorphic connection in $\mathcal{H}^{1,0} \to S$.

§3. Let $G(k,H)$ be the Grassmannian of k-planes in a vector space $H$ and

$$0 \to \mathcal{S} \to \mathcal{H} \to \mathcal{Q} \to 0$$

the universal bundle sequence (bundles are identified with locally free sheaves). By a *local subvariety* $U$ in $G(k,H)$ we mean a subvariety in some open set in $G(k,H)$. For example, $U$ could itself be an open set. Given a local subvariety $U \subset G(k,H)$ and a section $\nu \in H^0(U,\mathcal{Q})$ we ask: *When is* $\nu$ *the projection of a* **constant** *section* $v \in H \subset \mathcal{H}$ ?

MOTIVATION. A normal function is a section $\nu$ of $\mathcal{F}^m\backslash\mathcal{H}/\mathcal{H}_Z$, and to say that $\nu$ is induced from a *horizontal* section $v \in \mathcal{H}$ has intrinsic meaning since $\nabla(\mathcal{H}_Z) = 0$. Suppose $\nu$ is locally induced from $v \in \mathcal{H}$ with $\nabla v = 0$ and $S$ is algebraic and the variation of Hodge structure has no trivial factors. Then there is a plausibility argument (but not yet a proof) that $\nu$ *has finite order* (cf. Griffiths [4]).

Returning to the general discussion, suppose first that $U \subset G(k,H)$ is an open set. Let $\mathfrak{D}(\mathcal{Q})$ be the (filtered) algebra of differential operators on $\mathcal{Q}$ over $U$. For any subsheaf $\mathcal{F} \subset \mathcal{Q}$ set

$$\mathcal{F}^{\perp} = \{P \in \mathfrak{D}(\mathcal{Q}) : Pf = 0 \text{ for all } f \in \mathcal{F}\}.$$

It is elementary that (under very mild conditions)

$$\mathcal{F} = \{f \in \mathcal{Q} : Pf = 0 \text{ for all } P \in \mathcal{F}^{\perp}\}.$$

DEFINITION. $\mathcal{F}$ is *involutive* in case $\mathcal{F}^{\perp} \subset \mathfrak{D}(\mathcal{Q})$ is generated by first-order operators.

THEOREM 4. *Consider the subsheaf* $H \subset \mathcal{Q}$ *over* $U$. (*This is not a sheaf of* $\mathcal{O}$*-modules.*) *Then* $H$ *is involutive.*

This has the following meaning: We may construct a first-order operator*

$$\nu \mapsto \delta\nu$$

such that if the variation of Hodge structure is "sufficiently twisted", then $\delta\nu = 0$ if and only if $\nu$ is induced from $v \in H \subset \mathcal{H}$. Thus we must

(i) construct $\delta\nu$,

(ii) compute $\delta\nu_{\mathcal{Z}}$ geometrically in case the normal function comes from a cycle (analogue of cup product with the Kodaira-Spencer class),

(iii) compute examples to see how much information $\delta\nu$ carries.

Steps (i) and (ii) are in moderate shape, but Step (iii) has just been begun. (Actually the formalism underlying (i) and (ii) needs work in order to isolate the essential points.)

---

*Important notational convention: The $\delta\nu$ here is meant to suggest "the differential of $\nu$" insofar as this makes sense. It should not be confused with the $\delta$ used in Chapter XIV.

§4. The construction of $\delta\nu$ that follows is based on the theory of exterior differential systems, which may be viewed as a nonlinear contravariant version of the theory of D-modules. For each $s \in S$ we consider the part

$$T_s(S) \otimes H_s^{m,m-1} \to H_s^{m-1,m} \simeq (H_s^{m,m-1})^*$$

of the associated infinitesimal variation of Hodge structure. In $\mathbf{P}T(S) \times_S \mathbf{P}\mathcal{H}^{m,m-1}$ we set

$$\Xi = \{(\xi,\omega) : \xi \in \mathbf{P}T_s(S), \omega \in \mathbf{P}H_s^{m,m-1}, \text{ and } \xi \cdot \omega = 0\}.$$

Over $\Xi$ we have $\mathfrak{L} = \pi_1^* \mathcal{O}(1) \otimes \pi_2^* \mathcal{O}(1)$ with fibers $\mathfrak{L}_{(\xi,\omega)} = \xi^* \otimes \omega^*$. For a normal function $\nu$ we will define

$$\delta\nu \in H^0(\Xi, \mathfrak{L}).$$

For this, lift $\nu$ to $v \in \mathcal{H}$ locally. Then

DEFINITION. $(\delta\nu)(\xi,\omega) = Q(\nabla_\xi v, \omega)$.

We shall check that this makes sense. First, by quasi-horizontality, $\nabla v \in \mathcal{F}^{m-1}$ and $(\mathcal{F}^{m-1})^\perp = \mathcal{F}^{m+1}$. Thus all that matters is the class of $\omega$ in $\mathcal{F}^m/\mathcal{F}^{m+1} \cong \mathcal{H}^{m,m-1}$. Secondly, given another lifting

$$\tilde{v} = v + f + \lambda, \quad f \in \mathcal{F}^m, \quad \lambda \in \mathcal{H}_{\mathbf{Z}},$$

we have

$$\nabla \tilde{v} = \nabla v + \nabla f$$

and

$$Q(\nabla_\xi f, \omega) = -Q(f, \xi \cdot \omega) = 0.$$

It is clear that to understand $\delta\nu$ we must understand things like

$$\Sigma = \{\theta \in \mathbf{P}H^1(X,\Theta) : \theta_m \in \mathrm{Hom}^{(s)}(H^{m,m-1}(X), H^{m-1,m}(X)) \text{ is singular}\}.$$

EXAMPLE 5. Let $X = C$ be a curve and $\Sigma \subset PH^1(X,\Theta) = PH^0(2K)^*$ as above. It can be shown that

$$\Sigma = \bigcup_{D \in |K|} \overline{\phi_{2K}(D)} .$$

Using this it has actually been possible to make some computations (cf. Griffiths [5]).

EXAMPLE 6. Let $C$ be a nonhyperelliptic curve of genus 4. We will identify $C$ with its canonical model, which is the intersection of a quadric $F$ and a cubic $V$ in $P^3$:

$$C = F \cap V .$$

The kernel of

$$\delta^* : S^2H^0(K) \to T^* = H^0(2K)$$

gives $F$. We may then ask: Is it possible to reconstruct $V$ from the infinitesimal variation of Hodge structure associated to $C$?

The two rulings of $F$ cut out two $g_3^1$'s $|D_1|$ and $|D_2|$ on $C$, and

$$\nu(C) = u(D_1) - u(D_2)$$

gives a normal function, defined up to $\pm 1$. But $\delta\nu = -\delta(-\nu)$, and it can be shown that for $C$ general there is a natural identification

$$V = \{\delta\nu = 0\} .$$

For details see Griffiths [5, §6 (d)].

§5. *Computation of* $\delta\nu_{\mathcal{Z}}$ *in the geometric case (Example 3)*

We must give $Q(\nabla_\xi \nu_{\mathcal{Z}}, \omega)$ at a point $(\xi,\omega) \in \Xi$. Rather than giving the formula we will give conditions for which $Q(\nabla_\xi \nu_{\mathcal{Z}}, \omega) = 0$.

NOTATION 7. Let $X = X_s$ and $Z = Z_s$. We write $H^{p,q} = H_v^{p,q}(X)$ for the *variable part* of $H^{p,q}(X)$. Let $|Z|$ be the support of $Z$ and consider

$$0 \to I_{|Z|} \otimes \Omega^m \to \Omega^m \to \Omega^m \otimes \mathcal{O}_{|Z|} \to 0 \quad (\Omega^k = \Omega^k_X) .$$

We will say that $(\omega) \supseteq |Z|$ in case $\omega$ belongs to the image of

$$H^{m-1}(I_{|Z|} \otimes \Omega^m) \to H^{m-1}(\Omega^m) .$$

REMARK 8. In practice we will frequently have a surjective map

$$\delta^{m-1} : S^{m-1}T \otimes H^{2m-1,0} \to H^{m,m-1} \to 0 .$$

For example, if $X^{2m-1} \subset Y^{2m}$ is a sufficiently ample divisor and our family is $|\mathcal{O}_Y(X)|$, this is the case (cf. below). Then $\omega$ is of the form

$$\omega = \delta(\xi_1, \cdots, \xi_{m-1})\eta , \qquad \eta \in H^{2m-1,0} = H^0(K_X)$$

and $(\omega) \supseteq |Z|$ in case $(\eta) \supseteq |Z|$ in the usual sense (this is more geometric than the above).

Suppose now that $(\omega) \supseteq |Z|$ and consider the diagram

$$\begin{array}{ccccccc} & & & & H^{m-1}(I_{|Z|} \otimes \Omega^m) & & \\ & & & & \downarrow \text{cup product with } \delta(\xi) & & \\ H^{m-1}(\Omega^{m-1}) & \to & H^{m-1}(\Omega^{m-1} \otimes \mathcal{O}_{|Z|}) & \to & H^m(I_{|Z|} \otimes \Omega^{m-1}) & \to & H^m(\Omega^{m-1}) . \end{array}$$

Since by hypothesis $\rho(\xi) \cdot \omega = 0$ in $H^m(\Omega^{m-1})$, we have

$$[\omega \cdot \xi] \in H^{m-1}(\Omega^{m-1} \otimes \mathcal{O}_{|Z|})/H^{m-1}(\Omega^{m-1}) .$$

We define a linear map

$$\lambda : H^{m-1}(\Omega^{m-1} \otimes \mathcal{O}_{|Z|})/H^{m-1}(\Omega^{m-1}) \to \mathbf{C}$$

by writing $Z = \Sigma\, n_i Z_i$ and setting

$$\lambda(\phi) = \Sigma\, n_i \int_{Z_i} \phi .$$

THEOREM 9. *If* $(\omega) \supseteq |Z|$ *and* $\lambda[\omega \cdot \xi] = 0$, *then*

$$\delta\nu_{\mathcal{Z}}(\omega, \xi) = 0 .$$

To apply this theorem we consider the following (the following is highly speculative).

CONSTRUCTION. Suppose given a smooth variety $Y^{2m}$, a very ample line bundle $\mathcal{O}_Y(1)$, and a primitive Hodge class $\lambda \in H^{m,m}_{prim}(Y) \cap H^{2m}(Y, Z)$. Let $\nu$ be the corresponding normal function (cf. Chapter XIV); thus for $X \in |\mathcal{O}_Y(1)|$ smooth the point

$$\nu(X) \in J(X)$$

is well defined.

To each smooth $X \in |\mathcal{O}_Y(1)|$ there is associated an infinitesimal variation of Hodge structure $\{H_Z, H^{p,q}, Q, T, \delta\}$ where $\{H_Z, H^{p,q}, Q\}$ is the variable part of $H^{2m-1}_{prim}(X)$ and where

$$T = H^0(\mathcal{O}_Y(1))$$

with $\delta$ being determined by the map $\rho$ in

$$H^0(\mathcal{O}_Y(1)) \xrightarrow{r} H^0(\mathcal{O}_X(1)) \xrightarrow{\Delta} H^1(\Theta_X), \qquad \rho = \Delta \circ r$$

(here r is restriction and $\Delta$ is the coboundary in the exact cohomology sequence of $0 \to \Theta_X \to \Theta_Y \otimes \mathcal{O}_X \to \mathcal{O}_X(1) \to 0$).

On the other hand, by residues

$$H^{2m-1-p,p} \cong H^0(K_Y(p+1))/U(p,X) \tag{10}$$

where $U(p,X) \subset H^0(K_Y(p+1))$ is a subspace depending on $X$. By making $\mathcal{O}_Y(1)$ sufficiently ample we may assume that the induced embedding $Y \subset P^N$ is projectively normal and we set

$$R = \bigoplus_{\ell \geq 0} H^0(\mathcal{O}_Y(\ell))$$

We may also assume that the *Arbarello-Sernesi module* (cf. [2])

$$M = \bigoplus_{p \geq 0} H^0(K_Y(p+1))$$

is generated as an R-module in its lowest degree. By (10)

$$\bigoplus_{p \geq 0} H^{2m-1-p,p} = M_X$$

is a quotient vector space of $M$. *Then the axioms of an infinitesimal variation of Hodge structure imply that* $M_X$ *is a quotient* R*-module of* M *where the pairing*

$$H^0(\mathcal{O}_Y(1)) \otimes M_X \to M_X$$

*is given by* $\delta$. For $\omega \in H^0(K_Y(1))$ and $P \in H^0(\mathcal{O}_Y(m))$ we may consider the residue $\mathrm{Res}_X(P\omega) \in H^{m-1,m}(X)$ of $P\omega \in H^0(K_Y(m+1))$. We set $D = (\omega)$ and assume that $\mathrm{Res}_X(P\omega) = 0$. Then there is defined

$$\{P\cdot\omega\} \in H^{m-1}(\Omega^{m-1} \otimes \mathcal{O}_D)/H^{m-1}(\Omega^{m-1})$$

by the construction preceding the above theorem (and replacing $|Z|$ by $D$). We set

$$\Sigma = \{(\omega,P,X) : \mathrm{Res}_X(P\omega) = 0, \{P\omega\} = 0\}.$$

Note that: $\Sigma$ *is defined Hodge-theoretically.*

Returning to our normal function $\nu$, we write $\nu = \nu(\lambda, \mathcal{O}_Y(1))$ to emphasize that it depends only on the data $Y, \lambda, \mathcal{O}_Y(1)$. We note that $\delta\nu$ may be defined on $\Sigma$. For example, suppose that $P = \xi_1 \cdots \xi_m$ where $\xi_i \in H^0(\mathcal{O}_Y(1))$. Then

$$\psi = \mathrm{Res}_X(\xi_1 \cdots \xi_{m-1}\omega) \in H^{m,m-1}$$

and $\delta(\xi_m)\cdot\psi = 0$ in $H^{m-1,m}$. Hence $\delta\nu(\psi,\xi_m)$ is defined, and it is easily seen to be symmetric in the $\xi_i$'s. We then set

$$I(\lambda,\mathcal{O}_Y(1)) = \{\omega \in |K_Y(1)| : \delta\nu(\omega,P) = 0 \text{ whenever } (\omega,P,X) \in \Sigma\}.$$

Clearly this subvariety of $|K_Y(1)|$ is defined purely Hodge theoretically.

On the other hand, suppose that $\nu = \nu_{\mathcal{Z}}$ for some codimension-m algebraic cycle $\mathcal{Z} \subset Y$ whose support we denote by $|\mathcal{Z}|$. Setting

$$I(\mathcal{Z},\mathcal{O}_Y(1)) = |K_Y(1) \otimes I_{|\mathcal{Z}|}|$$

the above theorem implies that

$$I(\mathcal{Z},\mathcal{O}_Y(1)) \subseteq I(\lambda,\mathcal{O}_Y(1)) .$$

Thus $I(\lambda,\mathcal{O}_Y(1))$ gives a first candidate for the equations that define $\mathcal{Z}$. However, since in general $\mathcal{Z}$ is not unique (think of the difference of two $P^{m+1}$'s in a quadric $Q^{4m+2} \subset P^{4m+3}$) we will not have $I(\mathcal{Z},\mathcal{O}_Y(1)) = I(\lambda,\mathcal{O}_Y(1))$. In addition it is clear that our choice of $\mathcal{O}_Y(1)$ is rather arbitrary. Denoting by $\mathrm{cl}(\mathcal{Z})$ the fundamental class of $\mathcal{Z}$, we have for any $\ell \geqq 1$

$$\bigcup_{\mathcal{Z} \text{ with } \mathrm{cl}(\mathcal{Z})=\lambda} I(\mathcal{Z},\mathcal{O}_Y(\ell)) \subseteq I(\lambda,\mathcal{O}_Y(\ell)) . \tag{11}$$

Note that the left-hand side is a subvariety of $|K_Y(\ell+1)|$ that is ruled by the linear subspaces $|I(\mathcal{Z},\mathcal{O}_Y(\ell)|$. An obvious question is whether or not we have equality in (11) for some $\ell$ such that the subvariety $I(\lambda,\mathcal{O}_Y(\ell))$ is ruled and the base locus of one of the rulings actually gives a cycle $\mathcal{Z}$ with $\mathrm{cl}(\mathcal{Z}) = \lambda$. Clearly this is enormously speculative, but in [5] it is verified that the construction actually does work in one special case of divisors on a surface.

## REFERENCES

[1] A. Altman and S. Kleiman, Foundations of the theory of Fano schemes, Comp. Math. 34 (1977), 3-47.

[2] E. Arbarello and E. Sernesi, Petri's approach to the study of the ideal associated to a special divisor, Invent. Math. 49(1978), 99-119.

[3] H. Clemens and P. Griffiths, The intermediate Jacobian of the cubic threefold, Ann. of Math. 95(1972), 281-356.

[4] P. Griffiths, On the differential equation satisfied by the normal function of algebraic cycles, Amer. J. Math 101 (1979), 94-131.

[5] ————, Infinitesimal variation of Hodge structures (III): determinantal varieties and the infinitesimal invariant of normal functions, to appear in Comp. Math.

[6] P. Griffiths and J. Harris, On the variety of special linear systems on a general algebraic curve, Duke Math. J. 47 (1980), 233-272.

[7] S. Kleiman and D. Laksov, Another proof of the existence of special divisors, Acta Math. 132 (1974), 163-176.

PHILLIP GRIFFITHS
MATHEMATICS DEPARTMENT
HARVARD UNIVERSITY
CAMBRIDGE, MA 02138

**Library of Congress Cataloging in Publication Data**
Main entry under title:

Topics in transcendental algebraic geometry.

1. Geometry, Algebraic—Addresses, essays, lectures.
2. Hodge theory—Addresses, essays, lectures.
3. Torelli theorem—Addresses, essays, lectures.
I. Griffiths, Phillip, 1938- .
QA564.T66 1983 512′.33 83-42593
ISBN 0-691-08335-5
ISBN 0-691-08339-8 (pbk.)

Phillip Griffiths is Professor of Mathematics at Harvard University.